Springer Series in
Electronics and Photonics   23

Springer Series in Electronics and Photonics

Editors: D.H. Auston   W. Engl   T. Sugano

Managing Editor: H.K.V. Lotsch

This series was originally published under the title
Springer Series in Electrophysics
and has been renamed starting with Volume 22.

Volumes 1–20 are listed on the back inside cover

Chen S. Tsai (Ed.)

Guided-Wave Acousto-Optics

Interactions, Devices, and Applications

With 167 Figures

Springer-Verlag
Berlin Heidelberg New York London
Paris Tokyo Hong Kong Barcelona

Professor Chen S. Tsai, Ph.D.

Department of Electrical and Computer Engineering
and Institute for Surface and Interface Science, University of California,
Irvine, CA 92717, USA

Series Editors:

Dr. David H. Auston

Columbia University, Dept. of Electrical Engineering, New York, NY 10027, USA

Professor Dr. Walter Engl

Institut für Theoretische Elektrotechnik, Rhein.-Westf. Technische Hochschule
Templergraben 55, D-5100 Aachen, Fed. Rep. of Germany

Professor Takuo Sugano

Department of Electronic Engineering, The Faculty of Engineering
The University of Tokyo, 7-3-1, Hongo, Bunkyo-ku, Tokyo 113, Japan

Managing Editor: Dr. Helmut K.V. Lotsch

Springer-Verlag, Tiergartenstrasse 17
D-6900 Heidelberg, Fed. Rep. of Germany

ISBN-13:978-3-642-75227-8   e-ISBN-13:978-3-642-75225-4
DOI: 10.1007/978-3-642-75225-4

Library of Congress Cataloging-in-Publication Data. Guided-wave acoustooptics : interactions, devices, and applications / Chen Tsai, ed. p. cm.–(Springer series in electronics and photonics ; v. 23) Includes bibliographical references. ISBN-13:978-3-642-75227-8 (U.S. : alk. paper) 1. Integrated optics–Congresses. 2. Acoustooptics–Congresses. 3. Wave guides–Congresses. I. Tsai, Chen S. II. Series. TA1660.G8 1990 621.382'8–dc20   89-26320

The text was prepared using the PS™ Technical Word Processor
2154/3150-543210 – Printed on acid-free paper

Preface

The field of integrated- or guided-wave optics has experienced significant and continuous growth since its inception in the late 1960s. There has been a considerable increase in research and development activity in this field worldwide and some significant advances in the realization of working integrated optic devices and modules have been made in recent years. In fact, there have already been some commercial manufacturing and technical applications of such devices and modules.

The guided-wave-acoustooptics involving Bragg interactions between guided optical waves and surface acoustic waves is one of the areas of integrated-optics that has reached some degree of scientific and technological maturity. This topical volume is devoted to an in-depth treatment of this emerging branch of science and technology. Presented in this volume are concise treatments on bulk-wave acoustooptics, guided-wave optics, and surface acoustic waves, and detailed studies of guided-wave acoustooptic Bragg diffraction in three promising material substrates, namely, $LiNbO_3$, ZnO/SiO_2, and GaAs, the resulting wideband modulators and deflectors, and applications. The chapters cover not only the basic principles and theoretical analysis, but also the design, fabrication, and measurement of the resulting devices and modules, and their applications.

The applications that are described in some detail include light beam modulation and deflection, multiport switching and frequency shifting in communications; spectral analysis, correlation, and convolution in RF signal processing; and matrix-vector and matrix-matrix multiplications in computing. Because of page constraints some of the basic topics regarding the theoretical and practical aspects of integrated optics and surface acoustic waves are not covered. For material covering such topics the readers are referred to the large number of references listed at the end of the introductory chapter. The present volume is aimed at a wide spectrum of readers including graduate students, research scientists and engineers, design and practising engineers as well as research and development program managers.

The editor wishes to express his sincere thanks to those who helped review the chapters and provided their valuable suggestions. The reviewers included Drs. Bob Adler, Fred Hickernell, Janet Jackle, Gordon Kino, Adrian Korpel, Fred Leonberger, Henry Taylor, Bob Wager, Dick Williamsons, and Dick White. The editor is particularly grateful to Bob Adler for his most thorough and critical reviews of Chapters 5 and 8, and for his many suggestions for their improvement. Finally, the continuing support and encouragement of Drs. Helmut Lotsch and Theo Tamir throughout the preparation of this volume are also greatly appreciated.

Irvine, California, September 1989   C.S. Tsai

Contents

Contributors

Alferness, Rod C.
  AT&T Bell Laboratories, Crawford Hill Laboratories, Box 400
  Holmdel, NJ 07733, USA

Hamilton, Michael C.
  Boeing Electronics, P.O. Box 24969, Seattle, WA 98124-6269, USA

Mikoshiba, Nobuo
  Professor of Acoustoelectronics, Res. Inst. of Electrical Communication
  Tohoku University, 980 Sendai, Japan

Reeder, Thomas M.
  Vice President of Marketing, EE SOF, Inc.
  Westlake Village, CA 91362, USA

Spezio, Anthony E.
  Naval Research Laboratory, Code 5724, 4555 Overlook Avenue
  Washington, DC 20375, USA

Tien, Ping K.
  AT&T Bell Laboratories, Crawfords Corner Rd., Box 400
  Holmdel, NJ 07733, USA

Tsai, Chen S.
  Professor of Electrical and Computer Engineering
  Department of Electrical and Computer Engineering
  and Institute for Surface and Interface Science
  University of California,
  Irvine, CA 92717, USA

Wade, Glen
  Professor of Electrical and Computer Engineering
  Department of Electrical and Computer Engineering
  University of California at Santa Barbara
  Santa Barbara, CA 93106, USA

1. Introduction

Chen S. Tsai

With 3 Figures

Acousto-optic interactions, defined as the interactions between optical (light) waves and acoustic (sound) waves, and henceforth abbreviated as acousto-optics, commonly refer to the influences of the latter upon the former. Significant influences can be expected under certain situations as the refractive-index gratings created by the acoustic waves will cause diffraction or refraction of an incident light wave. Diffraction of light by sound was first predicted by *Brillouin* in as early as 1922 [1.1], and was observed experimentally by *Debye* and *Sears*, and by *Lucas* and *Biquard* in 1932 [1.2]. A great number of studies [1.3] were carried out subsequently using incoherent light sources and relatively low frequency acoustic waves in mostly liquid media. Impacts of these early studies lie mainly in the basic understanding of the thermal and acoustic properties of the liquid and gaseous solutions, as well as the solid materials involved. A more complete listing of these earlier works will be given in Chap. 2.

The simultaneous availability of coherent light sources through the advent of the laser and the very high frequency acoustic waves through advancement in piezoelectric transducer technology stimulated revival of interest in the subject in the 1960's [1.4-14]. Like all previous studies, these later ones were limited to Bulk-Wave Acousto-Optics (BWAO) in which both the light and the sound propagate as unguided (free) waves in mostly solid media. Concomitant success in growth of new and superior solid materials [1.15] has also enabled realization of various types of bulk Acousto-Optic (AO) devices including modulators, scanners, deflectors, Q-switches, mode lockers, tunable filters, multiplexers/demultiplexers, spectrum analyzers, correlators and other signal processors [1.16]. Some of these bulk AO devices have now been deployed in a variety of commercial and military applications. In the meantime, since the early 1970's a large number of studies have been focused on Guided-Wave Acousto-Optics (GWAO) in which both the light and the sound are confined to a small depth in suitable solid substrates [1.17-21]. This focus on guided-wave acousto-optics was a natural outgrowth of the guided-wave optics science and technology [1.22-36] and the surface-acoustic-wave device technology [1.37] that had been undergoing intensive research and development a few years earlier. These latest studies on guided-wave acousto-optics have already generated many fruitful results [1.21]. For example, the resulting wide-band planar AO Bragg modulators and deflectors are now widely used in the development and realization of microoptic modules for processing of radar signals, e.g., the integrated-optic RF spectrum analyzers. It is to be noted that acousto-optics in which either the light

Springer Series in Electronics and Photonics, Vol. 23
Guided-Wave Acoustooptics   Editor: C. S. Tsai
© Springer-Verlag Berlin, Heidelberg 1990

propagating as an unguided-wave while the sound as a surface wave or the light propagating as a guided wave while the sound as a bulk or leaky wave has also been studied. However, due to page constraints, these subjects will not be treated in this volume.

The brief chronology given above has clearly spelled out the two major subareas of acousto-optics, namely, bulk-wave and guided-wave acousto-optics. This topical volume is devoted to an in-depth treatment on the second subarea. Thus, except for Chap.2 by G. Wade, which provides a selective review on the first subarea, only guided-wave acousto-optics will be treated in the remaining chapters.

As indicated previously, guided-wave acousto-optics is concerned with the interactions between Guided-Optical Waves (GOW) and Surface Acoustic Waves (SAW). The GOW has its energy concentrated in a thin layer (typically of a thickness up to a few wavelengths of the optical wave involved) beneath a substrate material, and thus constitutes the basis for the now well-established interdisciplinary field of guided-wave or integrated optics. Integrated optics concerns basic phenomena, new device concepts, miniaturization and integration of GOW-based components such as lasers, modulators, switches, lenses, prisms, couplers, detectors, etc., in a common waveguide substrate to perform a variety of scientific and engineering functions. Similar to the prevalent integrated electronic circuits, the ultimate integrated-optic circuits or modules are expected to possess significant advantages over their discrete bulk counterparts. Some of the advantages are the smaller size and the lighter weight, wider bandwidth, lesser electrical drive power requirement, greater signal accessibility, a higher degree of integratability, potential for batch fabrication, and an ultimate reduction in cost. Similarly, the SAW-device technology concerns excitation, propagation, control, and utilization of the acoustic waves that have their energies concentrated in a small depth (typically a fraction of the wavelength of the acoustic wave involved) beneath suitable solid substrates. The SAW-device technology has already reached a high degree of maturity. For example, a variety of SAW devices and modules such as IF filters, oscillators, correlators, and convolvers are being batch-fabricated and deployed in both commercial applications such as TV sets and military applications in advanced communication and radar systems. In fact, a variety of guided-wave optical devices and modules have been studied and developed using mostly the lithium niobate ($LiNbO_3$) substrate and increasingly the gallium arsenite (GaAs) substrate as well. Such microoptic devices and modules have demonstrated some of the aforementioned advantages and are expected to provide unique applications in future single-mode optical fiber communications [1.38], and integrated-optic signal processing [1.21,22] and computing systems [1.39-41].

The introductory treatment on bulk-wave acousto-optics given in Chap.2 serves a useful pedagogical purpose as it provides not only an historical perspective on the subject, but also the physical principles and heuristic views as well as some analytical approaches that are equally applicable to guided-wave acousto-optics. Following Chap.2 a selective review on the principles, fabrication techniques, and coupling and measurement schemes on planar and channel waveguides as well as the promising waveguide substrates for AO and Electro-Optic (EO) devices, again with emphasis on the $LiNbO_3$ substrate, is presented in Chap.3 by R.C. Alferness and P.K. Tien. Chapter 4 by T.M. Reeder provides a review on the basic characteristics of the interdigital finger electrode transducers for excitation of the SAW and the design and evaluation of the transducers, using an equivalent circuit model, that are required in the construction of guided-wave AO devices.

Since both GOWs and SAWs are confined to thin layers of comparable thickness, efficient and wide-band AO Bragg interactions can be expected. Such technical potential has already been explored in detail in $LiNbO_3$ and zinc oxide(ZnO)/thermally-oxidized silicon (Si) substrates, and is being explored for the GaAs substrates as well. Chapter 5 by C.S. Tsai and Chap.6 by N. Mikoshia are devoted to in-depth treatments of the first two substrate materials, respectively. A progress report on the most recent advancement in realization of GHz AO Bragg modulators and deflectors using the third substrate material, and related waveguide lenses, will be given in Chap.8 by C.S. Tsai. Also, as indicated previously, the resulting wideband guided-wave AO Bragg modulators and deflectors in $LiNbO_3$ are now used widely in the development and realization of integrated-optic RF spectrum analyzer and correlator modules for real-time processing of wideband radar signals. For example, Fig.1.1 shows a hybrid integrated optic RF spectrum analyzer module that has been undergoing worldwide engineering development for some time. Chapter 7 by M. Hamilton and S. Spezio provides a detailed treatment of this integrated optic module.

Serious attempts to realize a variety of other integrated AO device modules with applications to optical communications, computing, and signal processing are also being made. For example, successful fabrication of high performance microlenses and microlens arrays using the Titanium-Indiffusion Proton-Exchange (TIPE) technique has enabled realization of a variety of multichannal integrated AO, EO and AO-EO Bragg modulator modules in the $LiNbO_3$ channel-planar composite waveguides. These multichannel integrated optic device modules have been shown to take up a small substrate dimension (up to 2.0cm only) along the optical path and are also inherently of high modularity and versatility [1.39-41]. Specifically, these integrated optic device modules have been utilized successfully to perform matrix-vector and matrix-matrix multiplications, optical switching and frequency shifting, RF spectral analysis and correlation, and programmable digital correlation and filtering. Figures 1.2 and 3

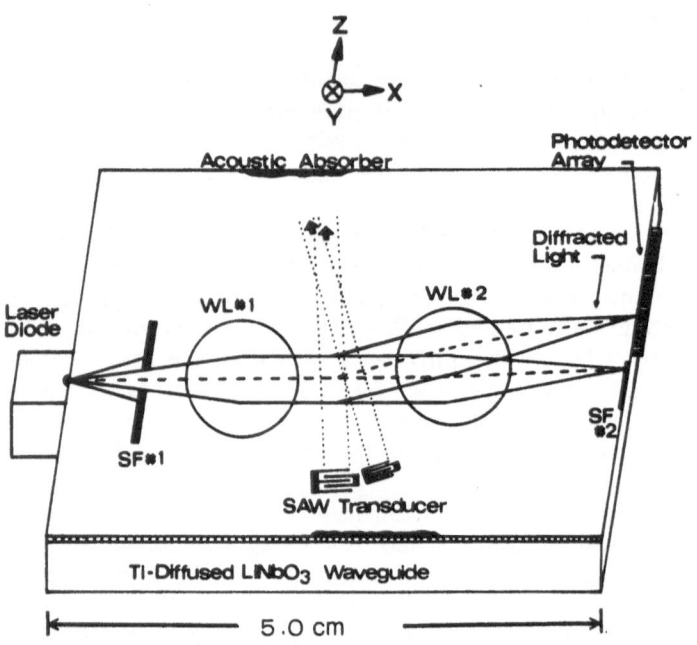

Z
⊗→X
Y

Acoustic Absorber

Photodetector Array

Diffracted Light

Laser Diode

WL#1

WL#2

SF#1

SAW Transducer

SF#2

Tl-Diffused LiNbO$_3$ Waveguide

|← 5.0 cm →|

Fig.1.1. A hybrid integrated-acoustooptic radio-frequency (RF) spectrum analyzer in a planar LiNbO$_3$ waveguide

Input Channel Waveguide Array

TIPE Microlenses Array

TIPE Integrating Lens

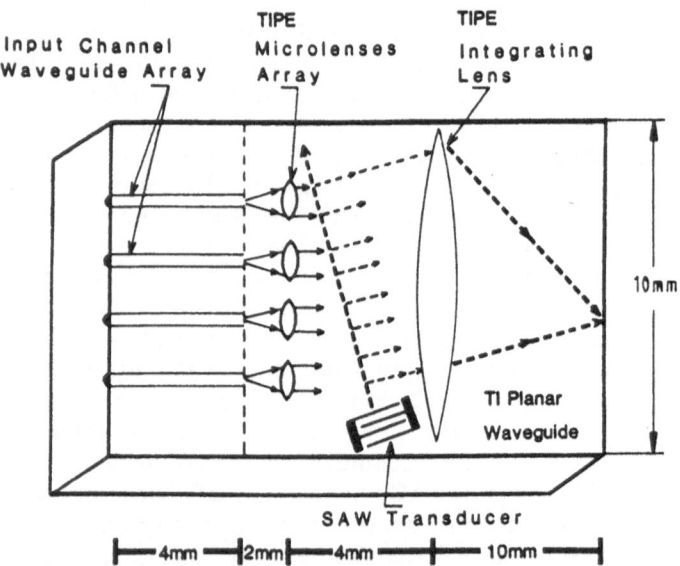

10 mm

Tl Planar Waveguide

SAW Transducer

|← 4mm →|← 2mm →|← 4mm →|← 10mm →|

Fig.1.2. A hybrid integrated-optic systolic-array computer in a channel-planar LiNbO$_3$ waveguide.

show, respectively, the multi-channel integrated AO and AO-EO Bragg modulator modules that were realized recently. Through the channel-waveguide and the linear microlens arrays, the very large channel capaci-

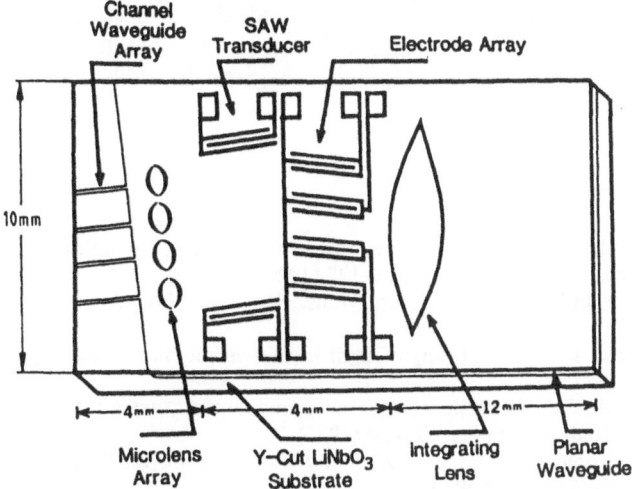

Fig.1.3. Multichannel integrated AO-EO modulator module in a channel-planar LiNbO₃ waveguide

ties that are inherent in the diode laser and the optical fiber as well as the photo-detector arrays may be conveniently exploited. Consequently, such integrated-optic device modules should facilitate realization of multi-channel optical computing as well as communication and RF signal processing systems. Thus the volume concludes with Chap.8 by C.S. Tsai that provides updated reports on the development and applications of these

Table 1.1. Applications of guided-wave acousto-optic Bragg modulators and deflectors

I. **Communications**
   Light-beam modulation and deflection
   Multiport switching
   Space-, Time- and wavelength-division multiplexing/demultiplexing
   Tunable optical wavelength filtering
   Optical-frequency shifting and heterodyne detection
   Optical interconnect

II. **Radio-Frequency Signal Processing**
   Spectral analysis or Fourier transform
   Pulse compression
   convolution
   Time- and space-integrating correlations
   Adaptive filtering
   Ambiguity function

III. **Computing**
   Matrix-vector multiplication
   Matrix-matrix multiplication
   Programmable analog and digital correlation/filtering
   Optical bistability

latest integrated AO device modules in LiNbO$_3$ and GaAs substrates together with those reported earlier. Finally, a list of applications that were demonstrated using the guided-wave AO devices is given in Table 1.1.

References

1.1   L. Brillouin: Diffusion de la lumière et des rayons X par un corps transparent homgène. Ann. Phys. (Paris) 9th Ser.,17, 88-122 (1922)
W.H. Bragg: X-ray and crystal structures, Trans. Roy. Soc. (London) A 215, 235 (1915)

1.2   P. Debye, F.W. Sears: On the scattering of light by supersonic. Proc. Nat. Acad. Sci. USA 18, 409-414 (1932)
R.Lucas, P. Biquard: Optical properties of solids and liquids under ultrasonic vibrations. J. Phys. Radium, 7th Ser., 3, 464 (1932)

1.3   A.B. Bhatia, W.J. Noble: Diffraction of light by ultrasonic waves. Proc. Roy. Soc. (London) A 220, 365-385 (1953)
M. Born, E. Wolf: *Principles of Optics*, 2nd ed. (Macmillan, New York 1964) and references therein
C.V. Raman, N.S. Nagendra Nath: The diffraction of light by high frequency sound waves. Proc. Indian Acad. Sci. Pt.I: 2, 406-412 (1935); Pt.II: 2, 413-420 (1935); Pt.III: 3, 75-84 (1936); Pt.IV: 3, 119-125 (1935); Pt.V: 3, 459-465 (1936)
J.C. Slater: Interaction of waves in crystals. Rev. Mod. Phys. 30, 197-222 (1958)

1.4   P.K. Tien: Bell Telephone Labs. Report, MM-61-124-8 (February 1961)

1.5   J. Melngailis, A.A.Maradudin, A. Seeger: Diffraction of light by ultrasound in anharmonic crystals. Phys. Rev. 131, 1972-1975 (1963)
W.R. Klein, B.D. Cook, W.G. Mayer: Light diffraction by ultrasonic gratings. Acoustica 15, 67-74 (1965)

1.6   H.V. Hance, J.K. Parks: Wideband modulation of a laser beam using Bragg-angle diffraction by amplitude modulated ultrasonic waves. J. Acoust. Soc. Am. 38, 14-23 (1965)
H.V. Hance: Light diffraction by ultrasonic waves as a multiple scattering process. Technical Report #6-74-64-35, July 1964, Lockheed Missiles and Space Co., Sunnyvale, CA

1.7   C.F. Quate, C.D.W. Wilkinson, D.K. Winslow: Interaction of light and microwave sound. Proc. IEEE 53, 1604-1623 (1965)

1.8   A.E. Siegman, C.F. Quate, J. Bjorkholm, C. Francois: Frequency translation of laser's output frequency by acoustic output coupling. Appl. Phys. Lett. 5, 1-3 (1964)
L.E. Hargrove, R.L. Fork, M.A. Pollack: Locking of the He-Ne Laser modes induced by synchronous intracavity modulation. Appl. Phys. Lett. 5, 4-5 (1964)
A.J. DeMaria, G.E. Danielson,Jr.: Internal laser modulation by acoustic lens-like effects. IEEE J. QE-2, 157-164 (1966)

1.9   M.G. Cohen, E.I. Gordon: Acoustic beam probing using optical techniques. Bell Syst. Tech. J. 44, 693-721 (1965)

1.10 A. Korpel, R.Adler, P. Desmares, W. Watson: A television display using acoustic deflection and modulation of coherent light. Proc. IEEE 54, 1429-1437 (1966)
A. Korpel: *Acoustooptics* (Dekker, New York 1988)

1.11 E.I. Gordon: A review of acoustooptical deflection and modulation. Proc. IEEE 54, 1391-1401 (1966)

1.12 R. Adler: Interactions of light and sound. IEEE Spectrum 4, 42-54 (May 1967)

1.13 R.W. Dixon: Acoustic diffraction of light in anisotropic media. IEEE J. QE-3, 85 (1967)

1.14 R.S. Chu, T. Tamir: Guided-wave theory of light diffraction by acoustic microwaves. IEEE Trans. MTT-17, 1002-1020 (1969)

1.15 R.W. Dixon: Photoelastic properties of selected materials and their relevance for applications of acoustic light modulators and scanners. J. Appl. Phys. 38, 5149-5153 (1967)
D.A. Pinnow: Guidelines for the selection of acoustooptic materials. IEEE J. QE-6, 223-238 (1970)

1.16 N. Uchida, N. Niizeki: Acoustooptic deflection materials and techniques. Proc. IEEE 61, 1073-1092 (1973)
I.C. Chang: Acoustooptic devices and applications. IEEE Trans. SU-23, 2-22 (1976)
Yu. V. Gulyaev, V.V. Proklov, G.N. Shkerdin: Diffraction of light by sound in solids. Sov. Phys. Usp. 21, 29-56 (1978)
A. Van der Lugt (ed.): Special section on acousto-optical signal processing. Proc. IEEE 69, 48-118 (1981)

1.17 L. Kuhn, M.D. Dakss, P.F. Heidrich, B.A. Scott: Deflection of an optical guided wave by a surface acoustic wave. Appl. Phys. Lett. 17, 265-267 (1970)

1.18 Y. Ohmachi: Acousto-optical light diffraction in thin films. J. Appl. Phys. 44, 3928-3932 (1973)

1.19 C.S. Tsai, Le T. Nguyen, S.K. Yao, M.A. Alhaider: High-performance guided-light beam device using two tilted surface acoustic waves. Appl. Phys. Lett. 26, 140-142 (1975)

1.20 R.V. Schmidt: Acoustooptic interactions between guided optical waves and acoustic surface waves. IEEE Trans. SU-23, 22-33 (1976)
R.V. Schmidt, I.P. Kaminov: Acoustooptic Bragg deflection in LiNbO$_3$ Ti-diffused waveguides. IEEE J. QE-11, 57-59 (1975)

1.21 C.S. Tsai: Guided-wave acoustooptic Bragg modulators for wideband integrated optics communications and signal processing. IEEE Trans. CAS-26, 1072-1098 (1979), and references therein

1.22 R. Schubert, J.H. Harris: Optical surface waves on thin films and their application to integrated data processors. IEEE Trans. MTT-16, 1048-1054 (1968)

1.23 P.K. Tien, R. Ulrich, R.J. Martin: Mode of propagating light waves in thin deposited semiconductor films. Appl. Phys. Lett. 14, 291-293 (1969)
P.K. Tien: Light waves in thin films and integrated optics. Appl. Opt. 10, 2395-2413 (1971)
P.K. Tien, R. Ulrich: Theory of prism-film coupler and thin-film lightguides. J. Opt. Soc. Am. 60, 1325-1337 (1970)
P.K. Tien: Integrated optics and new wave phenomena in optical waveguides. Rev. Mod. Phys. 49, 361-423 (1977)

1.24 S.E. Miller: A survey of integrated optics. IEEE J. QE-8, 199-205 (1972)

1.25 H.F. Taylor, A.Yariv: Guided-wave optics. Proc. IEEE 62, 1044-1060 (1974)
A. Yariv: Coupled-mode theory for guided-wave optics. IEEE J. QE-9, 919-933 (1973)

1.26 H. Kogelnik: An introduction to integrated optics. IEEE Trans. MTT-23, 2-16 (1975)

1.27 I.P. Kaminow: Optical waveguide modulators. IEEE Trans. MTT-23, 57-70 (1975)

1.28 R.C. Alferness: Guided-wave devices for optical communication. IEEE J. QE-17, 946-959 (1981)

1.29 T. Tamir (ed.): *Integrated Optics*, 2nd. ed., Topics Appl. Phys., Vol.7 (Springer, Berlin, Heidelberg 1979)
T. Tamir (ed.): *Guided-Wave Optoelectronics*, Springer Ser. Electr. Photonics, Vol.26 (Springer, Berlin, Heidelberg 1988)

1.30　A. Yariv: *Introduction to Optical Electronics*, 2nd ed. (Holt, Rinehart and Winston, New York 1976) Chap.13
A. Yariv: *Quantum Electronics*, 2nd. edn. (Wiley, New York 1975) Chap.19

1.31　R. Hunsperger: *Integrated Optics: Theory and Technology*, 2nd ed., Springer Ser. Opt. Sci, Vol.33 (Springer, Berlin, Heidelberg 1982)

1.32　H.P. Nolting, R. Ulrich (eds.): *Integrated Optics*, Springer Ser. Opt. Sci., Vol.48 (Springer, Berlin, Heidelberg 1985)

1.33　P.K. Tien, J.P. Gordon, J.R. Whinnery: Focusing of a light beam of Gaussian field distribution in a continuous and periodic lens-like media. Proc. IEEE 53, 129-136 (1965)

1.34　N.S. Kapany, J.J. Burke: *Optical Waveguides* (Academic, New York 1972)
D. Marcuse: *Theory of Dielectric Optical Waveguides* (Academic, New York 1974)
E.A.J. Marcatili: Dielectric rectangular waveguide and directional coupler for integrated optics. Bell Syst. Tech. J. 48, 2071 (1969)
H. Kogelnik: "Theory of Dielectric Waveguides" in *Integrated Optics*, 2nd. ed., ed. by T. Tamir, Topics Appl. Phys., Vol.7 (Springer, Berlin, Heidelberg 1979) Chap.2
H.G. Unger: *Planar Optical Waveguides and Fibres* (Clarendon, Oxford 1977)
A.W. Snyder, J.D. Love: *Optical Waveguide Theory* (Chapman and Hall, London 1983)
S. Ramo, J.R. Whinnery, T. Van Duzer: *Fields and Waves in Communication Electronics*, 2nd. ed. (Wiley, New York 1984)
T. Okoshi: *Planar Circuits for Microwaves and Lightwaves*, Springer Ser. Electrophys. Vol.18 (Springer, Berlin, Heidelberg 1985)

1.35　H.A. Haus: *Waves and Fields in Optoelectronics* (Prentice-Hall, Englewood Cliffs, NJ 1984)

1.36　A. Yariv, P. Yeh: *Optical Waves in Crystals* (Wiley, New York 1984)

1.37　R.W. Smith, H.M. Gerard, J.H. Collins, T.M. Reeder, H.J. Shaw: Design of surface wave delay lines with interdigital transducers. IEEE Trans. MTT-17, 865-873 (1969)
R.M. White: Surface elastic waves. Proc. IEEE 58, 1238-1276 (1970)
R.H. Tancrell, M.G. Holland: Acoustic surface wave filters. Proc. IEEE 59, 393-409 (1971)
G.S. Kino, R.S. Wagers: Theory of interdigital couplers on nonpiezoelectric substrates. J. Appl. Phys. 44, 1480-1488 (1973)
A.J. DeVries, R. Adler: Case history of a surface-wave TV IF filter for color television receivers. Proc. IEEE 64, 671-676 (1976)
H. Matthews (ed.): *Surface Wave Filter - Design, Construction, and Use* (Wiley, New York 1977)
A.A. Oliner (ed.): *Acoustic Surface Waves*, Topics Appl. Phys., Vol.24 (Springer, Berlin, Heidelberg 1979)
E. Dieulesaint: *Elastic Waves in Solids: Applications to Signal Processing* (Wiley, New York 1980)
E.A. Ash, E.G.S. Paige (eds.): *Rayleigh-Wave Theory and Application*, Springer Ser. Wave Phen., Vol.2 (Springer, Berlin, Heidelberg 1985)
B. A. Auld: *Acoustic Fields and Waves in Solids*, Vol.II (Wiley, New York 1973)
G.S. Kino: *Acoustic Waves - Devices, Imaging, and Analog Signal Processing* (Prentice-Hall, Reading, Mass. 1987)
C.K. Campbell: Applications of surface acoustic and shallow bulk acoustic wave devices. Proc. IEEE 77, 1453-1483 (1989)

1.38　D. Marcuse: *Light Transmission Optics* (Van Nostrand, Princeton, NJ 1972)
J.A. Arnaud: *Beam and Fiber Optics* (Academic, New York 1976)

C.K. Campbell: Applications of surface acoustic and shallow bulk acoustic wave devices. Proc. IEEE 77, 1453-1483 (1989)

1.38 D. Marcuse: *Light Transmission Optics* (Van Nostrand, Princeton, NJ 1972)
J.A. Arnaud: *Beam and Fiber Optics* (Academic, New York 1976)
M.K. Barnosk (ed.): *Fundamentals of Optical Fiber Communications* (Academic, New York 1976)
P.K. Cheo: *Fiber Optics Devices and Systems* (Prentice-Hall, Englewood Cliffs, NJ 1985)
D. Gloge (ed.): *Optical Fiber Technology* (IEEE Press, New York 1976)
C.K. Kao (ed.): *Optical Fiber Technology II* (IEEE Press, New York 1981)
C.K. Kao: *Optical Fiber Systems - Technology, Design, and Applications* (McGraw-Hill, New York 1982)
R.D. Maurer: Glass fibers for optical communications. Proc. IEEE 61, 452-462 (1973)
J.E. Midwinter: *Optical Fibers for Transmission* (Wiley, New York 1979)
S.E. Miller, E.A.J. Marcatili, T. Li: Research towards optical fiber transmission systems, Pt.I. Proc. IEEE 61, 1703-1726 (1973)
S.E. Miller, T. Li, E.A.J. Marcatili: Research towards optical fiber transmission systems, Pt.II. Proc. IEEE 61, 1726-1751 (1973)
D.B. Ostrowsky (ed.): *Fiber and Integrated Optics* (Plenum, New York 1978)
S.D. Personick: *Optical Fiber Transmission Systems* (Plenum, New York 1981)
E. Snitzer: Cylindrical dielectric waveguide modes. J. Opt. Soc. Am. 51, 491-498 (1961)
Y. Suematsu, K.I. Iga: *Introduction to Optical Fiber Communications* (Wiley, New York 1982)

1.39 C.S. Tsai, D.Y. Zang, P. Le: Guided-wave acoustooptic Bragg diffraction in a $LiNbO_3$ channel-planar waveguide with application to optical computing. Appl. Phys. Lett. 47, 549-551 (1985)
C.M. Verber, R.P. Kenan, J.R. Busch: Correlator based on an integrated optical spatial light modulator. Appl. Opt. 20, 1626-1629 (1981)

1.40 C.S. Tsai: "$LiNbO_3$-Based Integrated-Optic Device Modules for Communication, Computing, and Signal Processing" in *1986 Conference on Lasers and Electro-Optics, Technical Digest*, IEEE Cat. No.86CH2274-9, 44-46 (1986)
C.S. Tsai: "Titanium-Indiffused Proton-Exchanged Microlens-Based Integrated Optic Bragg Modulator Modules for Optical Computing" in: *Optical and Hybrid Computing*, ed. by H.H. Szu, SPIE. 634, 409-421 (1987)
D.Y. Zang, C.S. Tsai: Titanium-indiffused proton-exchanged waveguide lenses in $LiNbO_3$ for optical information processing. Appl. Opt. 25, 2264-2271 (1986)

1.41 C.S. Tsai, D.Y. Zang, P.Le: Multichannel integrated optic device modules in $LiNbO_3$ for digital data processing. 1988 Topical Meeting on Integrated and Guided-Wave Optics (Santa Fe, NM) Techn. Digest, pp.TuA5-1 to 4
C.S. Tsai, D.Y. Zang, P. Le: High-packing density multichannel integrated-optic modules in $LiNbO_3$ for a programmable correlation of binary sequences. Opt. Lett. 14, 889-891 (1989)
C.S. Tsai: Integrated optic device modules in $LiNbO_3$ for computing and signal processing. J. Mod. Opt. 35, 965-977 (1988)

2. Bulk-Wave Acousto-Optic Bragg Diffraction

Glen Wade

With 12 Figures

> "mais la lumière diffusée dans une direction qui fasse un angle 2θ avec le rayon incident proviendra uniquement de l'onde élastique dont la direction de propagation soit la bissectrice de l'angle 2θ des duex rayons, et dont la longeur d'onde Λ satisfasse à la formule $(2\Lambda\sin\theta=\lambda/n)$."
>
> Leon Brillouin, 1922 [2.1]

Diffraction of light by a narrow beam of high-frequency sound is called Brillouin scattering after the man who predicted it in 1922. If the sound beam is wide enough and the light incident upon it is at the appropriate angle, the diffraction which then takes place is most generally referred to as Bragg diffraction in analogy to the selective reflection of X rays by the lattice planes of crystals first described by W.H. Bragg in 1913. The ultrasonic phenomenon was verified experimentally in 1932 by P. Debye and F.W. Sears in the U.S.A. and by R. Lucas and P. Biquard in France.

Prior to the invention of the laser, this phenomenon generated interest mainly in the academic community. But it did have a few practical applications such as determining the velocity of sound, modulating the intensity and phase of light, imaging acoustic fields, correlating signals on optical beams, and analyzing spectra instantaneously. The advent of coherent light and the advances in high-frequency acoustics stimulated new interest in this field as its potential for new applications was recognized. Examples of these new applications include the visualizing of internal body structures for medical diagnosis; the nondestructive testing of materials; the modulating, detecting and frequency-shifting of light; the processing of signals; the probing of acoustic fields; and the convolving, pulse compressing, high-speed multi-port beam switching, time multiplexing and demultiplexing of optical pulse trains.

This chapter presents a tutorial review of the nature of bulk-wave acousto-optic Bragg diffraction and gives a number of applications. Since many of the physical concepts and the analytical approaches of bulk-wave acousto-optic Bragg diffraction are also applicable to guided-wave Bragg diffraction, this chapter also serves as an introduction to subsequent chapters. The first section describes a simple experiment which conveys a qualitative perspective of how Bragg diffraction works. The second section briefly reviews the history of the investigation of light-sound interaction which has led up to the development of Bragg diffraction and its many applications. The third section gives a heuristic picture of what

Springer Series in Electronics and Photonics, Vol. 23

Guided-Wave Acoustooptics   Editor: C. S. Tsai

© Springer-Verlag Berlin, Heidelberg 1990

takes place in Bragg diffraction. Such a picture is helpful in understanding the nature of the phenomenon and can be used effectively as a means of arriving at correct notions and relationships.

Bragg diffraction of light from sound can be looked at from the point of view of parametric excitation, and the fourth section does just that. A closely related but independent point of view is that obtained from quantum mechanics, and the fifth section describes Bragg diffraction in terms of photon-phonon interaction. The most rigorous analysis of the phenomenon presented in this chapter comes in the sixth section on wave theory. A sample calculation of Bragg diffraction effects is given in Sect.2.7.

Up to that point, the considerations have involved only conventional Bragg diffraction in which the incident light and the scattered light show up on opposite sides of the region occupied by the sound. This is the transmission case. However, an important case of interaction exists in which the incident and the scattered light both appear on the same side of the sound region. This, of course, is the reflection case and is described in Sect.2.8.

In all of the preceding sections, the acousto-optic interaction has been assumed to occur in such a way that the diffraction process does not change the polarization of the light being diffracted. Birefringence therefore plays no part in any of the effects thus far considered. However, Bragg diffraction may occur in an anisotropic medium. When this is the case, birefringence is a factor and the phenomenon is more complicated. Section 2.9 treats birefringent Bragg diffraction.

The question of what the proper combinations of material parameters are for optimum device performance is considered next. What characteristics should a good acousto-optic material possess? This subject is treated in Sect.10.

From all of these considerations certain conclusions become apparent and are presented in Sect.2.11.

2.1 A Simple Experiment

It is easy to perform an experiment that will tell a great deal about the interaction between light and sound, and the acousto-optic phenomenon known as Bragg diffraction. Take a box with transparent sides and fill it with water. Locate an ultrasonic transducer at one end, as shown in Fig. 2.1. Place an anechoic wall at the other end. Feed electrical power to the transducer and generate longitudinal compression waves in the water. These are acoustic waves which will travel through the water to be absorbed at the opposite end of the box by the anechoic wall.

Drive the transducer at a frequency of several Megahertz. Since acoustic waves propagate in water with a velocity of about 1500 meters per second, their wavelength Λ will be 1500/f meters, f being the ultrasonic frequency. The variations in compression in the water produce a

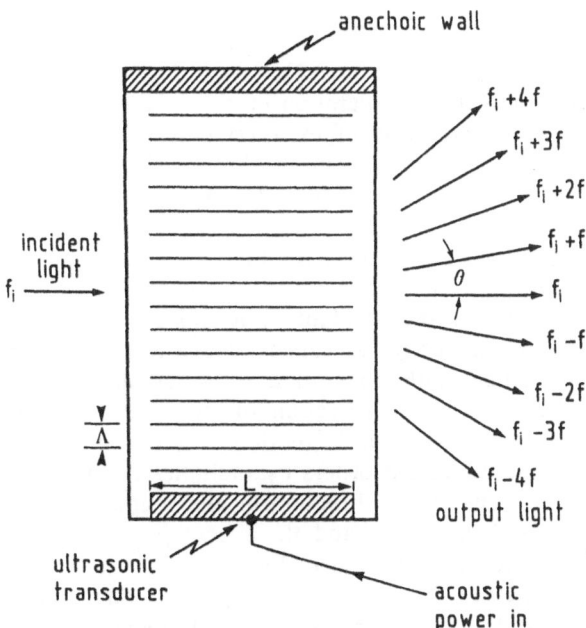

Fig.2.1. Experimental arrangement for observing Debye-Sears and Bragg diffraction of light by ultrasonic waves, where f_i is the light frequency and f is the sound frequency. For simplicity, only the gross effects are indicated. The effect of light refraction as it enters and leaves the medium of the sound propagation is not shown

corresponding variation of the optical refractive index Δn according to the Lorentz-Lorenz formula [2.2-4].

Now insert a beam of coherent light so that it passes through the sound, as shown in the figure. What effect will the sound have on the light? We would logically expect that the variation of refractive index would produce systematic scattering or diffraction. In the very simplest case where the acoustic waves form a narrow column and the light is incident upon them in a direction perpendicular to the column, it is easy to imagine that the light sees a moving "phase grating". To be sure, the phase grating in this case is a simple sinusoidal one, assuming that the conversion from electrical energy to refractive-index variation is linear, as it usually is to very good approximation when the acoustic power is low.

One substantial difference between this phase grating and an ordinary phase grating which is fabricated from solid, transparent material is that in this case the grating pattern propagates with the speed of sound. However, since the sound velocity is much, much smaller than the light velocity, we can see that the grating will not have moved very far during the travel time of the light. In fact, to a first approximation, it is reasonable to conclude that we can neglect the motion. If we do neglect it, we have precisely the case of a sinusoidal phase grating. The incident light of frequency f_i will be divided into a zero-order beam which passes straight through the medium and higher-order diffracted beams immerging at angles given by the familiar grating formula [2.5]

$$n\lambda = h\sin\theta \qquad\qquad (2.1)$$

where λ is the light wavelength; h the separation of the grating elements, θ the angle of diffraction, and n the order of the diffraction.

Let L be the width of the sound beam, Λ the wavelength of the sound, and λ the light wavelength measured in water. When the acoustic-wave column is narrow enough and the other parameters are such that the term $2\pi\lambda L/\Lambda^2$ is much, much less than unity, the diffraction is said to be in the "Raman-Nath" regime [2.6-9]. In this case a large number of diffracted orders can be produced if the incident light and the diffracting sound are of sufficiently high intensity. So-called "Debye-Sears" diffraction is a special case of Raman-Nath diffraction in which the incident light is directed perpendicularly to the direction of propagation of the acoustic waves. The fact that the sound is in motion causes a shifting of the frequencies of the various diffracted orders by mulitples of the sound frequency. Those which are bent in the direction of the sound propagation have higher frequencies than before, and those bent in the opposite direction have lower frequencies, as indicated in Fig.2.1.

When the quantity $2\pi\lambda L/\Lambda^2$ is much, much greater than unity, "Bragg" diffraction can be obtained under the right conditions. Assume that we go smoothly from the Raman-Nath regime to the Bragg regime by simply increasing the width of the acoustic beam. As long as the beam is narrow enough and $2\pi\lambda L/\Lambda^2$ is very much less than unity, we can get many diffraction orders, as indicated in Fig.2.1. If we progressively increase the width of the acoustic beam until we eventually reach the condition for which $2\pi\lambda L/\Lambda^2$ is much, much greater than unity, all but the zero-order light will eventually disappear, and we perceive no diffraction at all. When this happens, assume that we reorient the acoustic cell, rotating it slowly in the clockwise direction. If we turn it through an angle $\theta/2$ such that

$$\sin\theta/2 = \lambda/2\Lambda , \qquad\qquad (2.2)$$

we will observe that one of the diffracted orders reappears, namely the first down-directed order that we had previously obtained with the narrow acoustic beam. The frequency of this order is (f_i-f) as before. Thus two beams of light now show up, one constituting zero-order light with no frequency shift and the other, first-order downshifted light with frequency (f_i-f).

If we now slowly rotate the acoustic cell in the counterclockwise direction going back through its original position and beyond, we will at first see the diffracted order diminish in intensity and disapper. As we continue the rotation we will eventually observe the formation of a new diffraction order, this one pointing upwards with precisely the same orientation as the first up-shifted order in the narrow-beam experiment. The angle through which the cell has been rotated from its original position is again given by $\theta/2$ where $\sin\theta/2 = \lambda/2\Lambda$. The angle of diffraction for both the upshifted and downshifted orders is θ, and the diffraction is called Bragg diffraction.

These are simple experiments involving bulk waves. The consequences and ramifications of such experiments are what we will be discussing in this chapter. Note that here we are dealing with monochromatic, unidirectional (collimated) incident light and an acoustic beam consisting of plane waves at a single temporal frequency.

2.2 Historical Perspective

The diffraction of light by ultrasound is frequently referred to as Brillouin scattering after the man who first predicted the effect in 1922 [2.1]. However, it is interesting to note that one facet of the basic phenomenon, namely light diffraction by a phase grating, had been investigated and reported by *Lord Rayleigh* before the turn of the century [2.10]. Another important aspect of the analytical treatment was pointed out by *W.H. Bragg* [2.11] in 1915. *Bragg* showed that a single component of the diffracted waves is produced by a single plane wave or Fourier component of refractive-index variation. This method of Fourier analysis was extended and brought into practical use by *Duane* [2.12] in 1925. The approach was originally employed for X-ray diffraction from 3-dimensional crystals, but it is of great utility and provides excellent insight in the case of bulk diffraction of light by ultrasound.

As indicated in the title of this chapter, the particular type of diffraction we are concerned with here is generally referred to as Bragg diffraction. This is because it is greatly reminiscent of the above-mentioned selective diffraction of X rays from crystals, a phenomenon first described by *Bragg* in 1913 [2.13]. However, as opposed to X-ray diffraction, there is only one critical angle (the Bragg angle) in the acoustic case. When X rays are incident upon a distribution of atoms in a crystal, the spatial variation of the structure encountered by the X rays is not at all sinusoidal. But the acoustically-induced density variation in a transparent liquid or crystal is very nearly sinusoidal, and a sinusoidal variation gives rise to a single critical angle.

After Brillouin had predicted that light could be diffracted from sound, direct experimental verification was slow in coming. It was ten years before *Debye* and *Sears* [2.14] at MIT, and *Lucas* and *Biquard* [2.15] in France performed experiments which decisively confirmed the existence of the effect.

Debye and *Sears* directed a beam of light upon a column of sound generated by a quartz transducer. They then rotated the column continuously in one direction in order to observe the variation in the pattern of the transmitted light. Their experiment was similar to the one of Sect.2.1. In their paper they stated that, with normal incidence and "under favorable conditions as to the intensity of the (acoustic) vibration, more than 10 spectra to the right and to the left (of the zero-order light) have been obtained."

In an interesting paragraph of their paper, they showed that by measuring θ, the Bragg angle, and knowing λ, the wavelength of the light being used, they could calculate Λ, the wavelength of the sound. They had only to multiply the calculated wavelength by the frequency of the sound to obtain the sound velocity. They pointed out that this approach constitutes a "very simple method for the determination of the velocity of sound." They calculated the acoustic velocity at 1.7 and 16.5 MHz in both toluene and carbon tetrachloride and, observing no substantial differences at the two frequencies, concluded that "up to now no indication of a change in velocity with frequency has been observed."

As *Debye* and *Sears* correctly observed, it is easy to measure the phase velocity of sound in this fashion and the method has had frequent use ever since. The sound is generated by an oscillator of known frequency, and the wavelength is determined from Bragg's law so that the velocity can be found. The sound intensity can easily be great enough so that, with normal incidence of the light upon the column of sound, a large number of diffraction orders can be observed and the Bragg angle can easily be measured.

Although direct confirmation of Brillouin's predictions was first made by *Debye* and *Sears*, and by *Lucas* and *Biquard* in 1932, indirect confirmation had been obtained by *Gross* a couple of years earlier [2.16]. Thermal sound, naturally existing in a liquid, can diffract light just as readily as the artificially-generated sound of *Debye* and *Sears*, and of *Lucas* and *Biquard*. The molecules of any substance are in thermal motion and such motion can be looked upon as being caused by thermal sound waves. These waves have all possible propagation directions and a very large spectral range of frequencies. Any sound wave produces a variation of density, and therefore a variation of refractive index, and is capable of diffracting light. By shining monochromatic light into a sound cell, Brillouin scattering of that light can be observed even though the cell is not equipped with an electro-acoustic transducer for generating sound. The light will encounter components in the thermally-produced sound for which the Bragg condition is met and diffraction will result. The various orders of the diffracted light will be shifted in frequency as previously described. By examining the spectrum of the scattered light, the fact that Brillouin scattering has taken place can be verified. In 1930 *Gross*, at the Optical Institute of Leningrad, reported that he had been able to photograph several spectral components of light created by this scattering process [2.16]. A short time later *Meyer* and *Ramon* [2.17] published photographs of the effect.

It is interesting to note that two decades before *Debye* and *Sears* had directly confirmed Brillouin's predictions with their experiment, *Debye* wrote a paper [2.18] that contributed to an understanding of *Gross'* indirect confirmation. In the paper *Debye* postulated the existence of thermal sound to explain the variation of the specific heat of solids at low temperature. Eighteen years later when *Gross* saw spectral components due to Brillouin-scattered light, he correctly regarded them "as (being) a demonstration of the reality of Debye's waves."

Brillouin, in his original paper, assumed interaction between a sinusoidal plane wave of light and a sinusoidal plane wave of sound. He predicted that diffraction would occur at only one angle, the Bragg angle. In the first experiments of *Debye* and *Sears*, and of *Lucas* and *Biquard*, many orders of diffracted light were generated. This suggested the existence of a number of critical angles, not just one. *Debye* and *Sears*, observing the higher orders, noted that "the theory (predicted) only the first-order spectrum to the right and the first-order to the left." They then stated that there were higher harmonics in the sound and that "this departure from sinusoidal character... (accounted) for the existence of the higher-order spectra in the light."

Lucas and *Biquard* had a somewhat different explanation. Using ray optics, they calculated trajectories showing that the distribution of the light leaving the sound was spacially periodic. They concluded that this would result in an effect similar to that caused by a grating. Since many orders of diffracted light are observed from a grating, they felt that one should expect the same for light diffracted from an acoustic beam.

Brillouin took a second look at acousto-optic interaction [2.19] and pointed out that higher orders occurred only when the acoustic intensity was high. His original theory (predicting only one diffraction order) was a small-signal or weak-scattering theory and did not apply in the case of intense sound. In his second article, he interpreted the existence of the multiple orders as being due to multiple scattering of the diffracted light. He did not work out the large-signal or strong-scattering analysis in any detail, but he did show that the solution could be represented by a Mathieu equation and that the light leaving the cell could be expected to contain the many orders observed in the experiments.

Two years later, *Raman* and *Nath* in India, having approached the problem analytically from a grating point of view, published theoretical work starting from the analysis of *Lord Rayleigh* already referred to [2.10]. Their theory correctly predicted the temporal modulation of the light emerging from the sound, demonstrating that the transmitted light could be regarded as a phase-modulated carrier with sidebands spaced by f, the sound frequency. They showed that the carrier and sideband amplitudes were functions of the maximum phase excursion and had a Bessel-function dependence on the intensity of the sound.

Raman and *Nath* reported their results in a series of five articles published in quick succession [2.20-24]. In the first, they assumed the light to be incident perpendicularly on the sound column and to emerge from it with a wave-front of corrugated form. In subsequent articles they treated the case of oblique incidence and took into account the time variation of the refractive index which produces the frequency shift in the diffracted light. They derived an infinite set of coupled differential equations giving the spatial behavior of the various diffracted orders traversing the sound beam.

Raman and *Nath* were followed by many other researchers who used widely differing methods to derive somewhat similar sets of equations [2.7, 25-31]. Most of the approaches were more complicated than the

17

Raman-Nath formulation, but they all gave more or less similar results. None of them was specifically designed to investigate the interaction of arbitrary sound and light fields. Nearly all of them used interaction models in which the sound consisted of a single plane wave of finite width, and the light, a single plane wave of infinite width.

In all of the cases considered by the early workers, the interaction between the light and the sound was assumed to occur either in an isotropic medium or in an anisotropic medium in such a way that the diffraction process did not change the polarization of the light being diffracted, and hence the birefringence of the medium did not play a part. The first person to publish an analysis which was general enough to include all diffraction processes occuring in optically anisotropic media was *Dixon* [2.32]. He showed that under some conditions the effects of birefringence are so significant that normal Bragg conditions are not even approximately satisfied. Several interesting phenomena take place in anisotropic material which have no correspondence to those occurring in isotropic materials.

There have been a number of practical applications of ultrasonic light diffraction. One of the first was that of *Debye* and *Sears* already referred to. As we have seen, they measured the diffraction angle of the different orders with respect to the undiffracted light. Knowing the wavelength λ of the light, they obtained the sound wavelength Λ from (2.2). The sound frequency was determined "with an ordinary wavemeter". By multiplying the frequency by the wavelength, they obtained "a very simple method for the determination of the velocity of sound."

Another practical application was demonstrated a short time later when the Debye-Sears effect produced intensity modulation of light. The principle is easy to understand. Imagine that in Fig.2.1 the light emerging from the acoustic cell is gathered by a converging lens and, in the absence of any sound, is focused onto a small intercepting mask. When sound is introduced into the liquid, part of the light is diffracted and does not get intercepted by the mask. The optical intensity of the diffracted light is a function of the acoustic power.

Thus sound can intensity-modulate light. Because sound changes the optical refractive index of the medium through which it passes, it can also phase-modulate light (for example, in the case of a narrow light beam propagating for a short distance along a path parallel to well-separated sound wavefronts [2.29]). As we have seen, sound can deflect light. It can also bend light rays and therefore focus them [2.33, 34]. Sound can shift the frequency of light and change the way it propagates [2.35]. Sound can obviously be used in a variety of ways to process light. *Okolicsanyi* [2.36] was one of the pioneers in the 1930's to suggest a number of uses for sound-cells including employment as a unique type of light modulator designed specifically for use in the Scophony television projector [2.36, 37]. Ultrasonic light diffraction was later used in Schlieren-type images of acoustic fields, optical correlators, and instantaneous spectrum analyzers.

The emergence of these applications predated the advent of lasers with their large quantities of coherent light. The availability of lasers stimulated new interest in the field and gave rise to many new applications which had been impossible or impractical to implement, including optical-beam control and variable delay. Lasers were used in experiments to generate sound [2.38] and to investigate thermal-phonon processes. They were employed to probe acoustic fields to determine their far-field diffraction patterns [2.39] and to prodvide a mechanism for ultrasonic imaging [2.40-43]. Sound cells were used to shift the frequency of lasers and to deflect their beams for purposes of signal processing and display. Excellent discussions of a number of acousto-optic devices and applications have been given in [2.44].

In many of the early devices the interaction was strictly with bulk waves, but in more recent devices it has been with surface waves. Bragg diffraction of guided optical waves can readily be obtained with surface acoustic waves. This makes it possible to design and fabricate very wide-band guided-wave thin-film integrated chips for optical processing [2.45]. The development portends a wide range of uses for devices designed to meet specific needs. The processes can be made to be linear, and this fact yields one of the most attractive features of such processors - the ability to handle and separately identify simultaneous signals in dense environments. Using these techniques, spectrum analyzers, convolvers, and pulse compressors have so far been built. The applications include high-speed multi-port beam switches and deflectors for systems employing optical fibers, spectrum analyzers and other kinds of processors for wide-band RF signals, high-speed optical pulse modulators, and time multiplexors and demultiplexors of optical pulse trains having high data-rate capability. This approach to spectrum analysis not only gives small size and cost, but provides excellent probability of detection, high sensitivity for both cw and pulsed signals, wide-band operation without scanning, good frequency resolution, coherent conversion, and ability to sort simultaneous signals whether cw or pulsed. A substantial portion of the remainder of this book including its Chaps.5-8, will deal with devices and applications of this type of guided-wave acousto-optics.

2.3 Heuristic Picture

Before considering rigorous ways of looking at Bragg diffraction, we will examine an uncomplicated picture of what is going on. The picture is easy to describe and sketch and will give a heuristic view that is helpful in understanding the nature of the phenomenon. The picture can be used effectively as a means of arriving at correct notions and relationships. It is easily remembered and provides a simple way of thinking about Bragg diffraction so that many of the general properties become evident. Directions of propagation and frequency shifts in Bragg diffraction are indicated in Fig.2.2; the uncomplicated heuristic picture, in Fig.2.3.

The diffraction process generates a new beam of light shifted in both frequency and direction from those of the incident beam. Equation (2.2)

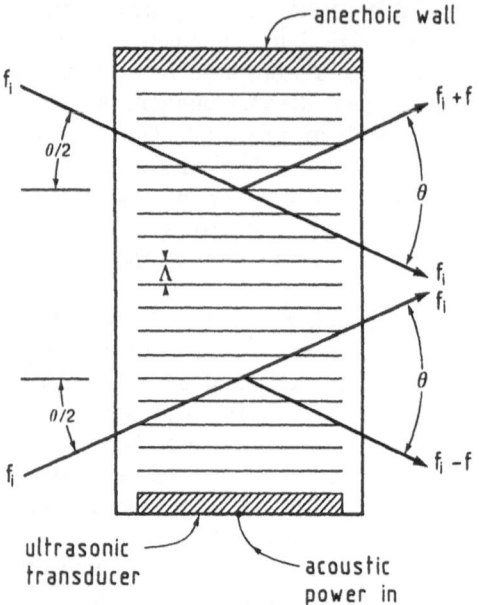

Fig.2.2. Upshifted and downshifted Bragg diffraction of light from sound. Angles and frequencies associated with the various beams are indicated. As in Fig.2.1, the bending effects due to light traversing media with different refractive indices are ignored

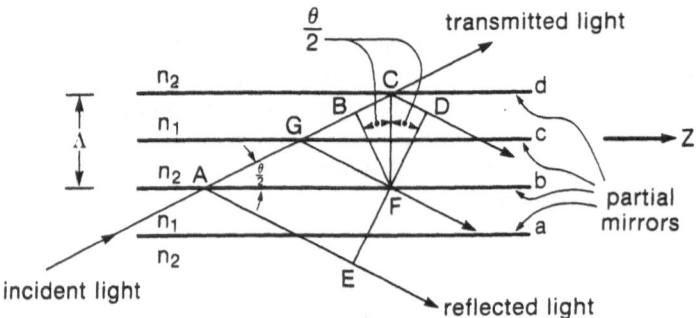

Fig.2.3. Simple picture depicting geometrical relationships for constructive reinforcement of light reflected from a stack of partial mirrors. The mirrors stem from an assumed abrupt variation in refractive index due to the sound. For simplicity, only four partial mirrors (a,b,c and d) and only three downshifted reflected rays (CD, GF and AE) are shown. Refractive indices n_1 and n_2 are indicated

gives the condition that must hold for this to happen: the angle between the direction of the light propagation and the sound wave fronts is just the arcsine of one-half the ratio of the light wavelength to the sound wavelength. This condition is true for both the incident and the generated beams. The reason is seen from the following heuristic argument based on the picture in Fig.2.3. As the light enters the medium traversed by the acoustic waves, it encounters a moving variation of refractive index. The actual variation is smooth and of sinusoidal form but for our simple picture we assume it to be abrupt and of rectangular form. The overall effect is as if the light were passing through a stack of uniform, transparent dielectric slabs, each a half sound wavelength thick and moving upward in Fig.2.2 at the sound velocity. For such a situation the light would experience a series of partial reflections at the periodically spaced planes where the abrupt changes take place. The refractive index throughout each slab is uniform, but the light in going from one slab to an adjacent one would encounter a slight change in refractive index. If the slabs were numbered consecutively from bottom to top, the even-numbered slabs would all have the same refractive index. So would the odd-numbered slabs, but their refractive index would differ by a small amount from that of the even-numbered slabs. An incident beam of light in passing through slab after slab would be slightly reflected from each interface. The stack of slabs can therefore be thought of as a stack of partial mirrors spaced one-half an acoustic wavelength apart. All the small reflections are in the same direction. If the conditions are such that the small reflections all add constructively (i.e., in phase) in the direction of their travel, there will be a cumulative effect resulting in the generation of a new beam of light, the Bragg-diffracted beam.

Figure 2.3 shows the correct geometry for this situation. The line BF is perpendicular to the ray representing the incident light and the line DE is perpendicular to the three rays representing the reflected light. BF and DE are therefore loci of points of constant phase for the incident and reflected beams respectively. Let us assume that $n_2 > n_1$ where n_1 and n_2 are the two refractive indices involved. If this is so, waves reflected from the mirrors at b and d will experience no change of phase due to reflection, but waves reflected from the mirror at c will be changed by 180°.[1] We can derive the Bragg condition very easily by noting that the angle of reflection is equal to the angle of incidence and that for the ray along AE to reinforce the ray along CD, we have

$$\overline{AB} + \overline{BC} + \overline{CD} = N\lambda + \overline{AE} \qquad (2.3)$$

where N is an integer. Since $\overline{AB} = \overline{AE}$, Eq.(2.3) holds when

$$\overline{BC} + \overline{CD} = 2\Lambda\sin(\theta/2) = N\lambda . \qquad (2.4)$$

Except for the factor N, the above expression is the same as (2.2). For the ray along FG to reinforce the ray along CD (and hence the ray along AE), we have

---

[1] Here the phase referred to is that of the H field. The stated conditions apply when H lies in the plane of incidence. If the fields are polarized so that E is in the plane of incidence and if the angle of incidence (90°-θ/2) is greater than Brewster's angle, the H-field phases are reversed. The phases are also reversed if the E field is used as the reference instead of the H field. These matters can alter the derivation slightly but the end results are the same as far as Bragg diffraction is concerned.

$$\overline{GB} + \overline{BC} + \overline{CD} = M\lambda + \overline{GF} - \lambda/2 \tag{2.5}$$

where M is an integer. The last term on the right takes into account the 180^0 change in phase resulting from the reflection. Since $\overline{GB} + \overline{BC} = \overline{GF}$, the equation holds when

$$\overline{CD} = M\lambda - \lambda/2 = \Lambda\sin(\theta/2) \tag{2.6}$$

which again is related to, but not quite the same as, (2.2). By equating (2.4 and 6) we can see that N = 2M-1. The rays along GF, AE and CD will all reinforce each other if N is odd. The heuristic picture is not accurate enough to specify the Bragg condition exactly, with its single critical angle. To show that N can only be unity we would have to consider a sinusoidal variation in refractive index, not the rectangular one assumed here.

For equally-phased rays such as those through points D and E, any integral multiple N produces constructive reinforcement. When we add oppositely-phased rays like the one through point F, it becomes necessary to exclude even multiples. If instead of $\Lambda/2$ slabs we had used much finer slabs to obtain a better approximation of a sinusoidal sound wave, more and more of the multiples would cancel; finally, with a purely sinusoidal distributon of n, only the case of N = 1 would remain.

If it is true that a sinusoidal wave in n corresponds to one and only one possible angle at which light of a given wavelength can be diffracted, then it is hardly surprising that a rectangular wave in n, which can be regarded as a superposition of odd harmonic sinusoidal components, should be capable of diffracting light at the angles appropriate to each component. In practice, of course, rectangular or even approximately rectangular sound waves are almost impossible to generate, because of the enormous transducer band width that would be needed; sinusoidal or nearly sinusoidal sound waves are much more common. For sinusoidal waves, if the Bragg condition does not hold, there will be destructive interference between the scattered light components and a Bragg-diffracted beam will not be generated.

Equation (2.2) can be represented nicely by a diagram of wave vectors. This is apparent from the following

$$\sin\frac{\theta}{2} = \frac{\lambda}{2\Lambda} = \frac{(2\pi/\Lambda)/2}{2\pi/\lambda} = \frac{K/2}{k} \tag{2.7}$$

where K denotes the magnitude of the wave vector for the sound, and k the magnitude of the wave vector for both the incident and the reflected light. We regard as negligible the tiny changes in light wave number associated with the slight variation in refractive index as the light goes from slab to slab and also the changes due to upshifts and downshifts in frequency since $f \ll f_i$. For the purposes of this diagram we can safely assume $|k_i| = |k_-| = |k|$. If we place end-to-end the wave vectors associated with the light and sound of Fig.2.3, we obtain the upper diagram of

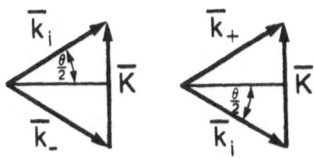

Fig.2.4. Wave-vector diagrams for downshifted Bragg diffraction (*upper diagram*) and upshifted Bragg diffraction (*lower diagram*). In each case k_i is the wave vector for the incident light and K, for the sound. k_- and k_+ are the wave vectors for the downshifted and upshifted light, respectively[*]

Fig.2.4. The resulting configuration is the trigonometric equivalent of (2.2) and is valid for downshifted Bragg diffraction. The lower diagram of Fig.2.4 shows the corresponding configuration for the upshifted case.

All along we have assumed that the angle of reflection equals the angle of incidence. This seems intuitively obvious so long as we have sharp boundaries between slabs of different n from which reflection take place. But for the case of a sinusoidal variation of n where sharp boundaries do not exist, does the reflection symmetry condition still hold? The wave vector diagrams of Fig.2.4 suggest that it should: there is no way we can draw these diagrams with two k vectors of equal length, except as isosceles triangles with the vector K as their base. But can we show with the help of Fig.2.3 that this must be so?

The single incident ray (Fig.2.3) intercepts the plane boundaries between slabs at specific points, such as A, G, C. Each of these points must now be assumed to diffract light in all directions, and for any chosen direction of the incident ray there exists some other direction - generally not symmetrical - where constructive interference occurs for the diffracted light.

But a single ray is a mathematical abstraction. Actually, the incident beam must be many wavelengths wide, i.e. made up of many parallel rays, if it is not to spread rapidly by purely optical diffraction. It can be shown (by trigonometry) that constructive interference between the components diffracted from many parallel incident rays is not possible unless the angles of incidence and exit are equal [Ref.2.46, Fig.8]. This is the physical counterpart of the geometrical observation that the wave vector triangle is always isosceles.

So much for the wave numbers. Now consider the frequencies. The upshifts and downshifts are due entirely to the Doppler effect. The partial mirrors are all in motion and their movement causes a change in the frequency of the reflected waves. To quantify the change, consider again the simple picture of Fig.2.3. A Doppler shift is produced when the length of the path traversed by a beam of waves is extended or shortened as the waves are being reflected. From Fig.2.3, if we imagine the partial mirrors to be moving upward, we can see that the path length of the waves for the portion of the incident beam that gets reflected is increasing while that for the portion not reflected is unchanged. Hence there will be a

---

[*] In the figures of this chapter vector quantities are indicated by overscores.

Doppler shift for the reflected beam but not for the incident beam. Consider the wave fronts passing through the line BF of Fig.2.3, being reflected at the moving mirror d, and then passing through the line DF. Let the phase of the incident wave traversing BF at a particular instant of time be

$$\phi_{BF} = 2\pi f_i t .$$ (2.8)

At that same instant the phase of the reflected wave passing through DF is somewhat smaller having been generated at an earlier time. It is given by

$$\phi_{DF} = 2\pi f_i (t-\tau)$$ (2.9)

where τ is the travel time for the waves going from BF to DF. Assuming the mirror motion is negligible during the wave travel time, we have for the instant of time depicted in the figure

$$\tau = \frac{\overline{BC} + \overline{CD}}{c} = \frac{2}{c} \overline{CF} \sin \frac{\theta}{2}$$ (2.10)

where c is the light velocity. Therefore at that instant

$$\phi_{DF} = 2\pi f_i \left[t - \frac{2}{c} \overline{CF} \sin \frac{\theta}{2} \right] .$$ (2.11a)

The frequency associated with any time-varying phase is given by $(2\pi)^{-1} d\phi/dt$. Hence the frequency for the incident wave as it crosses BF is f_i and for the reflected wave as it crosses DF is

$$f_{DF} = \frac{1}{2\pi} \frac{d}{dt}(\phi_{DF}) = \frac{1}{2\pi} \frac{d}{dt} \left[2\pi f_i \left[t - \frac{2}{c} \overline{CF} \sin \frac{\theta}{2} \right] \right]$$

$$= f_i \left[1 - \frac{2}{c} \frac{d}{dt}(\overline{CF}) \sin \frac{\theta}{2} \right] .$$ (2.11b)

To take the derivative we must be careful to properly interpret $\overline{CF}$. The point F identifies the crossing point of the two lines fixed in space, at one of which the phase of the incident wave is ϕ_{BF} and at the other, the phase of the reflected wave is ϕ_{DF}. F falls on moving mirror b only at the time instant depicted in the figure. At any later time the mirror will have advanced vertically but F remains motionless at the crossing point of the two fixed lines. C, on the other hand, must be regarded as moving since it is a point of reflection from d. With these definitions it is easy to see that $\overline{CF}$ is not constant with time. C moves away from F with the velocity of mirror d and therefore $d(\overline{CF})/dt = V$. Thus

24

$$f_{DF} = f_i - \frac{2f_i}{c} V \sin \frac{\theta}{2} \qquad (2.12)$$

where V is the velocity of the sound. Using (2.2) we obtain

$$f_{DF} = f_i - \frac{2}{\lambda} V \frac{\lambda}{2\Lambda} = f_i - f \qquad (2.13)$$

where f is the sound frequency and equal to V/Λ. Thus the downshifted light has the frequency $f_i - f$. By similar arguments we can easily show that the upshifted light has the frequency $f_i + f$. In both cases the change in frequency Δf_i is conveniently given by

$$\Delta f_i = \frac{2V \sin(\theta/2)}{c} f_i = f . \qquad (2.14)$$

The above discussion has been more mathematical than heuristic. However, the arguments can be cast into strictly heuristic terms.[2] Consider the moving partial mirror b in Fig.2.3. At the instant shown, it directs reflected light toward point E. One sound period later, it will have moved to position d and will direct reflected light to point D. We found previously that the light path to D is one light wavelength longer than the light path to E. We conclude: every time the column of mirrors (the sound wave) advances by one wavelength of sound, the light path associated with each mirror changes by one wavelength of light. If the sound frequency is f, each mirror moves through f acoustic wavelengths during each second. The light path, therefore, changes by f optical wavelengths per second, and as a result the optical frequency shifts by f Hz - up or down, depending on whether the light path associated with each mirror is getting shorter or longer.

One final bit of insight is obtainable from our heuristic picture. We derived the Bragg condition (2.2) and the frequency relationship (2.13) from simple geometric considerations. As we have seen from the previous discussion, this same simple geometry also shows that the incident beam becomes phase modulated. We can therefore invoke the principles of phase modulation and find out other characteristics of the interaction [2.46]. Doing this will tell us, for example, how much diffraction can take place. We so far do not have a functional relationship for the change of amplitude of the incident and downshifted beams or the transfer of power between them. To obtain such relationships we can observe from the heuristic picture of Fig.2.3 that another way to explain what is going on is to regard the dielectric slabs not as mirrors but as phase modulators of the incident and downshifted beams. In this way the slabs link the two beams togehter and produce a system of coupled waves. A close examina-

---

[2] The approach that follows was supplied by R. Adler who pointed out that the principle involved has broad application and fits general situations in which one wave is used to phase-modulate another.

tion of the figure indicates that Bragg diffraction contains all the necessary elements for coupling between the waves. Coupling can take place if a coupling mechanism exists (in this case, phase modulation due to the dielectric slabs), if the phase velocity of the waves is the same in the coupling direction (in this case, the direction associated with the modulation, that is, along the length of the slabs) and if the coupling is distributed over a region in space. All of these elements are present in Bragg diffraction and therefore we can treat it from a coupled-wave point of view.

If the variation of refractive index is small enough, the phase modulation caused by the slabs (Fig.2.3) will be very minute and we can use a simple first-order theory. Such a theory holds when the coupling between waves is so weak that we are justified in retaining only linear terms in the equations of motion and we therefore end up with linear equations. This is true essentially when the coupling changes the amplitude of both of the waves only slightly over the distance of one wavelength.

The dielectric slabs that produce the phase modulation (Fig.2.3) are moving. This means that the incident light as it travels from A to G is subject to periodic time variations in refractive index (the alternative point of view to the previously assumed periodic reflections from partial mirrors). The difference in phase between the light at G and the light at A is just

$$\psi = k_0 n \overline{AG} \quad \text{[radians]} \tag{2.15}$$

where k_0 is the vacuum propagation constant of the light, and n the refractive index. If, as the slabs move by, the total variation in the index is a small amount Δn, the light at G will experience a variation of phase by a small amount $\Delta\psi$ such that

$$\Delta\psi = k_0 \overline{AG} \, \Delta n = k_0 n \frac{\Delta n}{n} \overline{AG} \, . \tag{2.16}$$

Thus the wave of light passing through G is phase modulated. According to phase-modulation theory, this light can be regarded as consisting of a carrier and two side bands (provided $\Delta\psi$ is small in keeping with the weak-coupling assumption) [2.47,48]. The lower side band will have the frequency $f_i - f$, where f is the frequency of the periodic time variation in the refractive index. We can therefore identify this side band with the downshifted light of (2.13) which we discovered from examining the model based on partial mirrors. From the results of the previous considerations, we can expect this sideband beam to grow at the expense of the incident beam because of its favorable angular orientation. An upper side band with frequency $f_i + f$ is also generated, but it does not have favorable orientation for growth (if the slabs, and hence the partial mirrors, are moving upward in Fig.2.3). The upper sideband beam therefore will not grow. (If the motion is downward, it is the upper side band that grows, and not the lower). Phase-modulation theory [2.47,48] shows that the complex amplitude of the lower side band will be given by

26

$$\Delta U_-(z) = j\frac{1}{2}\Delta\psi U_i(z) = j\frac{1}{2}k_-\frac{\Delta n}{n}\overline{AG}\,U_i(z) \tag{2.17}$$

where z is the horizontal axis in Fig.2.3, with the left-to-right direction considered positive; $U_i(z)$ is the complex amplitude of the incident wave at G; and k_- the magnitude of the propagation constant of the down-shifted light, represents nk_0 from the previous equation. Let Δz be the projection of $\overline{AG}$ onto the axis. Then $\overline{AG} = \Delta z\sec(\theta/2)$, and we write

$$\frac{\Delta U_-(z)}{\Delta z} = j\frac{1}{2}k_-\frac{\Delta n}{n}\sec\frac{\theta}{2}U_i(z) \tag{2.18}$$

where $\Delta U_-(z)/\Delta z$ can be regarded as the rate of change with distance in the z direction of the complex amplitude of the lower side band. The variation in refractive index that causes $U_-(z)$ to be generated is abrupt and rectangular in this heuristic picture. The variation in the actual situation is smooth and sinusoidal. We will replace the above $\Delta U_-(z)/\Delta z$, a form quite in keeping with abruptness, with $dU_-(z)/dz$, a form more characteristic of smoothness. Thus, under the assumption that the two quantities are approximately equal, we obtain

$$\frac{dU_-(z)}{dz} = j\frac{1}{2}k_-\frac{\Delta n}{n}\sec\frac{\theta}{2}U_i(z)\ . \tag{2.19}$$

Equation (2.19) gives the rate of growth with distance of the downshifted side band. As it starts to grow, we might expect its increase in amplitude to be at the expense of a decrease in the amplitude of the incident wave. (Strictly speaking, energy conservation between the two light waves does not hold exactly because the acoustic wave is present and can be regarded as an energy sink or source depending upon whether the diffracted wave is downshifted or upshifted. This is a subtle point and is more easily explainded, as is done later, from a quantum-mechanical point-of-view). By referring to Fig.2.3, we can see that the mechanism resulting in generating $U_-(z)$ because $U_i(z)$ is phase modulated, exists also in the reverse direction. $U_-(z)$, as well as $U_i(z)$, is modulated by the moving slabs. Modulating $U_-(z)$ produces two side bands, the up-shifted one having the same frequency as $U_i(z)$. The partial mirror concept shows that reflections of $U_-(z)$ will be upward in Fig.2.3 in precisely the direction of propagation of $U_i(z)$. Using arguments similar to the ones already employed, we can show that these upward reflections initially will have the correct direction, frequency and phase to subtract from the incident wave. This is what causes the decrease in its amplitude. We have already seen that the downward reflections of the original incident wave from the partial mirrors initially have the right direction, frequency and phase to add to the downshifted wave. The nature of the coupling is the same in both directions and we can write an equation similar to (2.19) as a consequence of that fact. Thus

$$\frac{dU_i(z)}{dz} = j\frac{1}{2}k_i \frac{\Delta n}{n}\sec\frac{\theta}{2}\, U_-(z)\ . \tag{2.20}$$

Since k_i and k_- refer to the magnitudes of the propagation constants and are very nearly equal to each other, we can usually replace both by k.

Equations (2.19,20) constitute a pair of linear, first-order, coupled equations for the complex amplitudes of the traveling waves. Their solutions are simple sinusoids and indicate typical coupled-wave behavior:

$$U_i(z) = U_i(0)\cos\left[\frac{k}{2}\frac{\Delta n}{n}\sec\frac{\theta}{2}\,z\right]\ , \tag{2.21}$$

$$U_-(z) = j\sqrt{\frac{f_i-f}{f_i}}\; U_i(0)\sin\left[\frac{k}{2}\frac{\Delta n}{n}\sec\frac{\theta}{2}\,z\right]\ , \tag{2.22}$$

where we have replaced $(k_i k_-)^{1/2}$ by k in the argument of the sinusoidal terms in (2.21,22) and $(k_-/k_i)^{1/2}$ by $((f_i-f)/f_i)^{1/2}$ as a factor in (2.22). To these complex amplitudes we append the exponentials that specify the wave propagation. For the incident wave, we have

$$U_i(0)\cos\left[\frac{k}{2}\frac{\Delta n}{n}\sec\frac{\theta}{2}\,z\right] e^{-j(2\pi f_i t\, -\, \mathbf{k}_i\cdot\mathbf{r})} \tag{2.23}$$

and for the downshifted wave,

$$j\sqrt{\frac{f_i-f}{f_i}}\; U_i(0)\sin\left[\frac{k}{2}\frac{\Delta n}{n}\sec\frac{\theta}{2}\,z\right] e^{-j[2\pi(f_i-f)t\, -\, (\mathbf{k}_i-\mathbf{K})\cdot\mathbf{r}]}\ . \tag{2.24}$$

The power in the two waves is as follows

$$P_i(z) = \frac{U_i(z)\,U_i^{*}(z)}{2} = \frac{U_i^2(0)}{2}\cos^2\left[\frac{k}{2}\frac{\Delta n}{n}\sec\frac{\theta}{2}\,z\right] \tag{2.25}$$

and

$$P_-(z) = \frac{U_-(z)\,U_-^{*}(z)}{2} = \frac{U_i^2(0)}{2}\frac{f_i-f}{f_i}\sin^2\left[\frac{k}{2}\frac{\Delta n}{n}\sec\frac{\theta}{2}\,z\right]\ . \tag{2.26}$$

A plot of $U_i(z)$, $U_-(z)$, $P_i(z)$ and $P_-(z)$ is shown in Fig.2.5. At $(k/2)(\Delta n/n)\sec(\theta/2)z = \pi/2$, all the power, $U_i^2(0)/2$, initially brought into the system by the incident beam has been transfered out of it, $100(f_i-f)/f_i$ percent of the power showing up in the diffracted beam. Consideration of energy conservation suggests that the remaining power,

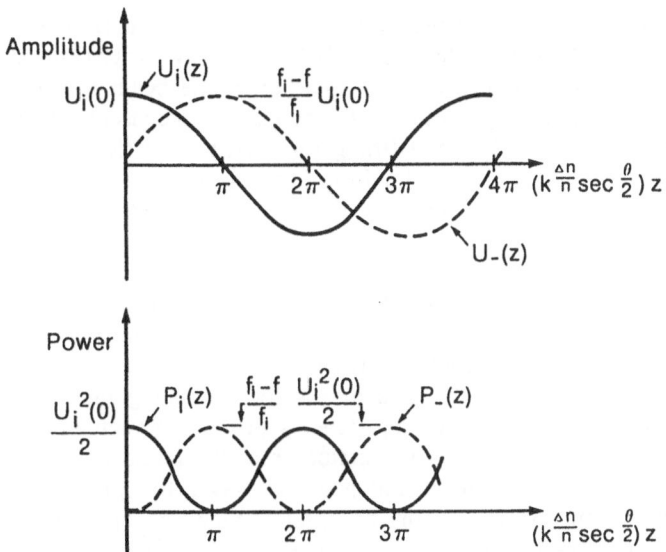

Fig.2.5. Plots of amplitude and power of the indicent and downshifted beams of Fig.2.3

or $100(f/f_i)$ percent of the initial power in the incident beam (an extremely minute quantity), has been transfered to the acoustic beam. As the beams continue to travel beyond that point to higher values of z, the energy transfer takes place in the opposite direction. At $(k/2)(\Delta n/n)\cdot \sec(\theta/2)z = \pi$, all the initial power is back in the incident beam and its amplitude is the same as at z = 0, although it has opposite phase. The two beams keep trading energy back and forth in a sinusoidal fashion, and with no loss in the system (a condition we have assumed), the transfer will continue unabated throughout the width of the sound beam.

Upshifted diffraction is equally well illustrated by what we have just considered since operation in the region $\pi/2 < (k/2)(\Delta n/n)\cdot (\sec\theta/2)z < \pi$ is in that category. The energy being transferred goes from an incident beam at a lower frequency to a diffracted beam at a higher frequency (by f) and in the transfer the sound acts as a slight source of energy instead of a slight sink.

This completes the heuristic picture. It is important to emphasize that the results
arrived at have not been obtained with rigor, but the approach provides valuable insight and, as we shall see when we perform a more rigorous analysis, it yields correct information.

2.4 Parametric Excitation and Bragg Diffraction

From heuristic considerations we have divised several fundamental relationships, including one for the frequency of the scattered sound (2.13) and another for its propagation-constant vector (Fig. 2.4). The frequency of the incident wave is f, and of the Bragg-diffracted wave, (f±F), where the plus or minus sign is chosen according to whether the beam has been upshifted (+) or downshifted (-). Similarly the propagation-constant vector of the incident wave is k_i, and of the Bragg-diffracted wave $(k_i \pm K)$, the sign again depending upon whether the beam has been upshifted (+) or downshifted (-). These two expressions are familiar in parametric interaction and are called the Tien relations [2.49, 50].

Consider briefly the question of parametric interaction with an input wave of frequency f_i and wave vector k_i. Theory tells us that under the proper conditions such interaction can be produced if a distributed reactive element of the medium in which the wave is propagating is caused to vary sinusoidally at a set frequency. The means for producing the variation is called a pump. The variation propagates with a wave vector called the pump wave vector. In the Bragg-diffraction case, the reactive element is the refractive index; the pump is the acoustic wave; the pump frequency f, and the pump wave vector K. Under conditions present in Bragg diffraction, parametric interaction does take place. It produces a new wave with frequency $(f_i \pm f)$ and wave vector $(k_i \pm K)$. Bragg diffraction of light from sound can be looked at simply as parametric interaction between the light and the sound. All the knowledge and techniques available for treating parametric interaction are immediately applicable to this kind of diffraction.

2.5 A Quantum-Mechanical Point-of-View

The frequency and wave-vector relations are easy to interpret quantum-mechanically. To show this, let us write these relations in equation form as follows:

$$f_\pm = f_i \pm f , \tag{2.27}$$

$$k_\pm = k_i \pm K . \tag{2.28}$$

We multiply (2.27) by Planck's constant h, and (2.28) by $\hbar = h/2\pi$ to obtain

$$hf_\pm = hf_i \pm hf , \tag{2.29}$$

$$\hbar k_\pm = \hbar k_i \pm \hbar K . \tag{2.30}$$

The terms of (2.29, 30) have the dimension of energy and momentum, respectively. The equations express conservation of these quantities, and a quantum-mechanical interpretation can be based on that fact.

The incident light conducts a swarm of photons into the interaction region and encounters a swarm of phonons conducted by the sound. The light is scattered by the sound. In the upshifted case, a portion of the incident wave with frequency f_i and wave vector $\mathbf{k}_i$ is changed to a scattered wave of frequency $(f_i + f)$ and wave vector $(\mathbf{k}_i + \mathbf{K})$. Each photon of the incident wave has energy hf_i and momentum $\hbar\mathbf{k}_i$, and each photon of the scattered wave has energy $h(f_i + f)$ and momentum $\hbar(\mathbf{k}_i + \mathbf{K})$. Thus the new (scattered) photons have an energy higher by hf and momentum different by $\hbar\mathbf{K}$. The scattering process can therefore be interpreted as having destroyed a photon of energy hf_i and momentum $\hbar\mathbf{k}_i$ and as having created a photon of energy $h(f_i + f)$ and momentum $\hbar(\mathbf{k}_i + \mathbf{K})$. Conservation of energy and momentum is achieved by the concomitant destruction of a phonon of energy hf and momentum $\hbar\mathbf{K}$. Thus the sound acts as an energy source. In this upshifted case, the quantities on the right-hand side of (2.29, 30) are for two input quantum-mechanical particles that are annihilated. The quantities on the left are for a single output particle that is created.

In the downshifted case, the scattering process is one in which a photon of the incident beam is destroyed by creating both a photon in the downshifted beam and a phonon in the sound beam. The sound in this case is an energy sink. These circumstances suggest that by means of Bragg diffraction, an incident beam of light can be used to amplify a beam of sound. Experimental verification of this type of amplification was demonstrated by *Korpel* et al. [2.38].

2.6 Wave Theory

We will now look at Bragg diffraction from the standpoint of wave theory and employ more rigor in analyzing the phenomenon than we have previously done. Our approach will be to use the d'Alembert wave equation and the approximations usually made in coupled-wave theory. This means, of course, that we will not attempt to be exact, but will simplify the analysis in justifiable ways as we proceed.

The situation is illustrated by the sketches and diagrams in Figs. 2.2 and 4. The wave vector $\mathbf{K}$ has an absolute magnitude given by $2\pi/\Lambda$ and is perpendicular to the moving planes of maximum index of refraction (the wave fronts) produced by the traveling sound. The variation of the refractive index between successive wave fronts is sinusoidal. The fact that there is variation, as we shall shortly see (and as we have already seen heuristically), results in coupling between two waves or modes of propagation in the medium. We can write the index of refraction as follows:

$$n(x, t) = n + \Delta n \cos(2\pi ft - Kx) \tag{2.31}$$

where x is the direction of propagation of the acoustic wave and n the refractive index of the medium. We can use (2.31) to enter the d'Alembert wave equation, which we express as follows

$$\nabla^2 U = \left(\frac{n(x,t)}{c}\right)^2 \frac{\partial^2 U}{\partial t^2} \qquad (2.32)$$

where U is a complex function of position and time whose real part is a scalar function that we may regard as representing either the electric or magnetic field strength of the light. At this point we invoke the condition for weak coupling, namely $\Delta n/n \ll 1$. In addition, we will assume that f $\ll f_i$ so that the temporal frequency of all light-wave components is very nearly f_i and we can write

$$\frac{\partial^2 U}{\partial t^2} \simeq -(2\pi f_i)^2 U . \qquad (2.33)$$

With (2.33) we can reduce the d'Alembert equation to the Helmholtz equation.

$$\nabla^2 U - \left(\frac{n(x,t)}{c}\right)^2 [-(2\pi f_i)^2]U = 0 ,$$

$$\nabla^2 U + k_0^2 n^2(x,t)U = 0 . \qquad (2.34)$$

There are numerous methods for solving (2.34). Since we already know a great deal about Bragg diffraction from previous heuristic considerations, one of the easiest ways for us to proceed is to make an educated guess at the form of the solution and use it to expand the above equation, throwing away the higher-order terms.

We start the conjecture by assuming the Bragg condition of Fig.2.4 with

$$k_i^2 = k_\pm^2 = (nk_0)^2 = k^2 \qquad (2.35)$$

where, as before, the subscript + is chosen to designate the upshifted beam, and - the downshifted beam. We guess the form of the solution for (2.34) to be

$$U = U_i(z)e^{-j(2\pi f_i t - \mathbf{k}_i \cdot \mathbf{r})} + U_\pm(z)e^{-j[2\pi(f_i \pm f)t - (\mathbf{k}_i \pm \mathbf{K}) \cdot \mathbf{r}]} \qquad (2.36)$$

where $U_i(z)$ and $U_\pm(z)$ are phasors. The presumed solution is consistent with the Tien relations and (2.27,28). If, as we have assumed, $\Delta n/n \ll 1$, the coupling between waves will be weak and the complex amplitudes, $U_i(z)$ and $U_\pm(z)$, will not change much over the distance of one wavelength. This permits us to neglect the second derivatives of these functions and $(\Delta n/n)^2$ compared with the remaining terms in the Helmholtz equation (that is, we can neglect $d^2 U_i(z)/dz^2$ and $d^2 U_\pm(z)/dz^2$ compared with $dU_i(z)/dz$ and $dU_\pm(z)/dz$, etc.). By doing this we obtain first-order coupled differential equations for the complex wave amplitudes.

From the previous heuristic considerations, we expect to find a typical coupled-wave situation between $U_i(z)$ and $U_\pm(z)$. If one of these waves is fed into the interaction region at the input, in general, we can expect to get both waves at the output. The presumed solution (2.36) is simply a summation of the two waves. If the coupling were reduced to zero (that is, if the amplitude of the acoustic wave and hence Δn were zero) the modes could exist separately. With no coupling, the two waves could propagate simultaneously in the same region and each would be completely unperturbed by the presence of the other. As written in (2.36), the waves are of planar form.

Neglecting the higher-order terms and noting that the Bragg condition permits simplifying the expression by cancelling some of the remaining terms and combining others, we obtain from the Helmholtz equation

$$2jk_{iz} \frac{d}{dz}U_i(z)\, e^{-j(2\pi f_i t - \mathbf{k}_i \cdot \mathbf{r})} + 2jk_{\pm z} \frac{d}{dz}U_\pm(z)\, e^{-j[2\pi(f_i \pm f)t - (\mathbf{k}_i \pm \mathbf{K})\cdot \mathbf{r}]}$$

$$+ \frac{\Delta n}{n}n^2 k_0^2\, U_i(z)\, (e^{-j[2\pi(f_i - f)t - (\mathbf{k}_i - \mathbf{K})\cdot \mathbf{r}]} + e^{-j[2\pi(f_i + f)t - (\mathbf{k}_i + \mathbf{K})\cdot \mathbf{r}]})$$

$$+ \frac{\Delta n}{n}n^2 k_0^2\, U_\pm(z)\, (e^{-j[2\pi(f_i \pm f - f)t - (\mathbf{k}_i \pm \mathbf{K} - \mathbf{K})\cdot \mathbf{r}]} + e^{-j[2\pi(f_i \pm f + f)t - (\mathbf{k}_i \pm \mathbf{K} + \mathbf{K})\cdot \mathbf{r}]}) = 0$$

$$(2.37)$$

where k_{iz} and $k_{\pm z}$ are the z components of $\mathbf{k}_i$ and $\mathbf{k}_\pm$, respectively. Perhaps the easiest way to proceed at this point is to decide whether we wish to consider an incident beam entering from above in Fig.2.2 or below. This will determine whether we have the upshifted or downshifted case and therefore which of the signs, plus or minus, is the appropriate one. Let us specify the downshifted case, and hence the minus sign. This is consistent with the situation considered in Sect.2.3 concerning the heuristic picture (Fig.2.3). Thus for each of the ± terms in the above equation, we choose the minus sign. This gives

$$\left[2jk_{iz} \frac{d}{dz}U_i(z) + \frac{\Delta n}{n} n^2 k_0^2\, U_-(z)\right] e^{-j(2\pi f_i t - \mathbf{k}_i \cdot \mathbf{r})}$$

$$+ \left[2k_{-z} \frac{d}{dz}U_-(z) + \frac{\Delta n}{n} n^2 k_0^2\, U_i(z)\right] e^{-j[2\pi(f_i - f)t - (\mathbf{k}_i - \mathbf{K})\cdot \mathbf{r}]}$$

$$+ \frac{\Delta n}{n}n^2 k_0^2 [U_i(z)e^{-j[2\pi(f_i + f)t\; (\mathbf{k}_i + \mathbf{K})\cdot \mathbf{r}]} + U_-(z)e^{-j[2\pi(f_i - 2f)t - (\mathbf{k}_i - 2\mathbf{K})\cdot \mathbf{r}]}] = 0 \;.$$

$$(2.38)$$

From this equation we can derive a pair of first-order coupled equations as follows: Multiply each term of the equation by $\exp(-jk_i \cdot r)$ and integrate over all x and y and over a distance in the z direction long enough so that many periods of sinusoidal variation are included, but short enough so that $dU_i(z)/dz$ and $U_-(z)$ do not change appreciably. All but the first two terms on the left vanish approximately.[3] Since the term on the right-hand side is zero, we must set the remaining integral equal to zero also. To do this we equate the integrand to zero and obtain

$$\frac{d}{dz} U_i(z) = j \frac{1}{2} k_i \frac{k_i}{k_{iz}} \frac{\Delta n}{n} U_-(z) \tag{2.39}$$

where we have let nk_0 be represented by k_i, the magnitude of the propagation constant in the exponential multiplying the term that produces this equation. In similar fashion, if we multiply by $\exp[-j(k_i - K) \cdot r]$ and integrate we obtain

$$\frac{d}{dz} U_-(z) = j \frac{1}{2} k_- \frac{k_-}{k_{-z}} \frac{\Delta n}{n} U_i(z) . \tag{2.40}$$

Since the factors k_i/k_{iz} and k_-/k_{-z} are both equal to $\sec(\theta/2)$, Eqs.(2.39,40) are the same as (2.19,20) which were obtained heuristically. The solutions to these equations are (2.21,22).

As discussed in Sect.2.1 describing a simple experiment, a relatively wide acoustic beam is required to establish the Bragg regime. If the beam is narrow, the diffraction is of the Raman-Nath type and a large number of diffracted orders can be produced. Why this is so is explained below on the basis of wave theory and Bragg diffraction.

In the preceding derivation, we assume plane-wave beams of light and sound, see (2.31,36). In practice, these beams are not strictly planar. They have finite width and therefore infinitely wide spectra of plane-wave components rather than a single component each. The effective angular width of one of the spectra (that is, the range of spatial frequencies over which the spectral values are relatively high) is proportional to the wavelength divided by the beam width. Because the acoustic wavelength is a couple of orders of magnitude greater than the light wavelength, it is frequently the case that the effective angular plane-wave spectrum of the sound beam is much broader than that of the light beam. In some applications, however, such as for acoustic light modulation, it is advantageous to make the two spectra approximately equal [2.51]. Nevertheless, respon-

---

[3] They vanish approximately because each of the various exponential factors $\exp[+j(k_i \cdot r)]$, $\exp[+j(k_i - K)r]$, $\exp[+j(k_i + K)r]$ and $\exp[+j(k_i + 2K)r]$ represent a sinusoid at a different spatial frequency. As such, each is orthogonal to the others. When we multiply by $\exp[-jk_i \cdot r]$ and integrate over all x and y and over the prescribed z, the orthogonality produces a near zero value for each integral except the one containing the factor $\exp(+jk_i \cdot r)$ (which cancels with the multiplying factor producing unity). Thus only one integral remains, the one containing the first two terms on the left.

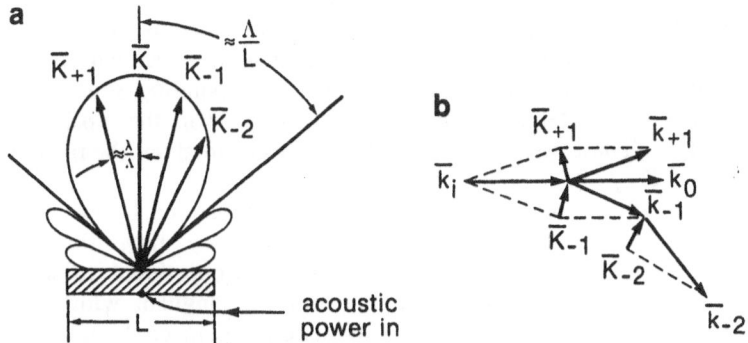

Fig.2.6a,b. Radiation pattern for the sound beam showing specific spectral components which produce first- and second-order down and upshifted light in Debye-Sears diffraction. In (a) various sound-wave vectors which diffract the light are shown. The wave-vector diagram for various orders of Bragg diffraction are shown in (b). For convenience, we have made the scale in (b) smaller than in (a)

sibility for the nature of the diffraction (Raman-Nath or Bragg) is customarily ascribed to the sound. For small acoustic beamwidth, the acoustic spectrum gives a broad range of relatively large plane-wave components for the light to interact with. This is the reason so many diffraction orders appear in Debye-Sears diffraction which, as has already been described in Sect.2.1, is a special case of Raman-Nath diffraction where the incident light enters the interaction region at right angles to the main component in the sound. Figure 2.6 illustrates the situation.

In Fig.2.6a the radiation pattern of the sound beam is shown with the central propagation constant vector $\mathbf{K}$ pointing vertically upward. The propagation constant vectors for three other components in the pattern are also shown. These components are responsible for producing first- and second-order downshifted and upshifted light in the Debye-Sears case. The interaction is as follows: The incident light enters the interaction region propagating in the horizontal direction. Its wave vector $\mathbf{k}_i$ is shown in Fig.2.6b with horizonal orientation. Part of the light passes through the cell undiffracted and produces a zero-order spot with the wave vector $\mathbf{k}_0$ = $\mathbf{k}_i$. However, a portion of the incident light meets the Bragg condition (2.2) with respect to the $\mathbf{K}_{-1}$ spectral component in the sound and is diffracted downward producing downnshifted light with wave vector $\mathbf{k}_{-1}$. As shown in Fig.2.6a, $\mathbf{K}_{-1}$ is a vector on the right-hand side of the radiation pattern. Since the pattern is symmetrical about the vertical axis, a corresponding vector exists on the left-hand side. In Fig.2.6a, it is called $\mathbf{K}_{+1}$. As shown, it makes the same angle with respect to the vertical axis as $\mathbf{K}_{-1}$ but is on the opposite side of the axis. $\mathbf{K}_{+1}$ also meets the Bragg condition with respect to $\mathbf{k}_i$ and produces upshifted light with vector $\mathbf{k}_{+1}$, as shown in Fig.2.6b. Thus first-order beams of downshifted and upshifted light are generated simultaneously.

Second-order beams can also be generated. The light vector $\mathbf{k}_{-1}$ meets the Bragg condition with respect to the sound vector $\mathbf{K}_{-2}$. First-order Bragg diffracted light encounters a sound beam component with which to interact producing a second-order of diffraction. The wave vector for this second-order, downshifted light is k_{-2}, as shown in the figure. For simplicity, we are not showing the corresponding sound and light vectors on the left-hand side of the diagram for the upshifted second-order light. These vectors exist and if we were to show them, they would be labeled K_{+2} and k_{+2}.

The Bragg condition tells us that the orders are separated by an angle λ/Λ (for small θ). As previously stated, diffraction theory shows that the angular width of the acoustic spatial spectrum is proportional to Λ/L. Hence by dividing the angular separation into the angular width we should get an indication of how many diffraction orders could possibly appear. The inverse of this quantity multiplied by 2π is given the symbol Q and is frequently used to specify the condition for Bragg diffraction (where there is no multiple scattering). Thus

$$Q = 2\pi \; \frac{\lambda/\Lambda}{\Lambda/L} = \frac{2\pi\lambda L}{\Lambda^2} \; . \tag{2.41}$$

When Q>>1, the system is in the Bragg regime and multiple scattering will not take place. For Q<<1, we have the Raman-Nath regime with its multiple scattering. These facts were stated in Sect.2.1.

2.7 A Sample Calculation

Equations (2.21,22) give the variation for the complex amplitudes associated with the incident and Bragg-diffracted light. Equations (2.25,26) describe the corresponding power variation. It is instructive to examine the practical consequences of these equations. Specifically let us consider how much acoustic power is required to cause a given fraction of the incident light to undergo Bragg diffraction in water.

The ratio $\Delta n/n$ is the key factor in this consideration and it is determined by the following equation [2.46]

$$\frac{\Delta n}{n} = \frac{np^2}{2} \left| \frac{\partial u}{\partial x} \right|_{max} \tag{2.42}$$

where u is the particle displacement in the x direction and hence $|\partial u/\partial x|_{max}$ is the amplitude of the acoustic strain. The quantity p is the elasto-optic coefficient.

The acoustic strain can be calculated from energy considerations. To do this, we focus our attention on a point particle in the water through which the sound propagates. With no sound present, the particle can be thought of as remaining at rest in its equilibrium position. When sound is introduced the particle oscillates back and forth through the equilibrium position with an instantaneous displacement of $u(x,t)$. To relate acoustic strain to acoustic power we start with an expression for the density of the energy stored in the water due to the sound that is present. The average value of the stored kinetic energy density is

$$W = \frac{1}{2} \rho \langle \dot{u} \rangle^2 \qquad (2.43)$$

where ρ is the mass of the water per unit volume, and $\dot{u}$, the partial derivative with respect to time of the particle displacement in the x direction and therefore the instantaneous particle velocity. < > denotes taking the time average. We emphasize that $\dot{u}$ is not the velocity of the sound wave. It is the instantaneous velocity, at a particular place, of individual little water particles that are moving backward and forward because sound is traveling through the water at that place. On the average, half the stored energy within any small region is potential energy. The makeup of the instantaneous stored energy changes sinusoidally with time between the kinetic and potential forms, its total density remaining constant at 2W.

Since u is the displacement due to the propagating sound, it can be regarded as a manifestation of the sound wave and therefore as propagating with the wave velocity. As the wave moves, its effects such as stress, strain, displacement, refractive-index variation, and particle velocity move along with it. In fact, each of these quantities can be looked upon as constituting the acoustic "wave". Therefore we can regard u as being a sinusoidal function of $(2\pi ft-Kx)$. This is consistent with (2.31) which gives the variation in refractive index as a sinusoidal function of the same quantity. Thus for the particle velocity in terms of acoustic strain we have

$$\dot{u} = \frac{\partial u}{\partial t} = -\left(\frac{2\pi f}{K}\right)\frac{\partial u}{\partial x} = -f\Lambda\frac{\partial u}{\partial x} \qquad (2.44)$$

from which we observe

$$\langle \dot{u} \rangle^2 = \frac{V^2}{2}\left|\frac{\partial u}{\partial x}\right|^2_{\mathrm{max}} \qquad (2.45)$$

where V is the acoustic wave velocity.

We may regard the acoustic power as stemming from the propagation of the stored energy. Along with stress, strain, displacement and other quantities, the stored energy advances with the wave velocity V. As we have seen, the oscillatory motion of the water molecules within a small region and the stress associated with their displacement from the equilibrium position produces the kinetic and potential energy that is stored there. This stress and motion propagate with the wave velocity. Every second, a continuous column V units long of the small regions, each with a stored energy density of 2W, passes through a fixed plane oriented at right angles to the direction of the propagation. Thus the density of the energy flow, or power, through the plane is given by

$$P_s = (2W)V = V\rho\langle\dot{u}\rangle^2 = \frac{\rho}{2}V^3 \left|\frac{\partial u}{\partial x}\right|^2_{max} .$$ (2.46)

From the preceding equations we obtain

$$\left|\frac{\partial u}{\partial x}\right|_{max} = \sqrt{\frac{2P_s}{\rho V^3}} \quad \text{and} \quad$$ (2.47)

$$\frac{\Delta n}{n} = \sqrt{\frac{2P_s}{\rho V^3}} \frac{n^2 p}{2} .$$ (2.48)

The argument of the sine and cosine terms of (2.21-26) can now be expressed in terms of the width of the sound beam, the acoustic power density, the wavelength of the light, and such medium properties as density, acoustic velocity, refractive index and elasto-optic coefficient.

Using (2.48) we can write

$$\frac{k}{2} \frac{\Delta n}{n} \sec\frac{\theta}{2} z = \frac{\pi}{\lambda} \sqrt{\frac{2P_s}{\rho V^3}} \frac{n^2 p}{2} \sec\frac{\theta}{2} z .$$ (2.49)

Because the light wavelength is usually much less than the sound wavelength, $\sec(\theta/2)$ is ordinarily very nearly unity and can be regarded as such to good approximation. If we multiply λ in the denominator by the refractive index, we obtain the free-space wavelength λ_0. The right-hand side of (2.49) becomes

$$\frac{\pi n^3 p}{\sqrt{2\rho V^3}} \frac{z}{\lambda_0} \sqrt{P_s} .$$ (2.50)

This is an expression we can easily deal with by obtaining the values for the physical constants from appropriate references. For water we have n = 1.33, $\rho = 10^3$ kg/m^3, V = $1.5\cdot10^3$ m/s and p = 0.31. If the light source is a helium-neon laser, λ becomes $0.63\cdot10^{-6}$ m. Using these values in the above expression, we get $1.4z\sqrt{P_s}$ as the argument of the sine and cosine terms of (2.21-26). From (2.25,26) we can write

$$\frac{P_-(z)}{P_i(0)} = \sin^2(1.4z\sqrt{P_s})$$ (2.51)

where we have neglected the factor $(f_i-f)/f_i$, assuming $f_i \gg f$. If the above ratio is very small compared to unity, we can replace the sine-squared term, to a good approximation, by its argument squared, obtaining

$$\frac{P_-(z)}{P_i(0)} = 1.96z^2 P_s \ . \tag{2.52}$$

We can use this equation to get an indication of how much acoustic power is required to produce Bragg diffraction. Consider specifically the question of diffracting 5% of the light from a helium-neon laser. The light is incident at the Bragg angle upon a column of acoustic waves put into the water by a transducer 4 cm wide and 2.5 cm high. From (2.52) we have

$$\frac{P_-(\ell)}{P_i(0)} = 1.96L^2 P_s = 1.96L^2 P_t / LH = 1.96(L/H)P_t \tag{2.53}$$

where L is the length of the transducer, H is the height of the transducer, and P_t is the total power in the sound beam. Thus

$$P_t = \frac{P_-(\ell)}{P_i(0)} \frac{1}{1.96} \frac{H}{L} = 0.05 \frac{1}{1.96} \frac{2.5}{4} = 16 \ mW \ . \tag{2.54}$$

Note that the aspect ratio H/L of the transducer, and not the area (HL), shows up in the final expression for the total acoustical power.

As we see, 16 mW of sound power in the water will produce a modest amount (5%) of diffraction. How about a large amount? If we set the argument of the sine-squared term in (2.26) equal to $\pi/2$, we can calculate how many Watts of sound power would be required to diffract all the incident light. With the same values for the physical constants as before, the calculation indicates that about 0.8 W would be necessary. Our simple theory has not taken into account losses which are always present. Because of loss, it is not possible to obtain 100% diffraction. By exercising proper care, however, 95% diffraction can be obtained.

2.8 The Reflection Case

Up to this point we have considered conventional Bragg diffraction in which the incident light and the scattered light show up on opposite sides of the region occupied by the sound. An important case of interaction between light and sound exists in which the incident and scattered light both appear on the same side of the sound region. Consider the geometry illustrated in Fig.2.7 where the light wave-vectors are shown pointing in roughly opposite directions. As far as the z-directed components are concerned, the incident and diffracted waves travel in exactly opposite directions. This is generally referred to as the reflection case [2.7]. If k_{iz} is regarded as positive, then we can designate k_{tz} as negative.

A specific situation of the reflection case is one in which the two light beams and the sound are all collinear. For this situation there is for-

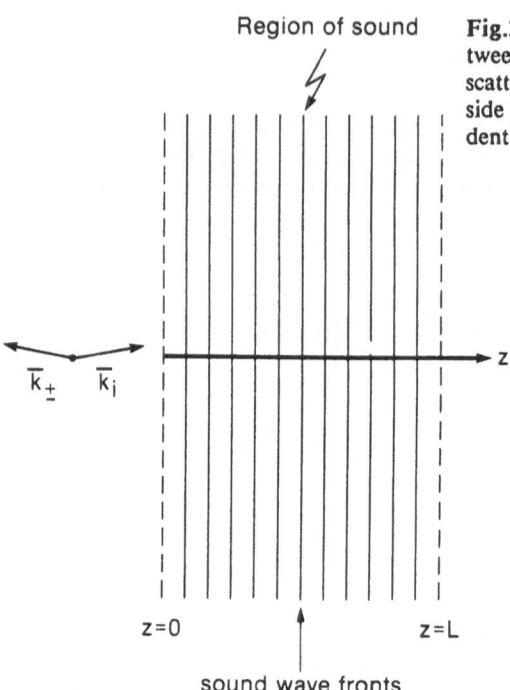

Region of sound

Fig.2.7. A geometry of interaction between light and sound in which the scattered light emerges from the same side of the sound region that the incident light enters

$\overline{k}_\pm$ $\overline{k}_i$

z

z=0   z=L

sound wave fronts

ward-traveling light (the incident beam), backward-traveling light (the diffracted beam) and either forward- or backward-traveling sound. If the sound is forward traveling, the scattered light is downshifted and we choose the minus sign as the subscript for $k_\pm$. If backward traveling, the light is upshifted and we choose the plus sign.

By following steps similar to the ones we have already taken in Sect.2.6 for the case of conventional Bragg diffraction, we can readily show that (2.39,40) hold for the general reflection case. The solutions to these equations will be different from those in conventional Bragg diffraction because $k_{\pm z}$ is negative and the boundary conditions are different. Specifically, we no longer have $U_\pm(0) = 0$ since the scattered light emerges at that boundary. Instead, we have $U_\pm(L) = 0$ since no back-scattered light can be generated by sound beyond $z = L$ where no sound exists. Solving (2.39,40) with these boundary conditions and employing a negative value for $k_{\pm z}$ gives

$$\mathbf{U}_i(z) = \frac{\cosh\left[\frac{k}{2}\frac{\Delta n}{n}\frac{k}{\sqrt{|k_{iz}k_{\pm z}|}}(L-z)\right]}{\cosh\left[\frac{k}{2}\frac{\Delta n}{n}\frac{k}{\sqrt{|k_{iz}k_{\pm z}|}}L\right]}\mathbf{U}_i(0) \qquad \text{and} \qquad (2.55)$$

$$U_{\pm}(z) = j \sqrt{\frac{f_i \pm f}{f_i}} \; \frac{\sinh\left[\dfrac{k}{2}\dfrac{\Delta n}{n}\dfrac{k}{\sqrt{|k_{iz}k_{\pm z}|}}(L-z)\right]}{\cosh\left[\dfrac{k}{2}\dfrac{\Delta n}{n}\dfrac{k}{\sqrt{|k_{iz}k_{\pm z}|}}L\right]} \; U_i(0) \; . \tag{2.56}$$

As before, the power in the two waves is given by $U_i(z)U_i^*(z)/2$ and $U_{\pm}(z)U_{\pm}^*(z)/2$. The ratio of the diffracted power to the incident power is

$$\frac{P_{\pm}(z)}{P_i(z)} = \frac{f_i \pm f}{f_i}\tanh^2\left[\frac{k}{2}\frac{\Delta n}{n}\frac{k}{\sqrt{k_{iz}k_{\pm z}}}(L-z)\right] . \tag{2.57}$$

Thus for very large L, the power ratio at z = 0 is

$$\frac{P_{\pm}(0)}{P_i(0)} = \frac{f_i \pm f}{f_i} \; . \tag{2.58}$$

This equation is consistent with the parametric interpretation of Bragg diffraction described in Sect.2.4. The expression is well known and is referred to as Manley-Rowe relation [2.52]. The equation is also consistent with the quantum-mechanical point of view treated in Sect.2.5, particularly in terms of conservation of energy and the annihilation or creation of photons.

A plot of $P_i(z)$ and $P_-(z)$ is shown in Fig.2.8 for the case of collinear beams. The plot is for the downshifted case and therefore the sound, as well as the incident light, is assumed to be moving in the positive z direction. The incident light, at frequency f_i, decays exponentially as it travels into the sound region. The diffracted light, at f_i-f, builds up exponenetially as it travels in the negative z direction.

Note that here we are considering light diffraction from a sound wave of large relative amplitude. One of the assumptions in the derivation

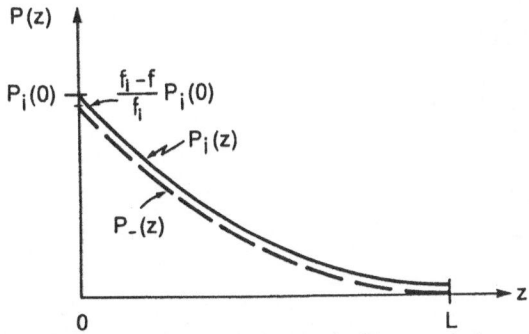

Fig.2.8. Plots of power in the incident and downshifted light beam of Fig.2.7. For this case, sound of large amplitude traveling in the positive z (*forward*) direction is assumed. $P_i(z)$ also flows forward, but $P_-(z)$ flows backward

of the above equations is that the sound amplitude at $z = L$ is the same as at $z = 0$. Since the sound in the down-shifted case is an energy sink, the constant-amplitude assumption is not valid unless the total energy absorbed by the sound is small compared with its initial energy. If the sound enters the interaction region with low amplitude and the input light enters with large amplitude, the behavior will not be as indicated by the above equations, and stimulated emission of phonons in the sound can take place as the quantum-mechanical point of view indicates.

2.9 Birefringent Bragg Diffraction

In the previous analyses, the acousto-optic interaction has been assumed to occur either in an isotropic medium or in an anisotropic medium in such a way that the diffraction process does not change the polarization of the light being diffracted. Birefringence has played no part in any of the phenomena associated with the diffraction we have so far considered. The type of diffraction we have studied is called isotropic [2.32,44]. It gives rise to the simple wave-vector diagrams we have seen in Fig.2.4 where the vectors for the incident and diffracted light are treated as being equal. Since these vectors have a common starting point, their end points fall on the arc of a single circle, as shown in Fig.2.9a.

It was mentioned in Sect.2.2 that *Dixon* [2.32] had analyzed the case of diffraction where birefringence does play a part. He showed that birefringent effects can be so significant that the normal Bragg conditions for the isotropic case may not even be approximatedly satisfied.

Consider Bragg diffraction in an anisotropic medium when the incident and diffracted polarizations differ. Assume, for example, that the incident beam is extraordinarily polarized and that the diffracted beam is produced with ordinary polarization. This situation would occur under the

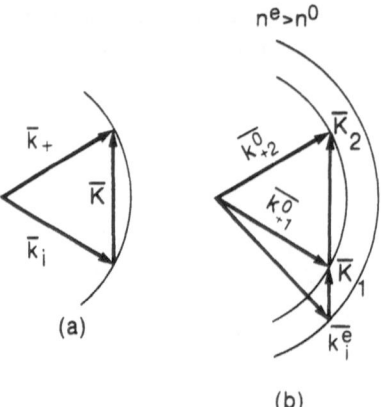

(a)

(b)

Fig.2.9. Wave vector diagrams for upshifted Bragg diffraction in (a) the isotropic case, and (b) a birefringent case where for simplicity the acoustic wave vector is shown extending vertically upward

proper circumstances in a positive uniaxial crystal where the incident beam and the acoustic beam are both propagating perpendicular to the optic axis. Wave-vector diagrams for this case when $n^e > n^o$ are shown in Fig.2.9b. The superscripts e and o refer to extraordinary and ordinary, respectively. As illustrated, there are two acoustic wave vectors (K_1 and K_2) in the direction indicated (vertical), and hence two different acoustic frequencies, for which the birefringent Bragg condition is met and wave-vector diagrams can be constructed. Since $k_i^e > k_+^o$, the end points of the vectors now fall on the arcs of two different circles.[4] By using the two circles, we can easily construct the appropriate wave-vector diagrams but they no longer lead to (2.2), the well-known condition for Bragg diffraction in the isotropic case. If the incident beam were ordinarily polarized, there would be only one acoustic wave vector in the vertical direction, and hence only one acoustic frequency, for which the birefringent Bragg condition is met. This becomes immediately apparent by considering the appropriate wave-vector diagram. For that situation, k_i falls on the inner arc. Only one value of a vertically-extending K can connect the point of intersection of k_i on the inner arc to a point on the outer arc.

Only one K value can exist also in the transmission case for collinear interaction. The wave-vector diagram for this situation is shown in Fig.2.10a. From the figure it is apparent that

$$K_{min} = k_i^e - k_+^o \tag{2.59}$$

from which the acoustic frequency is easily found to be

$$f_{min} = \frac{V}{\lambda_0} (n^e - n^o) . \tag{2.60}$$

K_{min} is the shortest vector that can be made to extend between the two arcs and therefore f_{min} is the lowest acoustic frequency for Bragg diffrac-

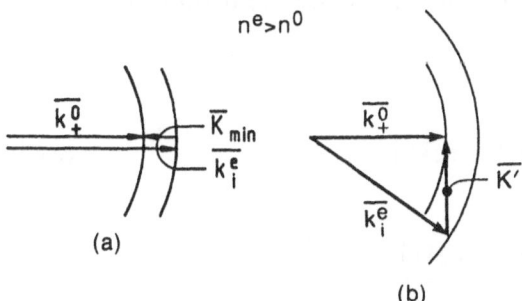

(a)

(b)

Fig.2.10. Wave vector diagrams for upshifted Bragg diffraction of (a) collinear birefringent interaction where, for simplicity, all the wave vectors are shown with horizontal orientation, and (b) tangential birefringent interaction for acoustic waves propagating vertically

---

[4] If the incident beam does not enter perpendicular to the optic axis, ellipses take the place of the circles, but the behavior remains qualitatively the same.

43

tion under these circumstances. There is no type of diffraction equivalent to this in the isotropic case.

Figure 2.10b shows another limiting case. This time $\mathbf{K'}$ is tangent to the inner arc for $k_+{}^o$. Thus we have

$$(K')^2 = (k_i{}^e)^2 - (k_+{}^o)^2 \tag{2.61}$$

from which we readily obtain

$$f' = \frac{V}{\lambda_0}\sqrt{(n^e)^2 - (n^o)^2} \ . \tag{2.62}$$

$\mathbf{K'}$ and f' are significant in that we can change their magnitudes substantially without bringing about much change in the angle required for Bragg diffraction between the incident light and the direction of sound propagation. For a given angle of incidence on the acoustic column, Bragg diffraction will take place over a wide range of acoustic frequencies and, as the acoustic frequency is changed, the diffracted light will experience a large change in direction. This is a condition of obvious importance in designing an optical scanner.

For simplicity, the acoustic vectors in all the wave-vector diagrams so far presented in this section have been shown pointing either horizontally or vertically. A more general situation is depicted in Fig.2.11; but even here, in order to draw the diagram at all, we have had to assume a set of specific conditions. For example, we assumed that an incident extraordinary wave diffracts into a more rapidly traveling ordinary wave. If the speed of travel were the other way around a different diagram would be needed. A number of other changes in the assumptions could be made and each would give rise to a somewhat different wave-vector digaram. In [2.32], *Dixon* compared possible wave-vector geometries for a number of these cases. He gave the following general expressions for the angles θ_i and $\theta_\pm$, where the angles and their positive values are defined in Fig.2.11.

$$\sin\theta_i = \frac{\lambda_0}{2n_i \Lambda}\left[1 + \frac{\Lambda^2}{\lambda_0{}^2}(n_i{}^2 - n_\pm{}^2)\right]$$

$$= \frac{\lambda_0}{2n_i V}\left[f + \frac{V^2}{f\lambda_0{}^2}(n_i{}^2 - n_\pm{}^2)\right] , \tag{2.63}$$

$$\sin\theta_\pm = \frac{\lambda_0}{2n_i \Lambda}\left[1 - \frac{\Lambda^2}{\lambda_0{}^2}(n_i{}^2 - n_\pm{}^2)\right]$$

$$= \frac{\lambda_0}{2n_i V}\left[f - \frac{V^2}{f\lambda_0{}^2}(n_i{}^2 - n_\pm{}^2)\right] . \tag{2.64}$$

The symbols are the same as before except that the superscripts for the refractive indices have been removed since the equations hold regardless of whether the incident beam is ordinarily or extraordinarily polarized as

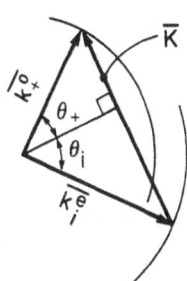

Fig.2.11. A more general geometry for the wave vector diagram of upshifted, birefringent Bragg diffraction

long as the diffracted beam has the other polarization. These equations are readily obtained by scrutinizing each of the geometries and writing down the appropriate trigonometric expressions, just as we did in deriving (2.60,62). Note that the last term on the right of each of these equations goes to zero if there is no birefringence (that is, if $n_i = n_+$). The equations then reduce to (2.2) (as they should) and we can identify the $\theta/2$ of isotropic Bragg diffraction with the above θ_i and θ_+.

Birefringent Bragg diffraction has advantages for optical beam deflection and other uses. An interesting consequence of the operation of the principles we have just considered is that multiple diffraction in the Raman-Nath regime, described for the isotropic case in Sects.2.1 and 6, cannot take place in the birefringent case. Birefringent diffraction in an anisotropic medium can occur for both transverse and longitudinal acoustic waves. This type of diffraction offers important options in designing devices for acousto-optic signal processing.

2.10 Acousto-Optic Materials

A chapter on bulk-wave Bragg diffraction would not be complete without considering the proper combinations of material parameters for optimum device performance. A good acousto-optic material should have fine optical quality and should be available in large enough size for fabricating a desired device. Both the optical and acoustic attenuation should be as low as possible.

In addition to the obvious considerations identified above, a good material should have high acousto-optic figures of merit. The first such figure, along with numbers for various materials, was published by *Smith* and *Korpel* in 1965 [2.53]. Since then many workers have found that figure and other figures to be of substantial value and several are in common use. A number of excellent references provide detailed explanations and derivations concerning them [2.9,44,51,53-56]. Having already gone through much of theory needed for their justification, we can now take a brief look at the rationale for these figures of merit.

From (2.25, 26, 49) we can write

$$\frac{P_\pm(z)}{P_i(0)} \simeq \sin^2\left[\frac{\pi}{\sqrt{2}} \frac{n^3 p}{\sqrt{\rho V^3}} \frac{z}{\lambda_0} \sec\frac{\theta}{2} \sqrt{P_s}\right].$$

(2.65)

The subscript $\pm$ appears here in place of the subscript - used in the previous equations because the principles involved apply equally well to upshifted and downshifted diffraction. The square of the argument of the sine-squared term on the right is called the scattering efficiency [2.51] and is given the symbol η. The term "scattering efficiency" is well chosen because, as we have already seen in Sect.2.7, the fraction of the incident beam that is scattered into the diffracted beam is closely approximated by η whenever $\eta \ll 1$. Thus for small η we have

$$\frac{P_\pm(z)}{P_i(0)} \simeq \sin^2\sqrt{\eta} \simeq \eta = \frac{\pi^2}{2}\left[\frac{n^6 p^2}{\rho V^3}\right]\frac{z^2}{\lambda_0^2}\left(\sec\frac{\theta}{2}\right)^2 P_s.$$

(2.66)

For high scattering efficiency, η should be as large as possible. The material parameters involved in η are all lumped together inside the brackets on the right-hand side of (2.66). They constitute the first figure of merit proposed [2.53]. The symbol M_2 is presently used to designate this figure. Thus

$$M_2 = \frac{n^6 p^2}{\rho V^3}.$$

(2.67)

Simply stated, M_2 relates the diffraction efficiency to the acoustic power for a given device geometry.

Equations (2.65,66) apply only when the Bragg condition is precisely met between the incoming planar light waves and the planar acoustic waves doing the diffracting. If this condition does not hold, the intensity of the scattered light is zero. As we discussed in Sect.2.6, light beams and acoustic beams are never strictly planar. Any actual beam will have a finite cross section and therefore an infinite spectrum of planar-wave components. Each of the components in the acoustic beam, if it meets the Bragg condition with respect to a component in the incident light beam, is capable of producing Bragg diffraction. Equations (2.65,66), which are based on strict planar geometry, are still useful but need to be modified to take into account the magnitude of the diffracting component in the acoustic beam. For an acoustic beam with rectangular cross section, the spatial spectrum has a sinc function variation which when used as a multiplier in (2.65,66) produces the modification required. Thus we have in place of (2.66)

$$\frac{P_\pm(z)}{P_i(0)} \simeq \frac{\pi^2}{2}M_2\frac{L^2}{\lambda_0^2}\left(\sec\frac{\theta}{2}\right)^2 P_s\left(\frac{\sin(KL\Delta\theta/2)}{KL\Delta\theta/2}\right)^2$$

(2.68)

where the bracketed function on the right is the appropriate multiplier. In that bracket, $\Delta\theta$ is the angle that the component doing the diffracting makes with the peak component in the central lobe of the acoustic beam, and L, the width of the acoustic beam.

Equation (2.68) was first derived by *Gordon* and has been justified rigorously in [2.51] where it is shown that the angular variation of the scattering interaction is completely independent of the position of the incident light beam along the sound beam. This is important because it permits choosing the most convenient place along the acoustic beam to make the evaluation. That place is often the position of the beam waist, real or virtual, where the cross section can be very nearly rectangular. For a flat transducer the beam waist occurs precisely at the transducer surface and has precisely the shape of the transducer (which we have assumed to be rectangular).

The sinc-squared function is down to one-half its peak value when

$$\frac{1}{2} KL\Delta\theta_{1/2} \simeq \pm\, 0.45\pi\,. \tag{2.69}$$

This gives the following expression for the total angular spread of the wave vectors in the sound beam between the half-power points for Bragg diffraction of the light:

$$2\,\Delta\theta_{1/2} \simeq 1.8\,\frac{\pi}{KL}\,. \tag{2.70}$$

The situation is illustrated in Fig.2.12. Note that because of the geometry involved, the light beam can be deflected over a total angle of $4\Delta\theta_{1/2}$ by sound-beam components between these half-power points. For planar light propagating in the k_i direction to be deflected over the above angular range, the frequency of the sound must be changed over a corresponding frequency range. Such light has a single spatial frequency com-

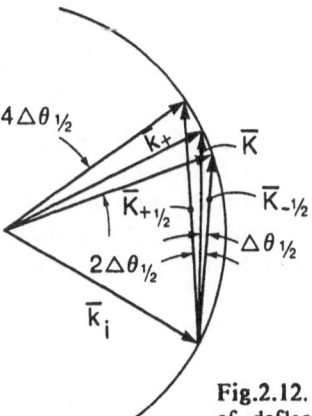

Fig.2.12. Light-beam deflection diagram illustrating the range of deflection angles between half-power points for a sound beam of finite width

ponent and the Bragg condition (2.2) is a precise requirement for it. To change θ, one must change Λ and hence f. Thus, (2.70) and Fig.2.12 relate directly to the frequency sensitivity of the sound in diffracting the light.

By using (2.70) and the Bragg condition, we can easily calculate the range of values in the sound frequency Δf required to produce the angular deflection range depicted in Fig.2.12, i.e.

$$\Delta \sin \frac{\theta}{2} = \Delta \left(\frac{K}{2k} \right) \qquad \text{or}$$

$$\Delta \left(\frac{\theta}{2} \right) = \pi \frac{\Delta f}{Vnk_0} \sec \frac{\theta}{2} = \frac{1.8\pi V}{2\pi f L} \ . \tag{2.71}$$

From this equation we obtain [2.51]

$$\Delta f = \frac{1.8 n k_0 V^2 \cos(\theta/2)}{2\pi f L} \ . \tag{2.72}$$

If we are interested only in diffracting large amounts of power from an incident light beam, a high diffraction efficiency η is important, and M_2 is the appropriate figure of merit. If, in addition, we are interested in deflecting the light beam over a large range of angles, then large Δf is also important. A new figure of merit can be obtained by forming the product of η and Δf. Hence from (2.66,67) we have

$$\eta \Delta f = 0.9 \pi^2 \ \frac{n^7 p^2}{\rho V} \ \frac{L \sec(\theta/2)}{\lambda_0^3 f} \ P_s \ . \tag{2.73}$$

The material parameters involved in this product are lumped together inside the brackets on the right and are given the second important figure of merit

$$M_1 = \frac{n^7 p^2}{\rho V} \ . \tag{2.74}$$

It can be used to optimize the efficiency-bandwidth product for a fixed amount of power in the incident beam throughout the frequency range.

Other figures of merit are also in use and can provide good information depending upon the special considerations appropriate to the application of interest. The product of the transducer length and acoustic power density LP_s appears in (2.73). Since

$$P_t = LHP_s \tag{2.75}$$

where P_t is the total power in the sound beam and H is the transducer height, we could have used P_t/H in (2.73) in place of LP_s. Frequently the value of H is constrained by fabrication limits or electrical impedance, but when it is not and may be made as small as the diameter of the incident light beam, *Dixon* has shown [2.54] that

$$M_3 = \frac{n^7 p^2}{\rho V^2} \tag{2.76}$$

is highly useful as a materials figure of merit.

There is a fourth figure of merit, M_4, employed for wide-band deflectors and modulators in which the power density P_s is the limiting factor [2.44]. For M_4 we have

$$M_4 = \frac{n^8 p^2 V}{\rho} \, . \tag{2.77}$$

Data and tables pertinent to these figures of merit appear in [2.29, 44, 55, 57, 58].

2.11 Conclusion

The question of Bragg-diffraction of light from ultrasound has had a long and interesting history. The pace of development in this area is greater now than ever before. One of the reasons for this is that bulk acousto-optic systems, because of their unique characteristics, have emerged as a major factor in the technology for advanced signal processing [2.44]. However, as important as the bulk techniques are turning out to be, some of the most exciting work employs surface acoustic waves [2.45] along with solid-state lasers and photodetector arrays. In this work the objective has been to fabricate subsystems in which lasers, lenses, interaction media and photodetectors are integrated together on a common substrate.

The progress and developments currently being realized portend a broad spectrum of uses for devices employing acousto-optic Bragg diffraction. A knowledge of the fundamental phenomena in bulk material is essential to understanding the various applications involving Bragg diffraction whether the interaction occurs deep within an acousto-optic specimen or on its surface. To describe those phenomena in tutorial fashion has been the objective of this chapter.

Acknowledgments. The author wishes to express appreciation to an anonymous reviewer of a draft of this chapter for a number of helpful comments and criticisms which pointed out weaknesses in various explanations and errors that were present in the draft. As a result, numerous changes were made in the manuscript. In addition, the author would like to acknowledge indebtedness to Robert Adler to whom the author sent copies of various drafts and from whom the author received extremely helpful discussions, insights and suggestions for changes. Dr. Adler supplied several additions to the text which are incorporated into Sect.23. The author would also like to thank M. Galindo and S. Corr for their patience and skill in typing the many drafts that were necessary for producing the chapter, and A. Meyyappan for much needed consultation on the general format and the overall appearance of the text.

References

2.1   L. Brillouin: Ann. Physique (Paris), 9th Ser. 17, 88 (1922)
2.2   H.A. Lorentz: Wiedman. Ann. 9, 641 (1880)
2.3   L. Lorenz: Wiedman. Ann. 11, 70 (1881)
2.4   M. Born, E. Wolf: *Principles of Optics*, 2nd ed. (MacMillan, New York 1964) p.87
2.5   G.R. Fowles: *Introduction to Modern Optics* (Holt, Rinehart and Winston, New York 1968) p.123
2.6   G.W. Willard: Criteria for normal and abnormal ultrasonic light diffraction effects, J. Acoust. Soc. Am. 21, 101 (1949)
2.7   C.F. Quate, C.D.W. Wilkinson, D.K. Winslow: Interaction of light and microwave sound, Proc. IEEE 53, 1604-1629 (1965)
2.8   W.R. Klein, B.D. Cook: Unified approach to ultrasonic light diffraction, IEEE Trans. SU-14, 723-733 (1967)
2.9   A. Korpel: Acousto-optics - a review of fundamentals, Proc. IEEE 69, 48-53 (1981)
2.10 Lord Rayleigh: *The Theory of Sound*, Vol.2, 2nd ed. (Dover, New York 1945) 272a, pp.89-96. This is a Dover republication of work first published in 1894
2.11 W.H. Bragg: X rays and crystal structures. Trans. Roy. Soc. (London) A 215, 253 (1915)
2.12 W. Duane: The calculation of the X-ray diffraction power at points in a crystal. Proc. Natl. Acad. of Sci. (USA) 11, 489 (1925)
2.13 W.H. Bragg: Die Reflexion von Röntgenstrahlen an Kristallen. Phys. Z. 14, 472 (1913)
      W.L. Bragg: The diffraction of short electromagnetic waves by a crystal. Proc. Cambridge Phil. Soc. 17, 43 (1913)
2.14 P. Debye, F.W. Sears: On the scattering of light by supersonic waves. Proc. Natl. Acad. of Sci. (USA) 18, 409-414 (1932)
2.15 R. Lucas, P. Biquard: Optical properties of solids and liquids under ultrasonic vibrations. J. Phys. Rad., 7th Ser 3, 464 (1932)
2.16 E. Gross: Z. Physik 63, 685 (1930); Naturwiss. 18, 718 (1930); Nature 126, 201, 400, 603 (1930)
2.17 E.H.L. Meyer, W. Ramm: Phys. Z. 33, 270 (1932)
2.18 P. Debye: Ann. Physik 39, 789 (1912)
2.19 L. Brillouin: *Diffraction de la Lumière par des Ultra-sons*, Actualités Scientifiques et Industrielles, No.59 (Hermann, Paris 1933)
2.20 C.V. Raman, N.S.N. Nath: The diffraction of light by high frequency sound waves: Pt.I, Proc. Ind. Acad. Sci. 2, 406-412 (1935)
2.21 C.V. Raman, N.S.N. Nath: The diffraction of light by high frequency sound waves: Pt.II, Proc. Ind. Acad. Sci. 2, 413-420 (1935)
2.22 C.V. Raman, N.S.N. Nath: The diffraction of light by high frequency sound waves: Pt.III, Proc. Ind. Acad. Sci. 3, 75-84 (1936)
2.23 C.V. Raman, N.S.N. Nath: The diffraction of light by high frequency sound waves: Part IV, Proc. Ind. Acad. Sci. 3, 119-125 (1936)
2.24 C.V. Raman, N.S.N. Nath: The diffraction of light by high frequency sound waves: Pt.V, Proc. Ind. Acad. Sci. 3, 459-465 (1936)

2.25 E. David: Scattering of light by weak supersonic radiation. Phys. Z. **38**, 587 (1937)

2.26 P. Phariseau: On the Diffraction of Light by Progressive Supersonic Waves. Proc. Ind. Acad. Sci. **A44**, 165 (1956)

2.27 J.C. Slater: Interaction of waves in crystals. Rev. Mod. Phys. **30**, 197-222 (1958)

2.28 Y.R. Shen, N. Bloembergen: Theory of stimulated Brillouin and Raman scattering. Phys. Rev. **137**, A1787 (1965)

2.29 A. Korpel: *Acousto-Optics*, Applied Solid-State Sci., Vol.3, ed. by R. Wolfe (Academic, New York 1972) Chap.2, pp.73-179

2.30 D. Marcuse: *Light Transmission Optics* (Van Nostrand, New York 1972) pp.61-81

2.31 M. Cardona (ed.): *Light Scattering in Solids I*, 2nd ed., Topics Appl. Phys., Vol.8 (Springer, Berlin, Heidelberg 1983)

2.32 R.W. Dixon: Acoustic diffraction of light in anisotropic media. IEEE J. QE-3, 85-93 (1967)

2.33 A.J. DeMaria, G.E. Danielson Jr.: IEEE J.QE-2, 157 (1966)

2.34 L.C. Foster, C.B. Crumley, R.L. Cohoon: A high resolution optical scatter using a travelling-wave acoustic lens. Appl. Opt. **9**, 2154 (1970)

2.35 M. Szustakowski, B. Swietlicki: Acousto-optic conversion of TE and TM modes in a diffusive planar waveguide. Arch. Acoust. **7**, 271-279 (1982)

2.36 F. Okolocsanyi: The wave slot, an optical television system. Wireless **14**, 527-536 (1937)

2.37 D.M. Robinson: The supersonic light control and its Application to television with special reference to the scophony television receiver. Proc. IRE **27**, 483-486 (1939)

2.38 A. Korpel, R. Adler, B. Alpiner: Direct observation of optically induced generation and amplification of sound. Appl. Phys. Lett. **5**, 86 (1964)
       D.E. Caddis, C.F. Quate, C.D.S. Wilkinson: Conversion of light to sound by electrostrictive mixing in solids. Appl. Phys. **8**, 309-311 (1966)

2.39 M.G. Cohen, E.I. Gordon: Acoustic beam probing using optical techniques. Bell System Tech. J. **44**, 693 (1965)

2.40 A. Korpel: Visualization of the cross-section of a sound beam by Bragg diffraction of light. Appl. Phys. Lett. **9**, 425 (1966)

2.41 H.V. Hance, J.K. Parks, C.S. Tsai: Optical imaging of a complex ultrasonic field by diffraction of a laser beam. J. Appl. Phys. **38**, 1981 (1967)
       C.S. Tsai, H.V. Hance: Optical imaging of the cross section of a microwave beam in rutile by Bragg diffraction of a laser beam. J. Acoust. Soc. Am. **42**, 1345-13347 (1967)
       J. Harlice, C.F. Quate, B. Richardson: IEEE Symp. Sonics Ultrasonics, Vancouver (1967) Paper 1-4

2.42 G. Wade, J. Landry, A.A. deSouza: Acoustic transparencies for optical imaging and ultrasonic diffraction. 1st Int'l Symp. on Acoust. Holography (1967)

2.43 M. Ahmed, G. Wade: Bragg-Diffraction Imaging, Proc. IEEE **67**, 587-603 (1979)

2.44 I.C. Chang: Acousto-optic devices and applications. IEEE Trans. SU-23, 2-22 (1976)
       A. Vanderlugt (Ed.): Special Section on Acousto-Optic Signal Processing., Proc. IEEE **69**, 48-92 (1981)

2.45 C.S. Tsai: Guided-wave acoustooptic Bragg modulators for wideband integrated optic communications and signal processing. IEEE Trans. CAS-26, 1072-1098 (1979)

2.46 R. Adler: Interaction between light and sound. IEEE Spectrum **4**, 42-54 (May 1967)

2.47 S. Goldman: *Frequency Analysis, Modulation and Noise* (McGraw-Hill, New York 1948) p.152

2.48 H.S. Black: *Modulation Theory* (Van Nostrand, New York 1953) p.184

2.49 P.K. Tien: Parametric amplification and frequency mixing in propagating circuits. J. Appl. Phys. **29**, 1347-1357 (1958)

2.50 W.H. Louisell: *Coupled-Mode and Parametric Electronics* (Wiley, New York 1960) Chap.5

2.51 E.I. Gordon: A review of acousto-optical deflection and modulation devices. Proc. IEEE **54**, 1391-1401 (1966)

2.52 J.M. Manley, H.E. Rowe: Some general properties of non-linear elements, Pt.I- General energy relations. Proc. IRE **44**, 904 (1956)

2.53 T.M. Smith, A. Korpel: Measurement of light-sound interaction efficiencies in solids. IEEE J. QE-1, 283-284 (1965)

2.54 R.W. Dixon: Photoelastic properties of selected materials and their relevance for applications to acoustic light modulators and scanners. J. Appl. Phys. **38**, 5149-5153 (1967)

2.55 D.A. Pinnow: Guidelines for the selection of acousto-optic materials. IEEE J. QE-6, 223-238 (1970)

2.56 N. Uchida, N. Niizeki: Acousto-optic deflection materials and techniques. Proc. IEEE **61**, 1073-1092 (1973)

2.57 M. Gottlieb, T.J. Isaacs, J.D. Feichtner, G.W. Roland: Acousto-optic properties of some chalcogenide crystals. J. Appl. Phys. **45**, 5145-5151 (1974)

2.58 E.H. Young, S.K. Yao: Design Considerations for Acousto-Optic Devices. Proc. IEEE **69**, 54-64 (1981)

3. Optical Waveguides - Theory and Technology

Rod C. Alferness and Ping K. Tien

With 22 Figures

This chapter provides a theoretical and technological overview of dielectric optical waveguides that are commonly used in guided-wave acoustooptic devices. The zigzag and potential-well models are employed to explain the waveguiding condition, concepts of effective waveguide index and waveguide modes. The mode equation is derived and solved for the slab waveguide and approximate methods, including the effective-index technique, to analyze channel waveguides are discussed. The approach is physical and tutorial rather than rigorous. The primary optical waveguide technologies including glass, semiconductor and ferroelectric crystals are outlined. Fabrication and some waveguide evaluation techniques are discussed as are methods of coupling light in and out of waveguides.

3.1 Background

The availability of low-loss optical fibers in the early seventies has brought optical communication into reality, and with it there are needs for integrable optical devices and standardized system hardware. To satisfy these demands, research in optics has entered into two new areas: (i) miniaturization of optical components, and (ii) integration of optical components among themselves such as in integrated optics or with electronics such as in optoelectronics. In both areas, optical waveguides play a central role.

Waveguides used in integrated optics are thin-film (or slab) waveguides (Fig.3.1a) and channel waveguides (Fig.3.1b). The simplest waveguide is the slab waveguide shown in the top of Fig.3.1a. It is a thin layer of dielectric film deposited on a suitable substrate. The refractive index of the film has to be larger than that of the substrate. When a high-index film is sandwiched between a low-index substrate and a low-index air space, the film acts as a focusing lens which keeps the light wave inside the film as well as guiding it along the film surface. For that reason, the film is called a light-guiding layer or simply an active layer. A typical light-guiding film in a slab waveguide is 1 μm thick and several centimeters wide, and the light beam inside the film has the shape of a wide ribbon. In fact, the width of the beam is so much larger than the thickness that variation of optical fields along the width of the beam may be neglected for lower-order modes. As a result, a simple two-dimensional wave equation may be used for slab waveguides. Unless otherwise specif-

Springer Series in Electronics and Photonics, Vol. 23   53
Guided-Wave Acoustooptics   Editor: C. S. Tsai
© Springer-Verlag Berlin, Heidelberg 1990

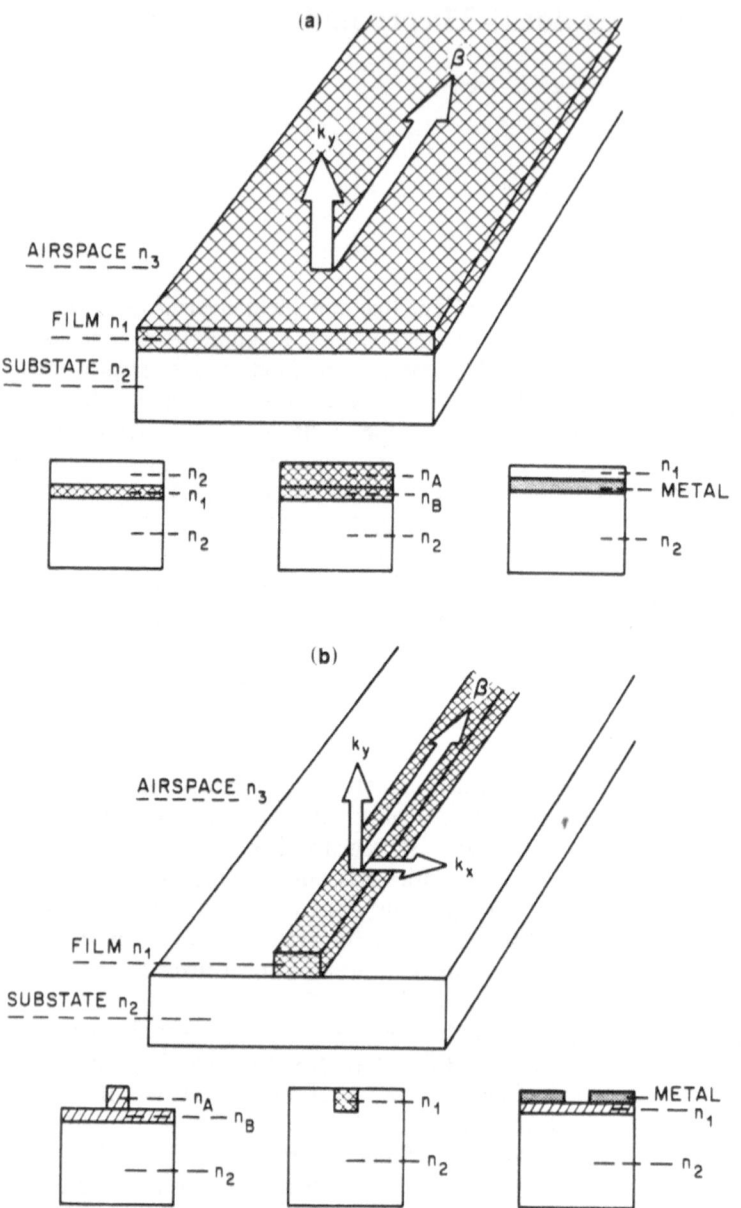

Fig.3.1. Schematic drawing of (a) slab or planar and (b) channel waveguides

ied, we will assume that light waves propagate in the z direction, the
width of the waveguide is placed along x, and the thickness along y.

However, for most device applications, a narrow and well-confined
light beam is required. In such cases, a channel waveguide is used which
in its simplest form is a narrow strip of thin high-index film deposited on
a low-index substrate, as shown in the top of Fig.3.1b. When a light wave

54

propagates in free space, the beam cross section continuously expands by diffraction. In channel waveguides, however, the light beam is confined to the same size regardless of propagation distance.

This capability of confining optical energy in a small cross section for a long distance is an important property of the channel waveguide. For the wavelength of interest in optical communication (0.8 to 1.6 μm in the infrared), a typical cross section of the channel waveguide is $1\,\mu$m x $1\,\mu$m, which is even smaller than the size of metal runners used in very-large-scale integrated electronic circuits. Some of the optical devices such as lasers and couplers have dimensions of hundreds of micrometers by hundreds of micrometers. They are only a few times larger than the basic inverters, NAND and NOR gates used in electronic circuits. Therefore, the same planar technology used for integrated electronics may be used for the fabrication of optical circuits, and in some cases, materials used for optics such as InGaAs, GaAs, InP, and InGaAsP may also be used for electronics.

Another important feature of the waveguide is the speed that an optical signal propagates in an optical circuit. The speed of electronic circuits is limited in large measure by the RC (resistance-capacitance product) constants of the interconnects. Consider the advanced one-micrometer-design rule, for a 2 mm long conducting path, the RC constants are tens of a nanosecond for the diffusion and polysilicon lines and a few nanoseconds for refractory metal. In the case of optical circuits, the speed is limited by optical dispersion which is only a fraction of a picosecond for 2 mm of light path. It is generally believed that the ultimate speed any electronic circuit may achieve is about tens of a picosecond, and that for speeds beyond this limit optoelectronic circuits have to be used.

Historically, the concept of dielectric waveguides was well understood from the microwave theory. Still, experiments involving light-wave propagation in a very thin layer of dielectric film was a scientific curiosity in the late sixties. It was then a great satisfaction when different waveguide modes were excited in a waveguide and were displayed as m-lines by a prism coupler [3.1]. In the meantime, important theories were written for a better understanding of the optics. These include calculation of field distribution in a channel waveguide by *Marcatili* [3.2], formulation of radiation modes by *Marcuse* [3.3] and the zigzag-wave model [3.4,5]. The concept of the potential well model was introduced in 1971 [3.4] and by that time, studies of complex waveguide structures for various applications such as those shown in Fig.3.1a,b were already started.

The zigzag-wave model provides a clear picture of wave propagation in the waveguide and has been used extensively in device analyses. For that reason, we will start our discussion from the zigzag-wave model, and then the potential-well model, mode equations, the effective index method, diffused waveguides, the WKB method, prism couplers, and m-line spectroscopy. In the discussion, we emphasize basic concepts of optics, and often alternative methods are presented to analyze the same problem in an attempt to aid the clarity of this presentation.

In the second part of this chapter (Sect. 3.9) we review the most important waveguide material systems and describe fabrication methods. In addition, the waveguide characteristics such as loss and mode size, and methods for coupling to optical waveguides are discussed.

3.2 Wave Equations and Wave Vectors

In integrated optics and optoelectronics we deal with planar technology, concerning ourselves mainly with wave propagation in rectangular coordinates. The wave equation involved is then

$$\left(\frac{\partial^2}{\partial x^2} + \frac{\partial^2}{\partial y^2} + \frac{\partial^2}{\partial z^2} \right) E = - k^2 n^2 E \ . \tag{3.1a}$$

Here, E is the optical field either electrical or magnetic; n is the refractive index of the medium; $k = \omega/c$; ω is the angular frequency of the optical wave; and c is the velocity of light in vacuum. The factor $\exp(j\omega t)$ is omitted in the above equation and will be in all our later discussions. A general solution of the above equation is a plane wave

$$E = A \exp(j\omega t + jk_x x + jk_y y + jk_z z) \ , \tag{3.2a}$$

where A is the amplitude of the wave. Substituting (3.2) into (3.1a), we find

$$(\partial^2 / \partial x^2) = - k_x^2 \ , \tag{3.3a}$$

$$(\partial^2 / \partial y^2) = - k_y^2 \ , \tag{3.3b}$$

$$(\partial^2 / \partial z^2) = - k_z^2 \ . \tag{3.3c}$$

In waveguide analysis, it is important to consider $(\partial^2 / \partial x^2)$ and $-k_x^2$ to be interchangeable, as are $(\partial^2 / \partial y^2)$ and $-k_y^2$, and $(\partial^2 / \partial z^2)$ and $-k_z^2$. Using these relations, the wave equation (3.1a) becomes

$$k_x^2 + k_y^2 + k_z^2 = n^2 k^2 \ . \tag{3.4a}$$

Taking $\hat{x}$, $\hat{y}$ and $\hat{z}$ as the unit vectors in the x, y and z direction, respectively, we can write (3.2a) as

$$E = A \exp(j\omega t + jn \cdot k) \tag{3.2b}$$

with the wave-vector projections (Fig. 3.1),

$$nk = k_x \hat{x} + k_y \hat{y} + k_z \hat{z} \ , \tag{3.5}$$

yielding the scalar relation (3.4a). Although (3.2-4) are mathematically equivalent, each of them accentuates a different aspect of the wave phenomenon. For example, (3.2a) emphasizes phases of the wave, important in the calculation when interferences of two waves are involved. On the other hand, (3.4a) is a relation of kinetic energy in the potential-well model, and (3.5) represents a plane wave by a wave vector which is a fundamental concept of the ray optics.

It is easy to see from (3.2) that if k_x, k_y or k_z are real, the optical field varies in x, y or z sinusoidally, otherwise it varies exponentially. For clarity, when k_x is imaginary, we replace k_x by jp_x and k_x^2 becomes $-p_x^2$, p_x being a positive real quantity. The same applies to k_y and k_z.

As mentioned earlier, we assume that light propagates along the z direction. Hence k_z must be real, and customarily we use β for k_z. Equation (3.1a) may then be written in the form

$$\left[\frac{\partial^2}{\partial x^2} + \frac{\partial^2}{\partial y^2}\right]E = -(n^2k^2 - \beta^2)E , \qquad \text{or} \qquad (3.1b)$$

$$(k_x^2 + k_y^2)E = +(n^2k^2 - \beta^2)E . \qquad (3.1c)$$

We notice from (3.1c) that both k_x and k_y can be real only when $nk > \beta$. Otherwise, either k_x or k_y or both k_x and k_y have to be imaginary. Consider a slab waveguide shown in Fig.3.2, which is devided in three regions: film (region I), substrate (region II), and air space (region III). They have the refractive indices n_1, n_2 and n_3, respectively. Because of

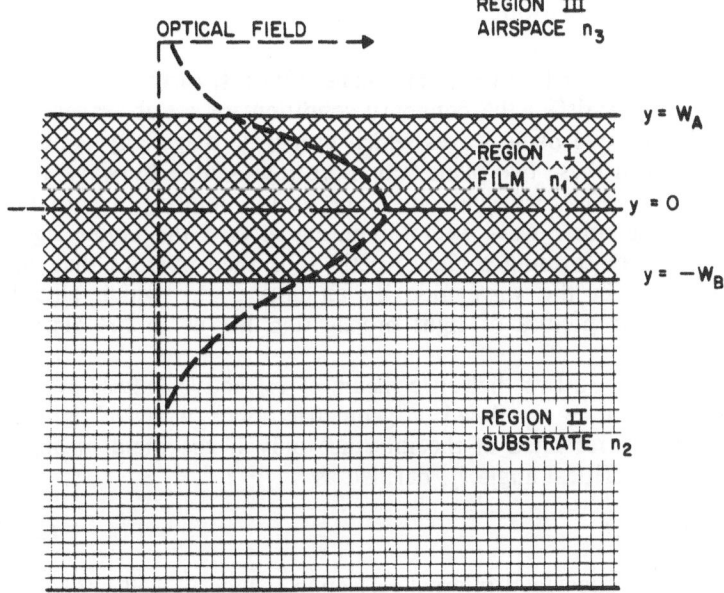

Fig.3.2. Optical field of planar dielectric waveguide

large width of the light beam compared with the thickness in the wave-guide, the variation of optical fields in x (along the width) may be neglected. Hence, $k_x = 0$ and we need to consider k_y only on the left side of (3.1c). As will be discussed in detail later, the field distribution in y is sinusoidal in the film and exponential in the substrate and air space. It is therefore necessary for $kn_1 > \beta$, and kn_2 and $kn_3 < \beta$. We have shown earlier in Sect.3.1 that when $n_1 > n_2$ and n_3, the film acts as a focusing lens and such lens action confines the optical energy inside the film. Here we argue that if the optical energy is confined in the film, fields must be sinusoidal in the film and exponentially decaying outside. This condition dictates the relation $kn_1 > kn_2$ and kn_3.

We can gain better insight into the above relation by considering the light-guiding film as a pipe inside which optical energy flows. The quantity β describes how the energy flows, or more precisely, how the wave propagates inside the pipe. If the optical fields were completely inside the pipe, the wave would propagate as though it were in an infinite medium of refractive index n_1 and β would be equal to kn_1. On the other hand, if the fields were completely outside the pipe, the wave would propagate as though it were in an infinite medium of refractive index n_2 and β would be equal to kn_2 (assuming here for the sake of convenience that $kn_2 = kn_3$). Now because part of the field is inside the pipe and part outside the pipe, β must assume a value between kn_1 and kn_2. It is clear that β is an eigenvalue of the wave and its value relative to kn_1 and kn_2 completely characterizes the propagation of the wave. The waveguiding action ceases to exist at $\beta = kn_2$, which is called the cutoff condition of the waveguide.

3.3 The Zigzag-Wave Model

Solutions to the wave equation (3.1a) are plane waves. It turns out that the wave solution which satisfies the boundary conditions of a slab waveguide is a superposition of two plane waves which propagate diagonally opposite to one another. When these two plane waves are represented by wave vectors discussed in Sect.3.2, we find one wave vector pointing toward the lower right and the other wave vector pointing toward the upper right, the two wave vectors forming a zigzag path. It is then obvious that one can describe waves in the waveguide by an optical ray which zigzags inside the film. This picture of light propagation led us to the zigzag-wave model.

Consider in Fig.3.3a two dielectric media of refractive indices n_1 and n_2, respectively, and $n_1 > n_2$. Based on geometric optics, a ray of light in medium n_1 is totally reflected at the interface only if the incident angle θ_1 is larger than the critical angle θ_c, defined as

$$\theta_c = \arcsin(n_2/n_1) . \qquad (3.6)$$

The phenomenon is called total internal reflection, which is well known in elementary optics. In a slab waveguide, the film of index n_1 is

Fig.3.3. Zigzag-wave model of optical waveguide

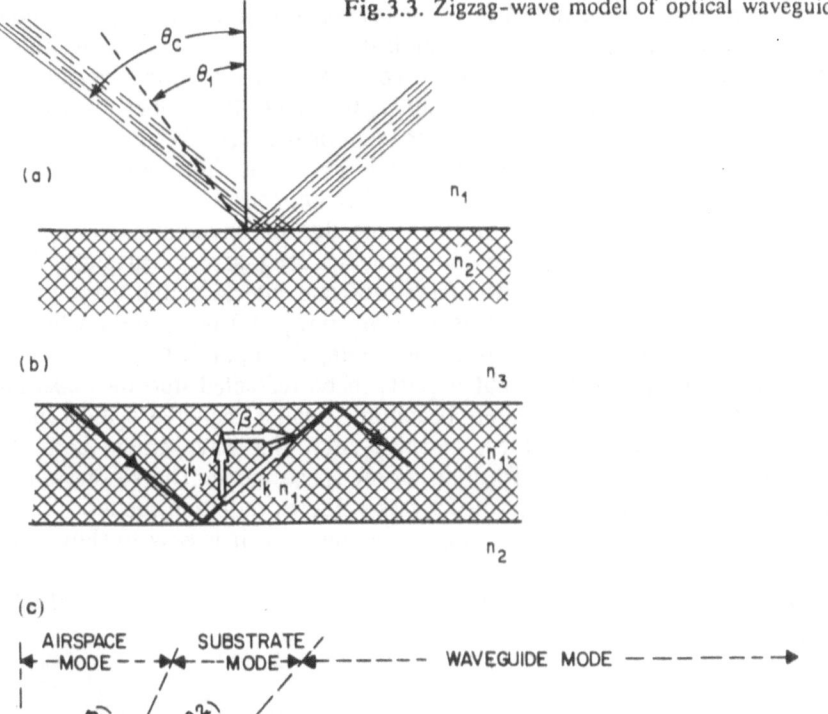

(a)

(b)

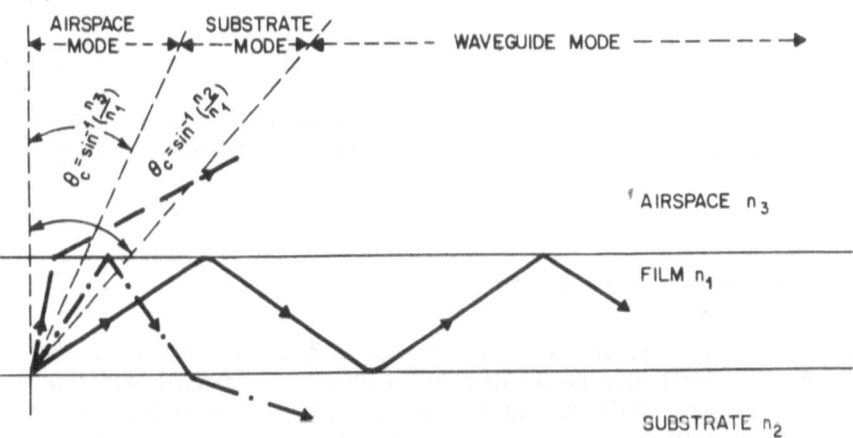

(c)

sandwiched between two low-index media n_2 and n_3. An optical ray which is totally reflected from the film-substrate interface will also be totally reflected at the film-air space interface. This ray of light will then be reflected back and forth between two interfaces, forming a zigzag wave. The optical energy is thus trapped inside the film as the wave propagates in the waveguide (Fig.3.3b).

Therefore, in ray optics, the wave in the waveguide does not travel straight in the z direction. Rather, it traces a zigzag path. There is a discrete set of zigzag angles corresponding to a discrete set of waveguide modes. On the other hand, in wave optics the zigzag wave may be considered as two planes superposed on one another in space. One wave travels toward the lower right, and the other toward the upper right. Both waves have the same k_z ($=\beta$), indicating a steady wave motion in the z direction. They have opposite values for k_y, indicating a standing wave along the y direction. It is clear that whether we consider a single optical ray or two superposed plane waves, the end result is the same. Ray optics are usually easier to handle mathematically and conceptionally.

In Fig.3.3c we start from an optical ray in the film which is nearly parallel to the interfaces, and examine what happens as the angle of incidence θ_1 is gradually reduced to zero. Note that the incident angle is measured from the normal to the interface. As long as $\theta_1 > \arcsin(n_2/n_1)$, which is the critical angle at the film-substrate interface, the wave in the film is a zigzag wave. As soon as $\theta_1 = \arcsin(n_2/n_1)$ or smaller, which is the cutoff condition of the waveguide mode, the light is still reflected at the film-air space interface, but it starts to be refracted into the substrate modes. As θ_1 is reduced further, becoming smaller than the critical angle at the film-air space interface, $\arcsin(n_3/n_1)$, the wave enters both the substrate and the air space, forming an air space mode. The substrate and air space modes have been defined by *Marcuse* as radiation modes.

Based on the wave vector diagram in Fig.3.3b, it is easy to show that

$$\beta = k_z = kn_1\sin\theta_1 , \qquad \text{and} \tag{3.7a}$$

$$k_y = kn_1\cos\theta_1 . \tag{3.7b}$$

It follows that when $\theta_1 = \arcsin(n_2/n_1)$ and $\arcsin(n_3/n_1)$, $\beta = kn_2$ and kn_3, respectively. These are relations of prime importance to waveguide optics.

3.4 The Potential-Well Model

The zigzag-wave model in the last section is built on the concept that a plane wave sheet may be divided into a number of parallel rays and we can use any one of the rays to represent the plane wave. In the potential-well model [3.4], we will introduce an entirely different concept, in which we consider a plane wave as a moving particle of momentum kn and kinetic energy k^2n^2.

According to the above concept, waves in media n_1, n_2 and n_3 have the kinetic energies $k^2n_1^2$, $k^2n_2^2$ and $k^2n_3^2$, respectively, where $n_1 > n_2 > n_3$. As a wave crosses the interface between media n_1 and n_2, it loses kinetic energy equal to $k^2(n_1^2-n_2^2)$. Similarly, it loses kinetic energy equal to $k^2(n_1^2-n_3^2)$ by entering into medium n_3. Because the total energy of the wave has to be conserved, the kinetic energy lost must be equal to the potential differences between the media n_2 and n_1 and those between n_3 and n_1, respectively. When these concepts are applied to a slab waveguide, we find the potential energy is the lowest in the guiding film, which is taken to be zero in Fig.3.4. The potential energy rises to $k^2(n_1^2-n_2^2)$ at film-substrate interface and to $k^2(n_1^2-n_3^2)$ at film-air

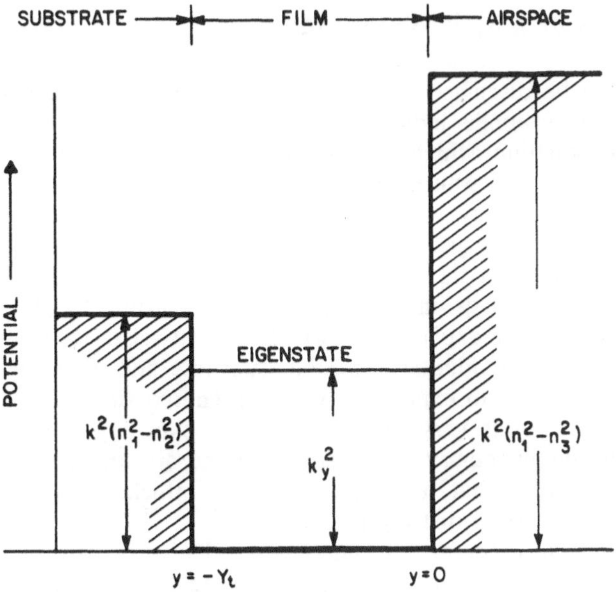

SUBSTRATE
FILM
AIRSPACE
POTENTIAL
EIGENSTATE
$k^2(n_1^2-n_2^2)$
k_y^2
$k^2(n_1^2-n_3^2)$
y = −Yt
y=0

Fig.3.4. The potential-well model

space interface. The result is a potential well (Fig.3.4) similar to the one studied in quantum mechanics.

As is well known in potential-well theory, there is a set of discrete energy states in the well. They are bound states of the particle which correspond to the waveguide modes or zigzag waves in optics. There are also continuous states above the well that correspond to the radiation modes in waveguide theory. According to (3.4a), kinetic energy involves x, y and z components which are $k_x{}^2$, $k_y{}^2$ and $k_z{}^2$, respectively. Since $k_z = \beta$, which is identical in film, substrate and air space, and since $k_x = 0$ in slab waveguides, the only component of the kinetic energy which varies from the film to substrate and to air space is $k_y{}^2$. Let us draw a set of bound-energy states in the well according to the potential-well theory. It is easy to verify that the vertical distance between the energy level of a bound state to the bottom of the well is $k_y{}^2$. The quantity $k_y{}^2$ is zero when the energy level meets the potential walls at the two interfaces and is negative by going into the walls. This again indicates that k_y is real in the well and imaginary outside, or the wave function is sinusoidal in the well and exponential outside the well. The wave function of a potential well is thus identical to the optical field distribution in the waveguide. In drawing an analogy between quantum mechanics and electromagnetic theory, one must exercise caution that boundary conditions of a potential well are continuity of the wave function and that of its derivative at potential walls, which are true for TE waves but are not true for TM waves. We will define TE and TM waves in the next section.

3.5 Mode Equations of a Slab Waveguide

The mode equations to be discussed in the next two sections are used to calculate eigenvalues β (or mode indices β/k), each of them defining a solution of the wave equation which satisfies boundary conditions of a given slab waveguide. These solutions are either waveguide modes or radiation modes. A simple slab waveguide (Fig.3.2) consists of three media: film, substrate and air space, and we will use subscripts 1, 2 and 3 to denote quantities in these media, and subscripts 12 and 13 to those at interfaces. Since waves propagage in the z direction, according to (3.2), fields vary in z in the form of $\exp(j\omega t - j\beta z)$ which is identical in all three media. The quantity β is then identical in film, substrate and air space. The fields vary in x according to $\exp(jk_x x)$ which is ignored since $k_x = 0$ in a slab waveguide. The fields vary in y according to $\exp(jk_y y)$, where k_y is real in the film and imaginary in the substrate and air space. For a given waveguide, we assume that n_1, n_2, n_3 and film thickness W are known. Below we write the field distributions separately in film, substrate and air space, and match them at the interfaces. The equation thus obtained is called the mode equation which may be used to solve for k_y. Once k_y is known, β may be determined from (3.4a). Although it is possible to derive mode equations based on the zigzag-wave model or the potential-well model, we prefer to work with Maxwell's equations in a traditional formulism of electromagnetic theory.

To avoid confusion, we will define all important parameters to be real and positive. Hence, if k_y is real, we use it as is; if k_y is imaginary, we substitute it by ip_y, where p_y is real and positive. We then have for (3.4a)

$$k_{y1}^2 = k^2 n_1^2 - \beta^2 , \tag{3.4b}$$

$$p_{y2}^2 = \beta^2 - k^2 n_2^2 , \tag{3.4c}$$

$$p_{y3}^2 = \beta^2 - k^2 n_3^2 . \tag{3.4d}$$

Using two curl equations of Maxwell's equations, we can express x and y components of the fields in terms of the z component as follows

$$E_x = -j \frac{1}{n^2 k^2 - \beta^2} \left[\beta \frac{\partial}{\partial x} E_z + \omega\mu \frac{\partial}{\partial y} H_z \right] , \tag{3.8a}$$

$$E_y = -j \frac{1}{n^2 k^2 - \beta^2} \left[\beta \frac{\partial}{\partial y} E_z - \omega\mu \frac{\partial}{\partial x} H_z \right] , \tag{3.8b}$$

$$H_x = -j \frac{1}{n^2 k^2 - \beta^2} \left[\beta \frac{\partial}{\partial x} H_z - \omega\epsilon n^2 \frac{\partial}{\partial y} H_z \right] , \tag{3.8c}$$

$$H_y = -j \frac{1}{n^2 k^2 - \beta^2} \left[\beta \frac{\partial}{\partial y} H_z + \omega \epsilon n^2 \frac{\partial}{\partial x} E_z \right] , \qquad (3.8d)$$

where μ and ϵn^2 are the permeability and dielectric constant of the medium, respectively. For MKS units, $\epsilon = 8.854 \cdot 10^{-12}$ Farad/meter and $\mu = 4\pi \cdot 10^{-7}$ Henry/meter. After eliminating all terms in (3.8) involving $(\partial/\partial x)$, which is zero, the result is two independent sets of equations. One set involves fields E_x, H_y and H_z only, which are called the TE wave. The other set involves fields H_x, E_y and E_z only, which are called the TM waves.

Taking a TE wave as an example, let the top surface of the film be at $y = W_A$ and the lower surface be at $y = -W_B$ so that the thickness of the film is $W = W_A + W_B$ (Fig.3.2). We can write

$$H_{z1} = A_1 \sin(k_{y1} y) , \qquad (3.9a)$$

$$H_{z2,3} = A_{2,3} \exp[-p_{y,2,3}(|y| - W_{A,B})] , \qquad (3.9b)$$

where the common factor $\exp(j\omega t - j\beta z)$ is omitted. Substituting (3.9) into (3.8) and remembering that $(\partial/\partial x) = 0$, we have

$$E_{x1} = -j \frac{1}{k^2 n_1^2 - \beta^2} [\omega\mu k_{y1} A_1 \cos(k_{y1} y)] , \qquad (3.10a)$$

$$E_{x2,3} = -j \frac{1}{k^2 n_{2,3}^2 - \beta^2} = \left[-\omega\mu p_{y2,3} A_{2,3} \exp\{-p_{y2,3}(|y| - W_{A,B})\} \right] . \qquad (3.10b)$$

By equating (E_{x1}/H_{z1}) to (E_{x2}/H_{z2}) at $y = W_A$, and (E_{x1}/H_{z1}) to (E_{x3}/H_{z3}) at $y = -W_B$, one obtains

$$\tan(k_{y1} W_A) = p_{y3}/k_{y1} , \qquad (3.11a)$$

$$\tan(k_{y1} W_B) = p_{y2}/k_{y1} . \qquad (3.11b)$$

A mode equation for the TE wave is obtained by combining (3.11a and b), namely

$$k_{y1} W = q\pi + \arctan(p_{y3}/k_{y1}) + \arctan(p_{y2}/k_{y1}) . \qquad (3.12)$$

In the same manner, we may derive a mode equation for the TM wave in a slab waveguide, which is

$$k_{y1} W = p\pi + \arctan(n_1^2 p_{y3}/n_3^2 k_{y1}) + \arctan(n_1^2 p_{y2}/n_2^2 k_{y1}). \qquad (3.13)$$

Here, p or q = 0,1,2,3... is the order of the waveguide mode. All arctangents have values between 0 and $\pi/2$. Equations (3.12, 13) are probably

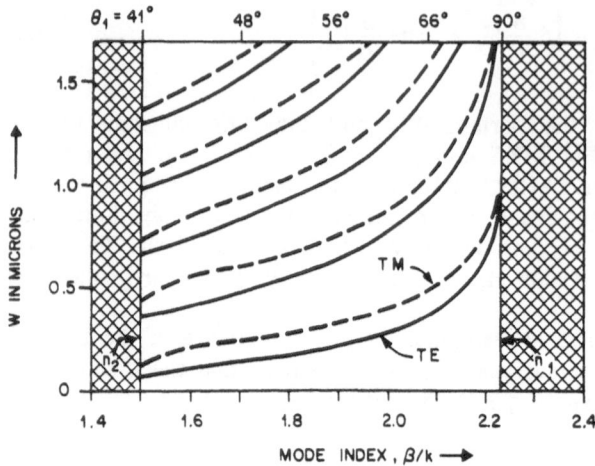

Fig.3.5. Mode index (β/r) versus waveguide thickness for ZnS on glass planar wave-guides for $\lambda = 1.06$ μm. Zigzag angles are shown for 1.5 μm thick waveguide

the most important equations in waveguide theory. For each value of p or q there is a waveguide mode. Fields of these modes are orthogonal to each other. Since the largest possible value of k_{y1} is $k(n_1{}^2-n_2{}^2)^{1/2}$ for both (3.12 and 13), we have p or q $< kW(n_1{}^2-n_2{}^2)^{1/2}/\pi$ and hence, the total number of waveguide modes that a waveguide can carry including TE and TM waves is less than $2 + [2kW(n_1{}^2-n_2{}^2)^{1/2}/\pi]$.

As a specific example, consider a waveguide formed by depositing 1.5 μm thick ZnS film on a glass substrate. Here, $n_1 = 2.2899$, $n_2 = 1.5040$ and $n_3 = 1.0$. Let the optical wavelength be 1.06 μm or k = $5.297 \cdot 10$ cm^{-1}. There are 5 TE modes at $\beta = 2.264$, 2.201, 2.086, 1.916 and 1.687 for mode orders q = 0, 1, 2, 3 and 4, respectively. Since $\beta =$ knsin(β_1) according to (3.6), the zigzag angle of these modes are $\theta_1 =$ 81.4, 74.0, 65,6, 56.8 and 47.4 degrees. These results are illustrated in Fig.3.5.

The terminology frequently used in waveguide theory is listed in Table 3.1 for reference.

Table 3.1. Terminology frequently used in waveguide theory

Waveguide-mode eigenvalue	β
Mode index	β/k
Mode order	p or q
Mode number	$(\beta^2-k^2n_2)^{1/2}(k_2n_1{}^2-k^2n_2)^{-1/2}$
Numerical aperture	$(n_1{}^2-n_2{}^2)^{1/2}$
V-number	$kW(n_1{}^2-n_2{}^2)^{1/2}$

3.6 Mode Equations of a Channel Waveguide

Before we discuss channel waveguides, let us recapitulate some essential features of a slab waveguide. Figure 3.6a shows a zigzag wave in a slab waveguide. The zigzag is the up-and-down wave motion shown in heavy dotted lines. The power flow or the Poynting vector is in the z direction. The fields responsible to this power flow are E_x and H_y for a TE wave and are H_x and E_y for a TM wave. Although the power flow is not dependent on H_z (TE wave) and E_z (TM wave), the zigzag angle is. Therefore, wave optics in the waveguide are completely characterized when these three field components are known.

In a channel waveguide, however, the situation is more complicated. The zigzag wave shown in Fig.3.6b involves both the up-down and sideways wave motions. More importantly, if we consider a TE wave, for example, because $(\partial/\partial x) \neq 0$, the fields contain components other than E_x, H_y and H_z such as E_y, H_x and E_z. Then both $E_x \cdot H_y$ and $E_y \cdot H_x$ could contribute to the Poynting vector in the z direction. To simplify this situation, we look for solutions in which $E_y = 0$ for TE waves (and $H_y = 0$ for TM waves) and argue that such solutions still form a complete set of waveguide modes. The above is at least intuitively true, although concrete proof is difficult. With $E_y = 0$, the Poynting vector in z consists of $(E_x \cdot H_y)$ only as though it were a slab waveguide. We call this type of solution TE-like waves, which behave very much like the TE waves of a slab waveguide.

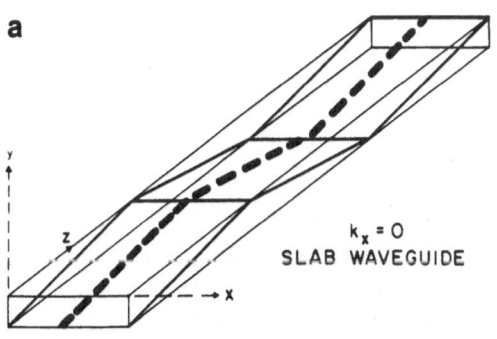

a

$k_x = 0$
SLAB WAVEGUIDE

Fig.3.6. Comparison of slab and channel waveguides. See text for details

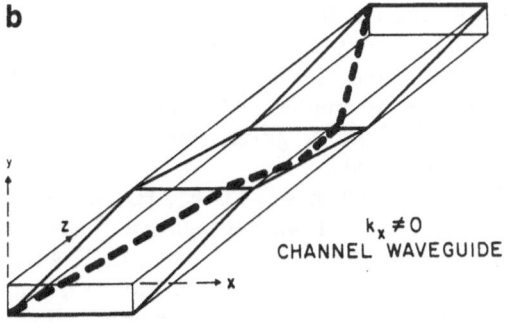

b

$k_x \neq 0$
CHANNEL WAVEGUIDE

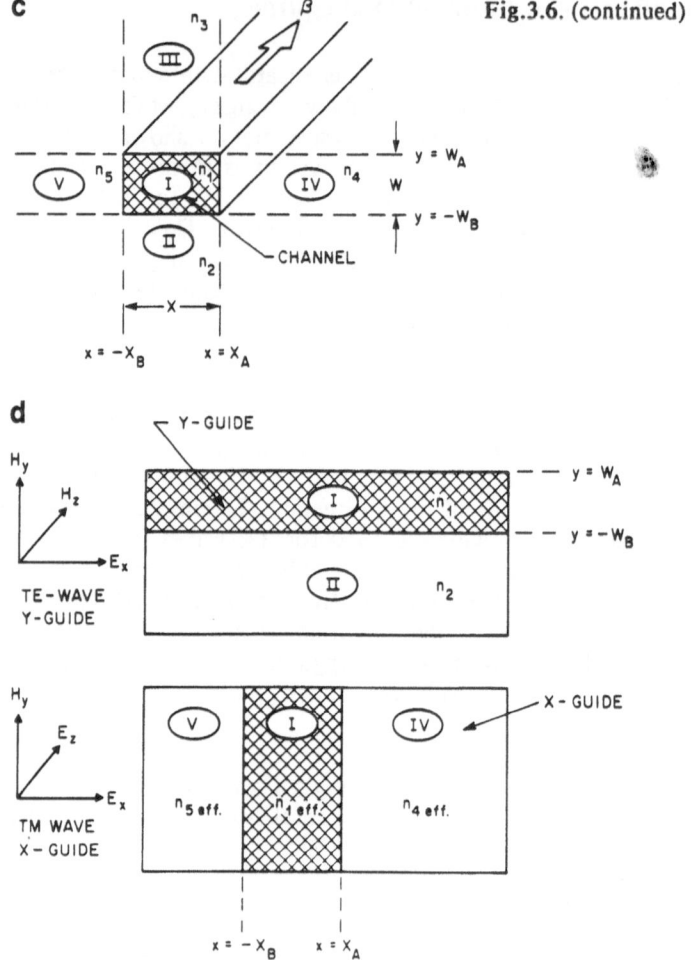

Fig.3.6. (continued)

c

n_3

β

III

$y = W_A$

V n_5 I n_1 IV n_4 W

$y = -W_B$

II

n_2 ——CHANNEL

|← x →|

$x = -X_B$ $x = X_A$

d

Y-GUIDE

H_y

H_z

$y = W_A$

I n_1

$y = -W_B$

E_x

TE-WAVE
Y-GUIDE

II n_2

H_y

X-GUIDE

E_z

V I IV

E_x

TM WAVE
X-GUIDE

$n_{5 \text{ eff.}}$ $n_{1 \text{ eff.}}$ $n_{4 \text{ eff.}}$

$x = -X_B$ $x = X_A$

The channel waveguide in Fig.3.6c is divided into five regions. Region I is the channel, and the regions immediately below, above, at right and at left of the channel are regions II, III, IV and V, respectively. Quantities in these regions are identified by subscripts 1, 2, 3, 4 and 5. The spaces at the four corners are ignored. If we view the waveguide as drawn in Fig.3.6c, regions I, II and III may be considered a slab waveguide. If we turn the diagram 90 degrees in the plane of the paper, regions I, IV and V may be identified as another slab waveguide. These two slab waveguides are called Y-guide and X-guide, respectively. They share region I which is the channel (Fig.3.6d).

Let the guide-Y carry a TE wave which has fields E_x, H_y and H_z in regions I, II and III. Let guide-X carry a TM wave which has fields E_x, H_y and E_z in regions I, IV and V. Interestingly, a TE wave in Y-guide and a TM wave in X-guide have E_x and H_y in common. More so, they are the field components responsible for the power flow in the z direction. We may then conclude that the TE-like wave of a channel wave-guide that we seek, is a combination of a TE wave in the Y-guide and a TM wave in the X-guide such that E_x and H_y determine the power flow. H_z determines the zigzag angle of the up-down wave motion, and E_z the zigzag angle of the left-right wave motion. We will show below such TE-like waves are indeed consistent with Maxwell's equations.

Let us write a wave equation for the channel, similarly to (3.1a),

$$\left(\frac{\partial^2}{\partial x^2} + \frac{\partial^2}{\partial y^2} + \frac{\partial^2}{\partial z^2}\right) E_1 = - k^2 n_1{}^2 E_1 \tag{3.1d}$$

or, similarly to (3.4a)

$$k_{x1}^2 + k_{y1}^2 + \beta^2 = k^2 n_1{}^2 , \tag{3.4b}$$

where n_1 is the refractive index of the channel. The channel has a width X and a thickness W, with X > W. Note that we have placed the longer dimension of the cross section along the x direction, and with this, we expect that $(\partial/\partial x)$ is smaller than $(\partial/\partial y)$ and may be ignored as far as the Y-guide is considered. With this approximation, the mode equation for the Y-guide is

$$k_{y1} W = q\pi + \arctan(p_{y3}/k_{y1}) + \arctan(p_{y2}/k_{y1}) , \tag{3.14}$$

which is identical to (3.12). After solving k_{y1} from (3.14) and substituting $(\partial^2/\partial y^2)$ by $(-k_{y1}{}^2)$ in (3.1d) for regions I, IV and V, we have

$$(-k_{x1}^2 - \beta^2)E_1 = - k^2[n_1{}^2 - (k_{y1}^2/k^2)]E_1 , \tag{3.15a}$$

$$(p_{x4}^2 - \beta_2)E_4 = - k^2[n_4{}^2 - (k_{y1}^2/k^2)]E_4 , \tag{3.15b}$$

$$(p_{x5}^2 - \beta^2)E_5 = - k^2[n_5{}^2 - (k_{y1}^2/k^2)]E_5 . \tag{3.15c}$$

They are the equations of the X-guide, which have the effective indices

$$n_{1eff} = \sqrt{n_1{}^2 - (k_{y1}^2/k^2)} , \tag{3.16a}$$

$$n_{4eff} = \sqrt{n_4{}^2 - (k_{y1}^2/k^2)} , \tag{3.16b}$$

$$n_{5eff} = \sqrt{n_5{}^2 - (k_{y1}^2/k^2)} . \tag{3.16c}$$

We can thus write immediately a mode equation for the X-guide, remembering that it is a TM wave

$$k_{x1} X = p\pi + \arctan \frac{n_{1eff}^2 P_{x4}}{n_{4eff} k_{x1}} + \arctan \frac{n_{1eff}^2 P_{x5}}{n_{5eff} k_{x1}} . \qquad (3.17)$$

Solving (3.14) for k_{y1} and (3.17) for k_{x1}, the eigenvalue β of the channel waveguide can be determined from (3.4b). We use the symbols TE_{qp} and TM_{qp} to identify modes of a channel waveguide. For each set of values of p and q, we have a TE-like wave. Following the same procedure outlined above, we can also obtain a TM-like wave for the same values of p and q. The maximum number of waveguide modes which a channel waveguide can carry is then

$$2 + [2k^2 WX(n_1{}^2 - n_2{}^2)(n_1{}^2 - n_4{}^2)/\pi^2]^{1/2} ,$$

where $n_2 > n_3$ and $n_4 > n_5$.

3.7 Effective-Index Method

The analysis given in Sect.3.6 is probably the best we can do for a channel waveguide without expanding the fields into a series of spatial harmonics. One drawback of the analysis is, however, that we completely ignore the four corner regions in Fig.3.6b, and the effect of these regions can be important in certain waveguide structures such as ridge waveguides and V-groove lasers. A modification of the above analysis which includes the effect of four corner regions is called the effective-index method [3.6]. The method has been used extensively in the analysis of lasers with excellent results. To illustrate this method below, we take a ridge waveguide as an example.

The ridge waveguide shown in Fig.3.7 consists of a high-index film of index n_1 deposited on a low-index substrate of index n_2. As usual, the airspace has the index n_3. The film has a small ridge at the center which confines a small part of optical energy in the ridge, leaving the larger part of the optical energy to the area closely below the ridge. The ridge waveguide is thus a channel waveguide although it has the appearance of a slab waveguide.

In the effective-index method, we divide the waveguide into three columns A, B and C, and the ridge occupies the center column B as shown in Fig.3.7. We consider the three columns of the ridge waveguide as three separate slab waveguides, or more precisely, three separate Y-guides. The quantities k_{y1}, k_{y4} and k_{y5} of the above guides may then be determined by applying (3.14) three times, and the results are used to calculate effective indices of regions I, IV and V as follows:

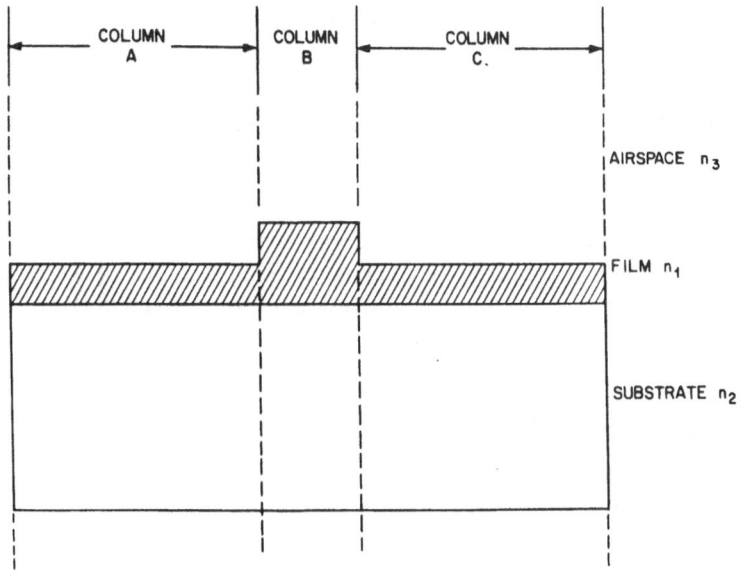

Fig.3.7. Effective-index analysis of ridge waveguide

$$n_{1eff} = \sqrt{n_1{}^2 - (n_{y1}^2/k^2)} \, , \qquad\qquad\qquad\qquad (3.16d)$$

$$n_{4eff} = \sqrt{n_4{}^2 - (n_{y4}^2/k^2)} \, , \qquad\qquad\qquad\qquad (3.16e)$$

$$n_{5eff} = \sqrt{n_5{}^2 - (n_{y5}^2/k^2)} \, . \qquad\qquad\qquad\qquad (3.16f)$$

Except for the k_y, equations in (3.16a-c) are identical to those in (3.16d-f). It is now a simple matter to follow the steps outlined in the last section for the calculation of k_{x1} from (3.17) and then using the values of k_{y1} and k_{x1} for the calculation of β from (3.15a).

It is clear that the only difference between the effective-index method and the analysis given in Sect.3.6 is the calculation of k_y, and thus the effective indices of regions IV and V. In the earlier analysis, only the Y-guide in the center column is considered and hence the effects of columns B and C are neglected. In the effective-index method, regions I, IV and V have different k_y they are k_{y4}, k_{y4} and k_{y5}, respectively. Since the k_y must be matched along the vertical edges of the channel, one may then ask how they could be different. The answer is that an exact analysis must include spatial harmonics for matching fields at the interfaces. When that is done, k_y for the leading term in the field expansion of region I could be slightly different from those of regions IV and V. For that reason, the effective-index method often gives more accurate results.

3.8 Diffused Waveguides and the WKB Method

The most commonly used technique to form waveguides is diffusion of titanium into a lithium niobate [3.7] substrate forming electro-optic and acousto-optic waveguides. The fabrication of these waveguides will be discussed in detail later. In a typical diffused waveguide, the refractive index is largest at the substrate surface ($y = 0$) and decreases monotonically below the surface (Fig.3.8). Let the index of the substrate be n_2 before diffusion, and its index at the surface be increased by an amount Δn after the diffusion. Assuming the thickness of the diffused layer to be d, we then have

$$\left(\frac{\partial^2}{\partial y^2} + \frac{\partial^2}{\partial z^2}\right)E = -k^2[n_2{}^2 + 2n_2\Delta n\exp(-|y|/d)]E \qquad (3.18)$$

for $y < 0$ in the substrate, and

$$\left(\frac{\partial^2}{\partial y^2} + \frac{\partial^2}{\partial z^2}\right)E = -k^2 n_{3E}^2 \qquad (3.19)$$

for $y > 0$ in the airspace which has a refractive index n_3. We have assumed an exponential diffusion profile to allow an analytical solution. Below we write the fields separately in the substrate and airspace, and obtain a mode equation by matching fields at $y = 0$.

An analytical solution of (3.1c) has been given by *Conwell* [3.8] in terms of Bessel functions. Considering again a TE wave, we may write

$$E_{x1} = A\, J_{2P_{y2}d}\left[2\sqrt{2n_2\Delta n}\ kd\ \exp(-|y|/2d)\right] \qquad (3.20a)$$

for $y < 0$, and

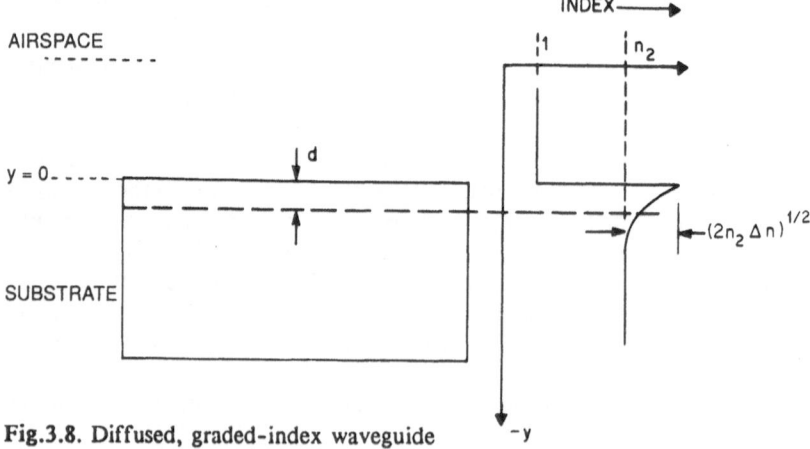

Fig.3.8. Diffused, graded-index waveguide

$$E_{x3} = \beta \exp(-p_{y3}y) \qquad (3.20b)$$

for $y > 0$, where the common factor $\exp(j\omega t - j\beta z)$ is omitted; A and B are amplitudes of the fields and $J_{2p_{y2}d}$ is the Bessel function of the first kind of $(2p_{y2}d)$th order. In addition we have from (3.3)

$$\beta^2 - p_{yj} = k^2 n_j{}^2 \qquad (3.21)$$

where $j = 2$ or 3. After obtaining H_{z1} and H_{z3} from E_{x1} and E_{x3} using (3.7) and equating the fields at $y = 0$, we obtain a mode equation in the following form:

$$\frac{J_{2p_{y2}d}(\xi) - J_{2p_{y2}d}(\xi)}{J_{2p_{y2}d}(\xi)} = - \frac{2p_{y3}}{k\sqrt{2n_2\,\Delta n}} \qquad (3.22)$$

where $\xi = 2(2n_2\Delta n)kd$. In practice, the right-hand member of (3.22) is a very large quantity and may be considered as infinity. Equation (3.22) is then simply

$$J_{2p_{y2}d}(\xi) = 0 \qquad (3.22a)$$

from which we can determine p_{y2}, and from p_{y2} we calculate β from (3.21). As an example, consider a waveguide: $n_2 = 2.46$, $n_3 = 1$, $d = 2.5$ μm and $(2n_2\Delta n) = 0.3$. The wavelength of the light is 0.633 μm. From the above data, we find $\xi = 27$. Substituting ξ into (3.22a), we find eight TE waveguide modes for $p_{y2} = 4.3$, 3.48, 2.82, 2.24, 1.7, 1.21, 0.76 and 0.34, respectively [3.8]. The corresponding value of β for the lowest-order mode is 2.509. The field distributions in the substrate for the lowest- and the fifth-order modes are plotted in Fig.3.9. It is interesting that fields of the higher-order modes extend inside the substrate far beyond the thickness of the diffused layer d, due to the exponential nature of index distribution.

Another method frequently used to study diffused waveguides is the use of the potential-well model. Based on (3.1c) we can draw the potential diagram shown in Fig.3.10. According to the well-known WKB approximation [3.9], the condition for the quantization of the momentum is

$$\int_{-y_t}^{0} k_{y2}(y)\,dy = q\pi + \Phi_{12} + \Phi_{13} , \qquad (3.23)$$

where $k_{y2}(y)$ is the momentum of the particle in the y direction in diffused waveguides and is a function of y,

$$[k_{y2}(y)]^2 + \beta^2 = k^2\,[n_2{}^2 + 2n_2\,\Delta n\exp(-|y|/d)] . \qquad (3.24)$$

71

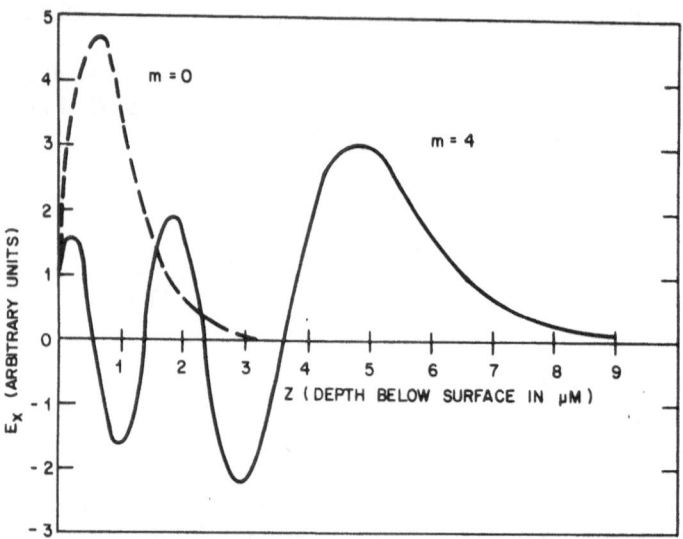

Fig.3.9. Field distribution of two modes of a diffused optical waveguide. See text for the waveguide parameters

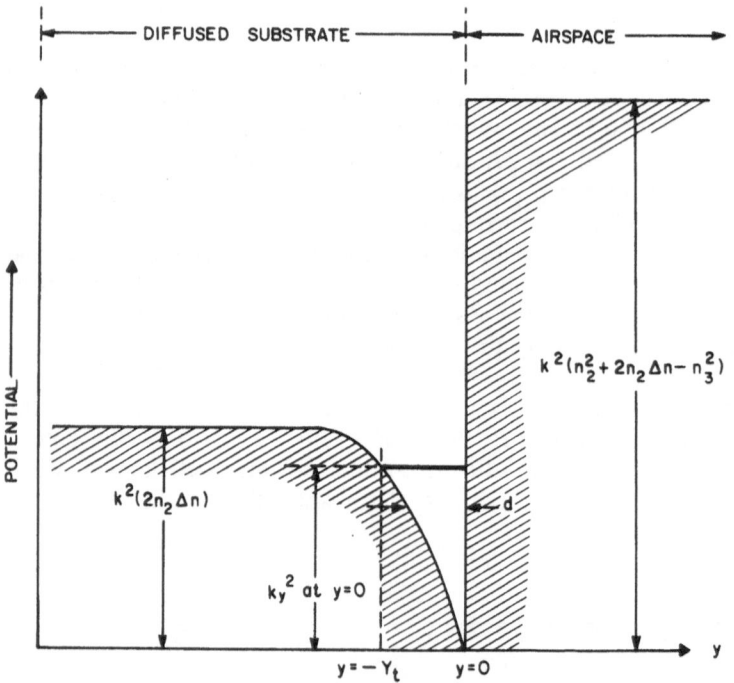

Fig.3.10. Potential-well model of diffused waveguide

We recall that $k_y^2(y)$ is the distance between the energy level of an eigenstate and the bottom of the potential well. The integral in (3.22) is carried out between two turning points $y = -y_t$ and $y = 0$ where the

72

Table 3.2. The mode indices computed by the WKB method are compared with those computed by the exact theory on E. Conwell. Case 1 is calculated for a diffused waveguide in which d = 2.23 μm, Δn = 0.0987, and n_0 0 2.177. Case 2 is for another diffused waveguide in which d = 0.931 μm, Δn = 0.043, n_0 = 2.177

m	Conwell's theory	WKB method
	Case 1: 9 modes	
0	2.2400	2.2399
1	2.2196	2.2192
2	2.2059	2.2057
3	2.1961	2.1959
4	2.1889	2.1888
5	2.1838	2.1835
6	2.1802	2.1800
7	2.1781	2.1778
8	2.1771	2.1771
	Case 2: 2 modes	
0	2.1897	2.1898
1	2.1789	2.1790

energy level meets the potential walls (Fig.3.10). The quantities Φ_{12} and Φ_{13} are the phases introduced to the wave function at the turning points. According to the standard WKB approximation, we may take Φ_{13} as $\pi/2$ and Φ_{12} as $\pi/4$. With these values understood, (3.22 and 24) may be used to solve for $k_{y2}(y)$ and β for each value of q which is the order of the mode. Table 3.2 lists the results computed by the WKB method as compared to those from Conwell's theory for two different diffused waveguides. The agreement between these two methods is better than one part in 10,000.

3.9 Waveguide Materials

Much research has been devoted to finding the optimum waveguide material system. Important substrate and waveguide characteristics include: low propagation loss, fabrication control of the waveguide width (W), depth (d) and waveguide-substrate index difference (Δn = n_1-n_2), good electro-optic and acousto-optic figures of merit, fabrication ease and reproducibility and stability with respect to environmental conditions, and time. Both the substrate absorption loss and scattering loss that may result due to waveguide formation are important in determining the total waveguide loss. The ability to guide both polarizations (TE and TM), to provide a relatively large Δn, and to conveniently produce channel waveguides are also desirable. Furthermore, the availability of large (e.g., 5 cm^2), defect-free substrate material is practically important.

Optical waveguides have been made in several important substrates using a variety of techniques. The most important substrates include glass, ferroelectric crystals, lithium niobate and lithium tantalate, as well as semiconductors, primarily GaAs and more recently InP. As described in detail above, to achieve planar (one-dimensional) waveguiding, the refractive index must be increased in the waveguide region relative to the surrounding region.

3.9.1 Glass

Early waveguides were formed by RF sputter-depositing Corning 7059 glass onto similar substrates [3.10]. The sputtered film has a refractive index about 0.1 larger than the substrate. Propagation losses of about 1dB/cm for $\lambda = 0.6328\ \mu m$ were reported.

Low-loss waveguides in glass can also be formed very conveniently by ion exchange in a molten solution [3.11]. Waveguide formation by Na$\longleftrightarrow$Ag ion exchange is shown schematically in Fig.3.11, but other ions, particularly potassium, may also be used. The sodium glass substrate is immersed in molten $AgNO_3$. The waveguide depth is determined by the melt temperature and immersion time while the value of Δn is determined by the sodium content of the glass and the Ag concentration in the melt. Generally, Δn values as high as $\Delta n = 0.1$ can be achieved. A typical melt temperature is 230°C. Smaller values of Δn can be achieved by diluting the melt with, for example, $NaNO_3$ [3.12]. However, very careful control of the relative Ag/Na concentration in the melt is essential for fabrication control over Δn.

Performing the ion exchange in an external applied electric field yields deeper waveguides [3.13, 14]. A double-exchange procedure, first in $AgNO_3$, then in pure $NaNO_3$ has been used to produce buried waveguides [3.13]. The second exchange, for a shorter time than the first, returns the refractive index near the surface to its bulk value. Strip waveguides of this type are ideally suited for fiber coupling, as described later.

3.9.2 Semiconductor

While waveguides have been reported in several semiconductor substrates, the most important systems are the ternary [3.15] GaAs/$Ga_x Al_{1-x}$As and the quaternary [3.16] InP/$In_y Ga_x As_{1-x} P_{1-y}$. Planar waveguides are formed on semiconductor substrates by lattice-matched growth of a thin epilayer. The epilayer can differ from the substrate either by doping or composition. The refractive index change due to the free carrier concentration N is [3.17]

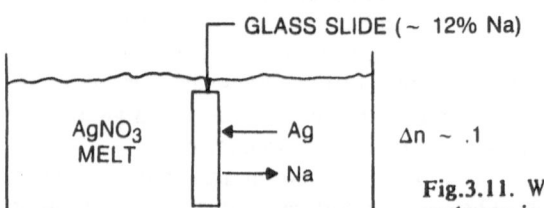

Fig.3.11. Waveguide formation by Ag ion exchange in sodium-based glass

74

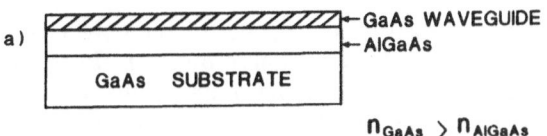

Fig.3.12. Planar, composition-based waveguide on (a) GaAs and (b) InP

$$\Delta n = - \frac{N\lambda^2 e^2}{8\pi\epsilon_0 n_s m^* c^2} , \qquad (3.25)$$

where λ is the free-space wavelength, e is the free carrier charge, ϵ_0 is the permittivity of free space, n_s is the bulk refractive index, m^* is the reduced mass, and c is the speed of light. Equation (3.25) indicates that to achieve waveguiding, the epilayer must have a lower doping level than the substrate. This is fortuitous because of the lower propagation loss in the lower doped region. Nevertheless, the optical mode's evanescent tail, which extends into the substrate, experiences the high free-carrier loss. The waveguide-substrate index change, achievable from the doping level difference only, is rather limited compared to that achievable by compositional change. In GaAs, for example, (3.25) indicates that for $\lambda = 1.06$ μm and $N = 10^{18}$/cm, this technique yields a Δn of about 0.006.

The heterostructure (Fig.3.12) in which the epilayer differs in composition from the substrate is the more desirable semiconductor waveguide. Such structures have, of course, been extensively studied for the development of semiconductor injection lasers and electro-optic devices as well. In the ternary system, GaAs has a refractive index larger than $Ga_2Al_{1-x}As$, which is proportional to x. Therefore, because the substrate material is GaAs, the simplest planar waveguide is a three-layer structure shown in Fig.3.12a. A waveguide-substrate index difference of $\Delta n = 0.1$ can be readily achieved. Correspondingly, a typical epitaxial thickness to produce single-mode waveguides is 1 μm or less. A typical planer epitaxial waveguide on InP is shown in Fig.3.12b. In this case the higher-index quaternary layer can be grown directly on the InP substrate. If a doped substrate is used, an intermediate buffer layer of undoped InP should be grown to eliminate unnecessary free carrier losses.

Epitaxial growth has been achieved with Liquid-Phase Epitaxy (LPE) [3.18], Vapor-Phase Epitaxy (VPE) [3.19] and Molecular-Beam Epitaxy (MBE) [3.20] epitaxy and with Metal-Organic Vapor-Phase Epitaxy (MOVPE) [3.21]. The first two techniques typically result in an approximately step index change, while the latter two can, in principle, provide a graded index. MBE and MOVPE also offer the advantage of finer film-thickness resolution, about 5 nm compared to about 50-100 nm for LPE.

The useful waveguide area is determined by the smaller of the defect-free substrate areas or the uniformity of the epitaxial growth. LPE-grown waveguides on GaAs are typically limited to about 1 cm² although it is not clear that larger areas of acceptable quality are impossible. The uniformity requirements to simply assure single-mode waveguides are not particularly demanding. According to (3.12, 13) the air clad step index waveguide can tolerate changes in depth of around plus or minus 30% and maintain single-mode operation. However, several acousto-optic and electro-optic devices are based upon either transverse or colinear phase-matched interaction, for example the acousto-optic deflector. Spatial variation of the effective index severly limits the performance of such devices. MBE and MOVPE offer advantages with respect to uniformity.

The propagation loss in waveguides on GaAs substrates, which were greater than 4 dB/cm for many years, have been dramatically reduced in the last several years, primarily due to to improved growth techniques. Presently, using MOVPE growth and careful waveguide design, waveguide losses as low as 0.2 dB/cm have been achieved for a wavelength of 1.3 μm [3.22]. Although GaAs waveguides have been explored principally for electro-optic device application, they have also been utilized for acousto-optic applications [3.23]. A progress report on the latter together with the most recent development on waveguide lenses in GaAs is given in Chap.8 by C.S. Tsai.

3.9.3 Lithium Niobate

The good electro-optic and acousto-optic properties of lithium niobate and lithium tantalate make these crystals excellent substrates for waveguide devices. A detailed treatment on resulting wideband acousto-optic waveguide devices and their applications is given in Chap.5 by C.S. Tsai. Waveguides have been formed in these crystals by outdiffusion, ion exchange, ion implantation and indiffusion. After a brief description of waveguides formed by the first three techniques we will discuss in detail diffused lithium niobate waveguides, which have been the most popular waveguides for device research.

The extraordinary refractive index n_e of $LiNbO_3$, a uniaxial crystal, increases with a decrease in the Li concentration. Thus a waveguiding layer can be formed by simply heating $LiNbO_3$ (and $LiTaO_3$ - most of the following discussion applies to either crystal but for simplicity we refer to lithium niobate only) to outdiffuse Li [3.24]. The waveguide depth depends upon the diffusion temperature and time. Although outdiffused waveguides are low-loss (<1dB/cm), the achievable index change is relatively small, $\Delta n \simeq 0.001$. As a result, relatively deep Li-deficient layers ($5-10\mu$m) are required. Typical diffusion parameters to form single-mode waveguides for $\lambda = 0.6328$ μm are a diffusion temperatue and time of 1000°C and about 10 h, respectively.

In addition to the rather low values of Δn, channel waveguides cannot be easily formed by outdiffusion. Furthermore, the ordinary refractive index n_o shows little dependence upon lithium concentration, so that light polarized along a direction affected by n_o is not guided. For example, for z-cut, y-propagating crystals, light polarized perpendicular to the substrate plane (TM-like polarization) can be guided, while light polarized in the plane (TE-like) cannot. A significant advantage of out-diffused $LiNbO_3$ waveguides is their enhanced resistance to optically induced refractive index change. Such photo-refractive changes in bulk $LiNbO_3$, frequently referred to as optical damage, can severely restrict the allowable optical power density for operation at visible wavelengths [3.25]. Several efficient waveguide devices including an early waveguide acousto-optic deflector [3.26] and electro-optic Bragg modulator have been demonstrated with outdiffused $LiNbO_3$ waveguides [3.27].

Ion exchange waveguides in $LiNbO_3$ were first made using the technique described for glass substrates: immersion in a molten bath of $AgNO_3$ [3.28]. Large index changes, Δn about 0.1, were achieved but only for the extraordinary index of x-cut crystals. Subsequently, waveguides were formed in $LiNbO_3$ and $LiTaO_3$, using either an $AgNO_3$ or a $ThNO_3$ melt [3.29]. However, subsequent efforts produced nonreproducible results and closer investigation indicated that little thallium, for example, entered the crystal [3.30]. Apparently the index change is caused by impurities, probably hydrogen, in the melt.

Much more controlled hydrogen ion (proton) exchange can be achieved using benzoic acid as the source [3.30]. When performed in pure benzoic acid, the extraordinary index increase is as large as $\Delta n = 0.12$ while the ordinary index decreases slightly. The index profile is step-like and planar waveguide losses of about 0.5 dB/cm have been measured. An advantage of this technique, in addition to the large index change, is the relatively simple fabrication process. The exchange apparatus is similar to that for ion exchange in glass, as shown in Fig.3.11. Typical exchange temperatures are in the range of 200-250°C.

Unfortunately, waveguides fabricated with a pure benzoic acid melt have exhibited problems including index instability in time [3.31] and reduced electro-optic [3.32] and acousto-optic [3.33] efficiency. Some of these problems have been overcome by using a diluted benzoic acid melt. In particular, dilution with lithium benzoate to provide excess lithium ions in the melt has led to improved index change stability and good electro-optic efficiency [3.34].

Research to improve the techniques for fabricating proton-exchange waveguides in lithium niobate is ongoing. Proton-exchange waveguides have several important features. These include the large index change, which potentially allows strong coupling in etched grating [3.35] or acousto-optic devices as well as improved waveguide lens [3.36] performance. Also important is their increased resistance to optical damage [3.37]. In addition, employed in conjunction with titanium indiffusion, proton exchange provides increased flexibility for design of passive devices such as polarizers [3.38], and efficient second-harmonic generators [3.39].

The most widely used waveguides for acousto-optic and electro-optic devices are made by diffusion of titanium into lithium niobate [3.7, 40].

The waveguides are fabricated by depositing a layer (thickness τ) of titanium on a polished, cleaned $LiNbO_3$ crystal and subsequently heating the crystal for several hours at a temperature of 950-1100°C. The produced graded refractive index depends upon the indiffused TiO_2 concentration. Typically, the titanium thickness is small compared to the diffusion depth and the concentration profile into the crystal can be written as

$$c(z) = \frac{2}{\sqrt{\pi}} \frac{\rho\tau}{D} \exp[-(z/D)^2] , \qquad (3.26a)$$

where ρ is the atomic density of the titanium film, and D is the diffusion depth

$$D = 2\sqrt{D_z}t . \qquad (3.26b)$$

D_z is the temperature-dependent diffusion coefficient, and t is the diffusion time,

$$D_z = (D_z)_0 \exp(-T_0/T) , \qquad (3.26c)$$

where $(D_z)_0$ and T_0 are material constants that also depend upon crystal orientation and stoichiometry [3.40].

The relationship between Δn and the indiffused concentration c depends upon whether the ordinary or the extraordinary index is used. Δn_e is linearly proportional to c for all values of interest. Δn_o is also linearly proportional to c for small c, but becomes sublinear at larger values of c [3.41]. For typical values of concentration required for guiding, $\Delta n_e > \Delta n_o$. In either case the peak index change, which occurs at the surface, depends upon τ/D. Peak Δn values as high as 0.01 to 0.02 have been achieved. Consequently the important waveguide parameters can be independently design-controlled via the fabrication parameters. For $LiNbO_3$ the diffusion can be performed below the Curie temperature (about 1150°C), eliminating the need to repole the crystal. The lower Curie temperature of $LiTaO_3$ ($T_c \simeq 850°C$) generally necessitates repoling the crystal after titanium indiffusion. As a result, $LiNbO_3$ has found more widespread use than $LiTaO_3$ for device applications. However, it should be noted that $LiTaO_3$ has the advantage of being much less susceptible to optical damage.

Single-mode $Ti:LiNbO_3$ waveguides can be formed over a rather wide range of fabrication parameters. Representative values for both visible ($\lambda = 0.6328\mu m$) and infrared ($\lambda = 1.06$ and $1.3\mu m$) operation can be found in the literature [3.43]. The particular diffusion parameters used also depend upon the application. For example, to reduce the drive power in active devices, a waveguide with the modal field concentrated near the surface is desirable to provide maximum interaction with the applied acoustic or electro-optic field. This requires strong guiding (large Δn) and relatively shallow diffusion depth that results from large τ, and small T and t. However, if these conditions are made too extreme, the diffusion

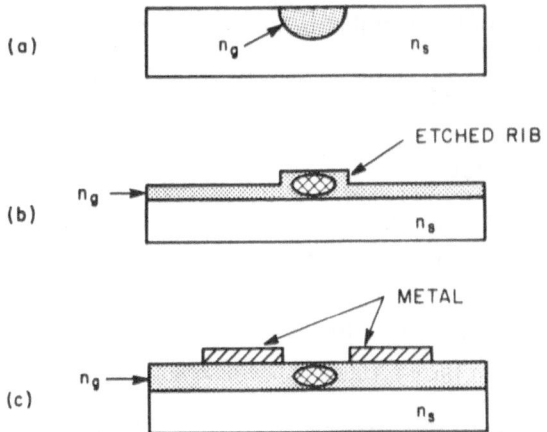

Fig.3.13. Channel waveguides (a) diffused, (b) ridge and (c) metal-clad

may be incomplete. The result is high loss due to poor guiding and scattering off the surface residue (undiffused surface TiO_2). Generally, there is sufficient variation from laboratory to laboratory (and perhaps between crystal suppliers) that one must experimentally determine the best fabrication parameters for one's own process and application.

Evaporation (both resistive and e-beam heating) and RF sputtering have been used for titanium deposition. The effective density may depend upon the deposition process. Therefore, for identical diffusion parameters the resulting peak refractive index change may be different for two titanium films of identical thickness deposited by different methods. In addition, to insure reproducible results it is preferable to use the same deposition rate by using, for example, a microprocessor-controlled feedback loop with an in-situ quartz monitor.

3.9.4 Channel Waveguides

For many waveguide device applications, particularly when coupling to optical fibers is necessary, channel waveguides are required. Several methods for obtaining channel waveguides are shown in Fig.3.13.

The simplest and most convenient channel waveguides are formed by photolithographic masking (Fig.3.13a). Indeed, the compatibility with masking is a significant advantage for a waveguide fabrication technique. Channel waveguides can be made in this way by ion exchange and metal indiffusion. For ion exchange waveguides, either an aluminum or silicon nitride negative pattern can be used to block exchange except in the open strip region. A boundary condition analysis of masked ion exchange indicates that a metal mask should give a sharp index change at the mark edge. The boundary is less abrupt for the dielectric mask [3.43].

For diffused waveguides the metal strip can be photolithographically delineated by etching or liftoff. The processing steps for fabricating strip $Ti:LiNbO_3$ waveguides by liftoff are shown in Fig.3.14. The mask to del-

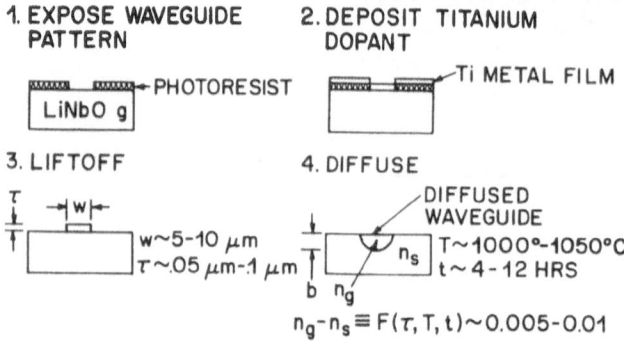

1. EXPOSE WAVEGUIDE PATTERN

2. DEPOSIT TITANIUM DOPANT

3. LIFTOFF

4. DIFFUSE

$n_g - n_s \equiv F(\tau, T, t) \sim 0.005 - 0.01$

Fig.3.14. Fabrication steps for titanium-diffused lithium niobate strip waveguides

ineate the strip or a generally more complicated device pattern is typically written by a computer-controlled scanning electron beam. This pattern is photographically reproduced onto the UV-sensitive photoresist spun onto the crystal (Fig.3.14). After metal deposition, the substrate is immersed in a solvent which dissolves the photoresist, removing the metal outside the channel. Good vertical photoresist sidewalls and substrate cleanliness are important for metal delineation. Using a flexible, conformable mask and vacuum draw for good substrate-mask contact, metal widths as narrow as 0.5 μm can be defined. The metal is then diffused into the crystal. The diffusion results in significant surface swelling. A photomicrograph of a pair of strip Ti:LiNbO$_3$ as viewed with Nomarski optics is shown in Fig.3.15.

Unless special measures are taken to avoid it, Li$_2$O outdiffusion can occur outside the strip region during titanium indiffusion [3.45]. This can result in undesirable planar guiding for the extraordinary polarization. Several methods have been used to suppress such planar guiding [3.46]. For example, by placing a source of Li$_2$O in the diffusion furnace, the net outdiffusion can be reduced sufficiently to eliminate surface guiding [3.47]. Another approach is to perform the diffusion in a closed container

Fig.3.15. Photomicrograph of pair of Ti:LiNbO$_3$ waveguides

[3.48]. Yet another technique, initially used to reduce optical damage, is to perform the diffusion in a water vapor-rich atmosphere [3.49,50]. No planar guiding is observed for strip waveguides indiffused in an atmosphere of argon bubbled through water with a flow rate of about 1 liter/min.

3.9.5 Ridge and Metal-Clad Channel Waveguides

Masking during waveguide formation is not always possible. For example, there is no convenient technique to pattern, to the small dimension required, an epitaxial layer during LPE growth. In such cases the ridge or rib waveguide (Fig.3.13b) is a practical means to achieve two-dimensional guiding. The ridge waveguide is formed by masked etching of a slab waveguide. The planar waveguide is first fabricated, as described above. Photoresist is then spun onto the substrate and the desired channel pattern is photolithographically reproduced. The photoresist pattern serves as a mask during etching in which the higher index layer is either totally or partially removed outside the channel region (Fig.3.13b). The reduced film thickness diminishes the effective index outside the channel, and the difference in effective indices between the channel and nonchannel regions produces lateral confinement. As discussed earlier, the degree of lateral confinement depends upon the effective index difference and thus upon Δn and the rib height. By etching away the entire epilayer, extremely well-confined lateral waveguiding can be achieved.

The etching can be performed either chemically or by physical etching processes such as ion-beam milling. Anisotropic etching techniques such as reactive-ion etching which combine the chemical and sputter etching processes have also proved to be very useful [3.51]. Important etching characteristics include good selectivity (substrate-mask etch ratio), anisotropic etching to avoid undercutting of the mask, and smooth, damage-free sidewalls. Minimizing sidewall irregularities, which is essential to achieve acceptable propagation loss, also requires excellent mask quality and photolithography.

A metal cladding reduces the effective index of the planar waveguide beneath it. Therefore by placing a metallic overlay outside the desired channel region (Fig.3.13c), a difference in effective index sufficient to provide lateral confinement can be produced [3.52]. Metallic overlays can also produce channel guiding by a strain-induced real index change [3.53].

3.9.6 Coupling to a Dielectric Waveguide

Light can be channeled into a waveguide from an external beam or another guiding medium (for instance, a fiber) by evanescent or endfire coupling techniques. Evanescent coupling methods involve distributed coupling over a finite interaction length and require matching the propagation constant of the waveguide to that of the external beam. The most important example is prism coupling. Endfire coupling can be achieved with a lens or by butt coupling with a fiber. Efficient endfire coupling

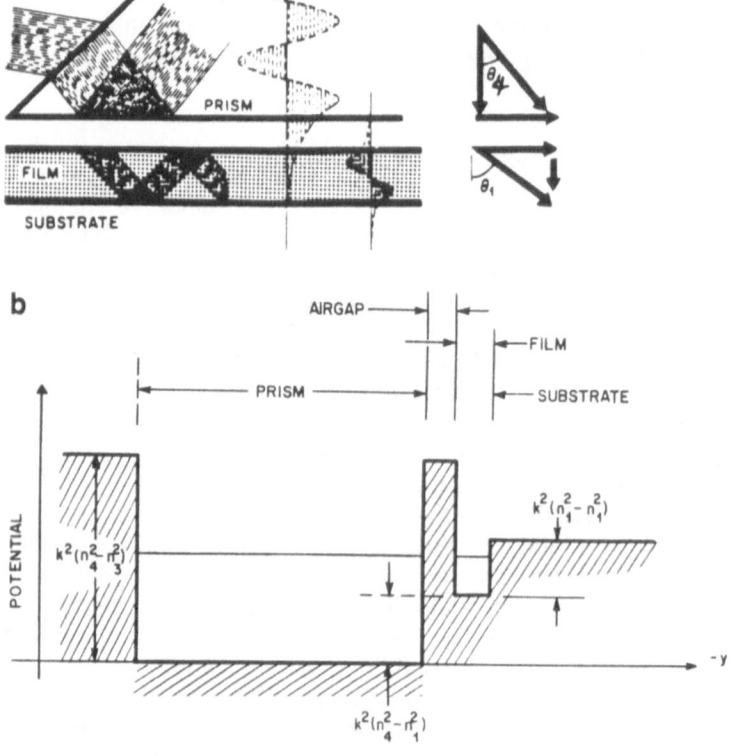

Fig.3.16. Prism coupling

requires a good matching of the coupled modes but eliminates the need for equal propagation constants of the coupled modes. A carefully polished endface is essential, however.

3.9.7 The Prism

In the prism coupling method [3.1], the laser light is coupled into a waveguide through the top surface of the film, which is usually flat and smooth. The laser beam can be hundreds of optical wavelengths in diameter and the position of the beam is also not critical. Moreover, by properly adjusting the direction of the laser beam as it is incident on the prism, any waveguide mode can be excited and each time only one mode is excited. The phenomenon which makes all this possible is optical tunneling.

Figure 3.16a illustrates the operation of the prism coupler. A rectangular prism of refractive index n_4 is pressed against the top surface of a waveguide, leaving a small air gap between them. The gap is usually on the order of one quarter of an optical wavelength. Assuming that the refractive index of the film is n_1, the operation of the prism coupler requires $n_4 > n_1$. Let the laser beam be incident on the base of the prism at an incident angle θ_4, the z component of wave vector in the prism is then $kn_4\sin\theta_4$. Suppose that this laser beam excites a light wave in the waveguide, which according to the zigzag-wave model has a z component wave vector

$$k_{z1} = \beta = kn_1\sin\theta_1 .\tag{3.27}$$

Since the wave in the prism is coupled to the waveguide through the top surface of the film, fields in the prism and those in the film must be matched along this surface. We have then

$$kn_1\sin\theta_1 = kn_4\sin\theta_4 ,\tag{3.28}$$

which is called the synchronous condition. For exciting a particular waveguide mode in the waveguide, the incident angle θ_4 must be adjusted so that the synchronous condition is satisfied for that mode. For that reason, only one waveguide mode is excited at one time. This is very important, since most optical devices work better when they act on a single mode of light wave propagation.

The principle of the prism coupler is optical tunneling. Based on the potential-well model, we may draw the potential diagram shown in Fig.3.16b. In this diagram, the prism is a very wide potential well in which energy levels are closely spaced and may be considered continuous. On the other hand, the waveguide is a narrow potential well which has a discrete set of eigenstates, each of them corresponding to a waveguide mode in electromagnetic theory. The two wells are separated by a potential wall which is the air gap. The wave in the prism must tunnel through this wall in order to be coupled to the waveguide.

According to the potential-well model, the bottom of the prism well is located $k^2(n_4{}^2-n_1{}^2)$ below that of the waveguide well. We recall that the vertical distance between an energy level and the bottom of the well as measured in the energy scale is $k_y{}^2$. Since the tunneling phenomenon requires eigenstates of the waves in the two wells to be on the same level and since

$$k_{yj} = kn_j\cos\theta_j ; \quad j = 1,4,...\tag{3.29}$$

we finally have the condition of optical tunneling as

$$k^2(n_4{}^2 - n_1{}^2) + k^2n_1{}^2\cos^2\theta_1 = k^2n_4{}^2\cos^2\theta_4 \tag{3.30}$$

which, of course, is identical to the synchronous condition (3.28).

The prism coupler is an indispensable tool for the study of waveguide structures and materials. In practice, the prism-waveguide assembly is mounted on a turntable so that angle θ_4 can be varied arbitrarily. When a set of waveguide modes is excited in a waveguide, one at a time, angles θ_4 thus obtained may be used to determine refractive indices of the film and substrate as well as thickness of the film. The accuracy of these measurements can easily reach a few parts in 10,000. Table 3.3 shows mode

Table 3.3. Comparison between observed and theoretically expected propagation constants of a ZnO film. Film thickness $W = 15881 \pm 60$ Å, refractive index of glass substrate $n_0 = 1.5127$ at 6328 Å and $n_0 = 1.5206$ at 4880 Å

λ	Polarisation	n_i	m	β/k		
				Observed	Theory	Difference
	TE	1.9732	0	a	1.9647	a
			1	1.9383	1.9389	-0.0006
			2	1.8961	1.8954	+0.0007
			3	1.8329	1.8332	-0.0003
			4	1.7518	1.7510	+0.0008
			5	1.6469	1.6473	-0.0004
			6	1.5248	1.5249	-0.0001
6328 Å	TM	1.19779	0	a	1.9686	a
			1	a	1.9404	a
			2	1.8933	1.8929	+0.0004
			3	1.8251	1.8249	+0.0002
			4	1.7353	1.7355	-0.0002
			5	1.6242	1.6246	-0.0004
	TE	2.0428	0	2.0360	2.0377	-0.0017
			1	2.0215	2.0223	-0.0008
			2	1.9973	1.9964	+0.0009
			3	1.9603	1.9596	+0.0007
			4	1.9117	1.9115	+0.0002
			5	1.8511	1.8513	-0.0002
			6	1.7786	1.7781	+0.0005
			7	1.6907	1.6910	-0.0003
			8	1.5898	1.5892	+0.0006
4880 Å	TM	2.0485	0	a	2.0430	a
			1	a	2.0265	a
			2	a	1.9987	a
			3	1.9598	1.9593	+0.0005
			4	1.9089	1.9078	+0.0011
			5	1.8432	1.8433	-0.0001
			6	1.7649	1.7652	-0.0003
			7	1.6727	1.6729	-0.0002
			8	1.5677	1.5687	-0.0010

[a] These modes could not be observed with the 26° rutile prism

indices of a ZnO/glass waveguide measured by a rutile prism coupler as they are compared with those calculated based on the values of refractive indices and film thickness measured independently by other methods.

We may also display waveguide modes as a set of bright lines (called m-lines) on a screen in an arrangement illustrated in Fig.3.17a. Here, two couplers are used: the left prism serves as an input coupler and the right one serves as the output coupler. Let a waveguide mode be excited in the waveguide. As light in this mode travels from left to right toward the

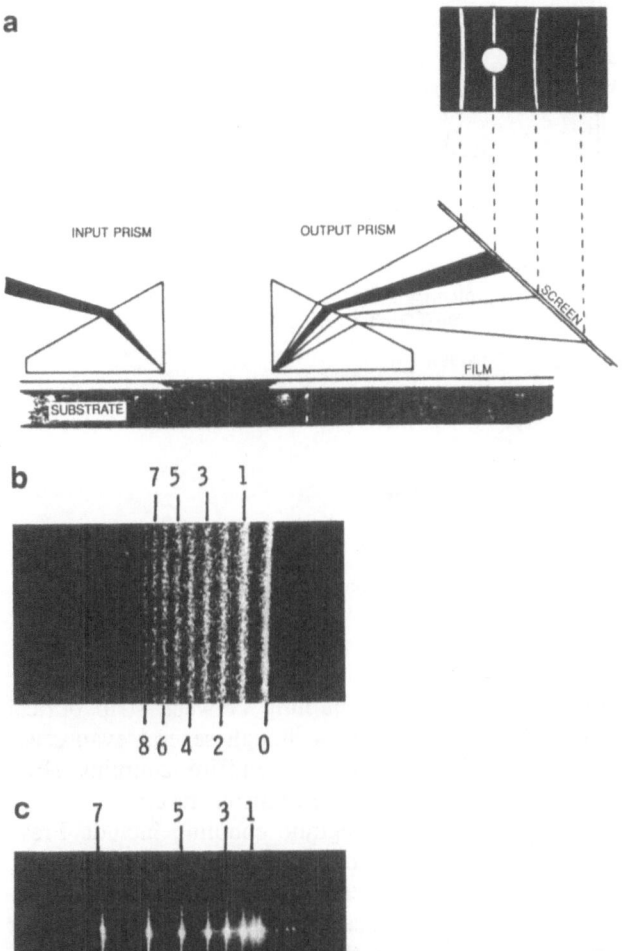

Fig.3.17. Waveguide modes displayed by prism output coupling (a) with photographs (b,c) of m lines

output coupler, part of the light is scattered into other waveguide modes due to roughness of film surface. Because of synchronous conditions, light in different waveguide modes is coupled out into space at different angles. When finally light from the output coupler is projected on a screen, we observe a bright spot and a set of bright lines, one line for each waveguide mode. They can be used to study the properties of the waveguide. For example, Fig.3.17b,c are photographs of m-lines obtained from a titanium-diffused lithium niobate waveguide and a ZnS/glass waveguide, respectively. It is interesting to note that the spacing between two neighboring m-lines decreases with the mode order in the diffused waveguide, whereas the reverse is true in the uniform-film waveguide. This sort of analysis is called m-line spectrosocopy.

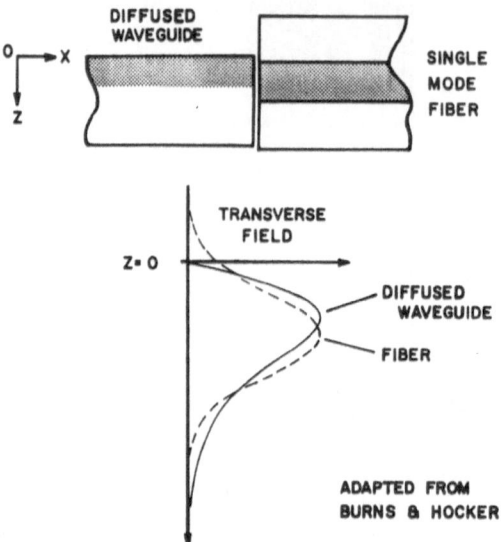

Fig.3.18. Fiber-coupling considerations for diffused waveguides

3.9.8 Fiber-Waveguide Coupling

For lightwave-communication application, coupling between strip optical waveguides and fiber is critically important. Although several evanescent coupling arrangements have been proposed, the endfire coupling (Fig. 3.18) approach has thus far produced the most promising results.

Loss components for endfire fiber-waveguide coupling include Fresnel loss and fiber-waveguide mode mismatch. The former can be reduced, if not totally eliminated, by using antireflection coatings or index-matching fluids. Mode mismatch can be reduced by fabricated waveguides whose modes closely approximate those of the fiber. The degree of mode mismatch is determined by the difference in the functional forms of the intensity profiles for the fiber and waveguide as well as by the difference in the characteristic sizes (Fig.3.19). Important for the first consideration is the fact that the fiber profile is symmetric while that of the waveguide is typically asymmetric because of the large index difference at the air cladding.

The calculated power-coupling coefficient between a circular fiber and a Ti:LiNbO$_3$ strip waveguide can be written approximately [3.53, 54] as

$$K = 0.93 \left[\frac{4}{[(w/a)^2 + \epsilon][(w/a)^2 + 1/\epsilon]} \right] \tag{3.31}$$

where a is the fiber mode diameter (1/e intensity), $w = (w_x w_y)^{1/2}$ is the waveguide geometric mean width, and ϵ ($=w_x/w_y$) is the eccentricity of the waveguide mode. The factor 0.93 in (3.31) represents the profile mis-

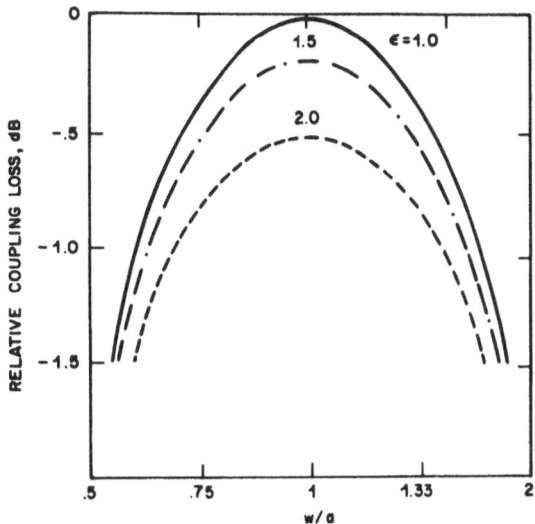

Fig.3.19. Calculated fiber-waveguide loss due to mode mismatch with mode ellipticity (ϵ) as a parameter

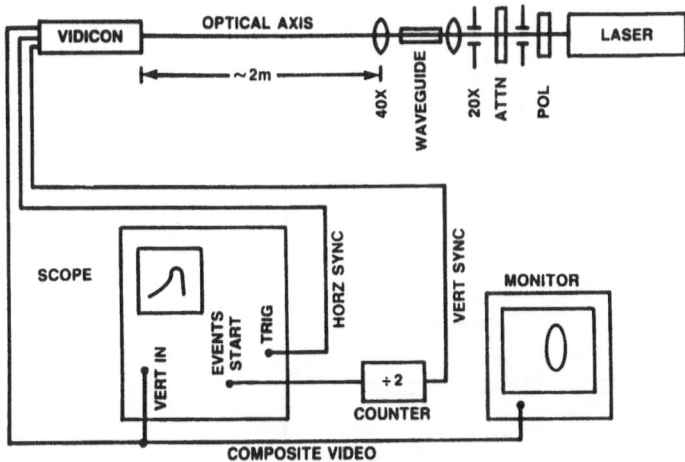

Fig.3.20. Experimental arrangement for measuring the spatial-mode profile of optical waveguides

match between fiber and Ti:LiNbO$_3$ waveguide modes. The dimensional dependence of the mode mismatch, the bracketed term in (3.31), is shown in Fig.3.19. The loss is actually quite tolerant of small dimension difference. To achieve good dimensional match, independent control of the waveguide width, depth and Δn via fabrication parameters is essential.

Good fiber-to-Ti:LiNbO$_3$ waveguide coupling can be achieved by determining the appropriate diffusion parameters to accomplish good mode matching. The fiber and waveguide modes can be examined and measured with the experimental setup shown in Fig.3.20. In Fig.3.21 the measured total fiber-waveguide-fiber insertion loss for about 1 cm long

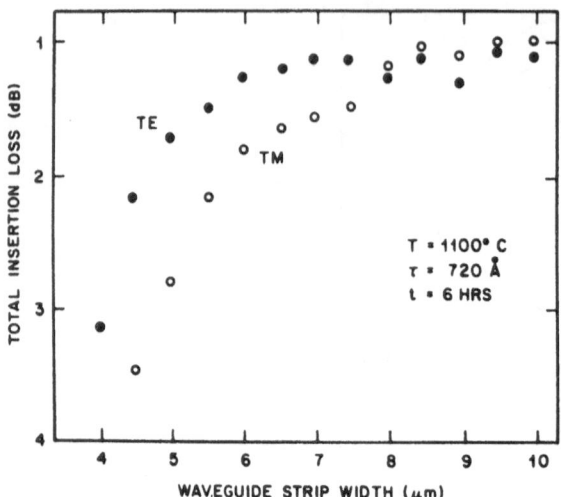

Fig.3.21. Measured total fiber-waveguide-fiber insertion loss for 1 cm long Ti:LiNbO₃ channel waveguide versus titanium metal-strip width

MODE SIZE VERSUS METAL STRIP WIDTH

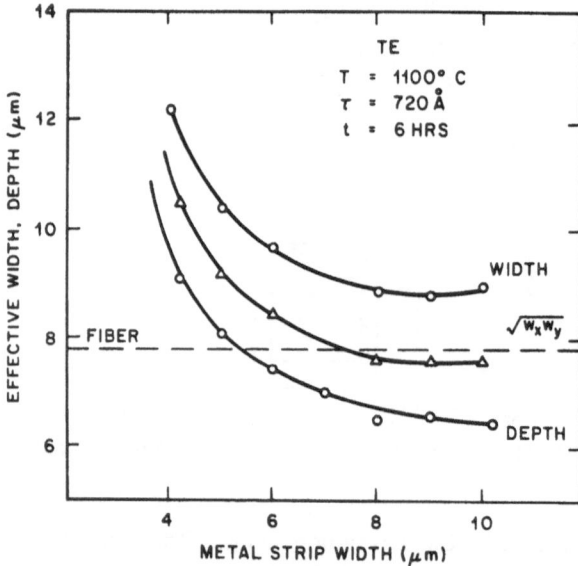

Fig.3.22. Measured waveguide optical mode width (full width at 1/e intensity) and depth for waveguide illustrated by Fig.3.21

Ti:LiNbO₃ waveguides made with strip width from 4–10 μm is shown. The diffusion parameters are those given and the optical wavelength is λ = 1.32 μm. Index matching fluid has been used at each interface. The waveguide mode width and depth for these waveguides for the TE polarization are shown in Fig.3.22. The fiber diameter is indicated. Very good match between the mean waveguide diameter and the fiber mode diameter has been achieved for strip widths between 7 and 10 μm with correspondingly low insertion loss. The results in Fig.3.21 include coupling as well as propagation loss in the waveguides. The latter can be determined by cutting the crystal and remeasuring the total insertion loss. The actual coupling loss determined in this way is about 0.35 dB/face which includes about 0.1 dB residual Fresnel loss due to imperfect index matching. The propagation loss is about 0.3 dB/cm [3.55]. For practical applications, fixed fiber pigtails are necessary.

References

3.1   P.K. Tien, R. Ulrich, R.J. Martin: Appl. Phys. Lett. 14, 291 (1969)
3.2   E.A.J. Marcatili: Bell System Techn. J. 48, 2071 (1969)
3.3   D. Marcuse: Bell System Techn. J. 48, 3187 (1969)
3.4   P.K. Tien: Appl. Opt. 10, 2395 (1971)
3.5   H.K.V. Lotsch: Optik 27, 239-254 (1968)
3.6   V. Ramaswamy: Bell System Tech. J. 53, 697 (1974)
3.7   R.V. Schmidt, I.P. Kaminow: Appl. Phys. Lett. 25, 458 (1974)
3.8   E.M. Conwell: Appl. Phys. Lett. 23, 328 (1973)
3.9   G.B. Hocker, W.K. Burns: IEEE J. QE-11, 270 (1975)
3.10 J.E. Goell, R.D. Standley: Bell System Tech. J. 48, 3445 (1969)
3.11 T.G. Giallorenzi, E.J. West, R. Kirk, R. Ginther, R.A. Andrews: Appl. Opt. 12 1240 (1973)
3.12 G. Stewart, P.J.R. Laybourn: IEEE J. QE-14, 930 (1978)
3.13 T. Izawa, H. Nakagome: Appl. Phys. Lett. 21, 584-486 (1972)
3.14 G.H. Chartier, P. Jaussaud, A.D. de Oliveira, O. Parriaux: Elect. Lett. 14, 132-134 (1978)
3.15 Cf. H. Inoue, K. Hiruma, K. Isida, T. Asai, H. Matsumura: IEEE J. LT-3, 1270 (1985)
3.16 Cf. P. Bachmann, A.J.N. Houghton: Elect. Lett. 18, 851 (1982)
3.17 Cf. E. Garmire: Semiconductor Components for Monolithic Integration, in *Integrated Optics*, 2nd. ed., ed. by T. Tamir, Topics Appl. Phys., Vol.7 (Springer, Berlin, Heidelberg 1979) Chap.6
3.18 K. Nakajima: Liquid-Phase Epitaxy, in *GaInAsP Alloy Semiconductors*, ed. by T.P. Pearsall (Wiley, New York 1982) Chap.2
3.19 G.H. Olsen: Vapor Phase Epitaxy in *GaInAsP Alloy Semiconductors*, ed. by T.P. Pearsall (Wiley, New York 1982) Chap.1
3.20 W.T. Tsang: MBE Growth Techniques, in *Semiconductor and Semi-metals*, ed. by R.K. Willardson, A.C. Beer, W.T. Tsang (Academic, New York 1985) Vol.22A, Chap.2
      K. Ploog, K. Graf: *Molecular Beam Epitaxy of III-V Compounds, A Comprehensive Bibliography 1958-1983* (Springer, Berlin, Heidelberg 1984)
      M.A. Herman, H. Sitter: *Molecular Beam Epitaxy*, Springer Ser. Mat. Sci., Vol.7 (Springer, Berlin, Heidelberg 1988)
3.21 G.B. Stringfellow: Organometallic Vapor-Phase Epitaxial Growth of III-IV Semiconductors, in *Semiconductor and Semimetals* 22A, Chap.3 (Academic, New York 1985)
3.22 K. Hiruma, H. Inoue, K. Ishida, H. Matsumura: Appl. Phys. Lett. 47, 186 (1985)
3.23 C.J. Lii, C.S. Tsai, C.C. Lee: IEEE J. QE-22, 868 (1986)
3.24 I.P. Kaminow, J.R. Carruthers: Appl. Phys. Lett 22, 326 (1973)

3.25 R.V. Schmidt, P.S. Cross, A.M. Glass: J. Appl. Phys. 51, 90 (1980)
3.26 R.V. Schmidt, I.P. Kaminow, J.R. Carruthers: Appl. Phys. Lett. 23, 417 (1973)
     C.S. Tsai, Le T. Nguyen, S.K. Yao, M.A. Alhaider: Appl. Phys. Lett. 26, 140 (1975)
3.27 R. Holman: Proc. Topical Meeting on Guided-Wave and Integrated Optics, Lake Tahoe, Nevada (1980)
3.28 M.L. Shah: Appl. Phys. Lett. 26, 652 (1975)
3.29 J.L. Jackel: Appl. Opt. 19, 1996 (1980)
3.30 J.L. Jackel, C.E. Rice, J.J. Veselka: Appl. Phys. Lett 41, 607 (1982)
3.31 A. Yi-Yan: Appl. Phys. Lett. 42, 633 (1983)
3.32 R.A. Becker: Appl. Phys. Lett. 43, 131 (1983)
3.33 R.L. Davis: Proc. SPIE 517, 74 (1984)
3.34 K.K. Wong: GEC J. Research 3, 243 (1985)
3.35 C. Warren, S. Forouhar, W.S.C. Chang, S.K. Yao: Appl. Phys. Lett. 43, 424 (1983)
3.36 D.Y. Zang: Opt. Commun. 47, 248 (1983)
     D.Y. Zang, C.S. Tsai: Appl. Phys. Lett. 46, 703 (1985)
3.37 J. Jackel, A.M. Glass, G.E. Peterson, C.E. Rice, D.H. Olson, J.J. Veslka: J. Appl. Pnys. 55, 269 (1984)
3.38 T. Findakly, B. Chen: Elect. Lett. 20, 128 (1984)
3.39 M. DeMichell, J. Botineau, S. Neveu, P. Sibillot, D.B. Ostrowsky, M. Papuchon: Opt. Lett. 8, 116 (1983)
3.40 R.C. Alferness: IEEE J. QE-17, 946 (1981)
3.41 M. Fukuma, J. Noda: Appl. Opt. 19, 591-496 (1980)
3.42 M. Minakata, S. Saito, M. Shibata, S. Miyazawa: J. Appl. Phys. 49, 4677 (1978)
3.43 R.C. Alferness, R.V. Schmidt, E.H. Turner: Appl. Opt. 18, 4012-4018 (1979)
     M. Fukuma, J. Noda: App. Opt. 19, 591-597 (1980)
3.44 C.D.W. Wilkinson, R. Walker: Elect. Lett. 17, 319 (1981)
3.45 T.R. Ranganath, S. Wang: IEEE J. QE-13, 290 (1977)
3.46 J.L. Jackel: J. Opt. Commun. 3, 82-85 (1983)
3.47 T.R. Ranganath, S. Wang: Appl. Phys. Lett. 30, 376 (1977)
3.48 R.J. Esdaie: Appl. Phys. Lett. 33, 733-735 (1978)
3.49 R.C. Alferness, L.L. Buhl: Opt. Lett. 5, 473-475 (1980)
3.50 J.L. Jackel, V. Ramaswamy, S.P. Lyman: Appl. Phys. Lett. 38, 509-511 (1981)
3.51 N. Rey Whetten (ed.): Plasma Etching and Reactive Ion Etching (Am. Phys. Soc., New York 1982)
3.52 J.C. Campbell, F.A. Blum, D.W. Shaw, K.C. Lawley: Appl. Phys. Lett. 29, 203-205 (1975)
3.53 T.M. Benson, T. Murotani, P.A. Houston, P.N. Robson: Elect. Lett. 17, 37 (1981)
3.54 W.K. Burns, G.B. Hocker: Appl. Opt. 16, 2048-2050 (1977)
3.55 R.C. Alferness, V.R. Ramaswamy, S.K. Korotky, M.D. Divino, L.L. Buhl: IEEE J. QE-18, 1807-1813 (1982)

4. Excitation of Surface-Acoustic Waves by Use of Interdigital Electrode Transducers

By Thomas M. Reeder

With 12 Figures

The efficiency of guided-wave acousto-optic diffraction is proportional to the total power flow P_a in the incident surface acoustic wave (SAW) beam. The efficiency and frequency response of guided-wave acousto-optic devices, therefore, are a strong function of the characteristics of the transducer used to excite the incident acoustic beam [4.1]. As described in Chap. 1, the interdigital-electrode SAW transducer [4.2,3], illustrated in Fig. 4.1, is utilized as the acoustic excitation source in all guided-wave acousto-optic devices under study today. The planar format and the relatively high electro-acoustic conversion efficiency of this transducer make it ideal for acousto-optic applications. Moreover, it is possible to fabricate multiple transducer arrays [4.1,4] which greatly extend the bandwidth of acousto-optic interaction by optimizing the spatial phase match of the acoustic and optic beams over a wide range of frequency.

SAW transducers have also been developed for other applications including radar correlation [4.5,6], highly selective IF filters for communication and television [4.7-9], and high-performance instrumentation [4.10]. Because of this extensive development, it is now possible to consider the synthesis of a transducer and associated electrical coupling network, which will provide near-optimum trade-off of acousto-optic efficiency bandwidth product.

In this chapter we review the basic characteristics of the interdigital SAW transducer with the aim of providing a simple, but quite useful equivalent circuit model for transducers utilized in acousto-optic devices. The model describes the electric admittance and electro-acoustic conversion efficiency for the general interdigital transducer configuration and may be used to estimate both electric-circuit driving-point requirements

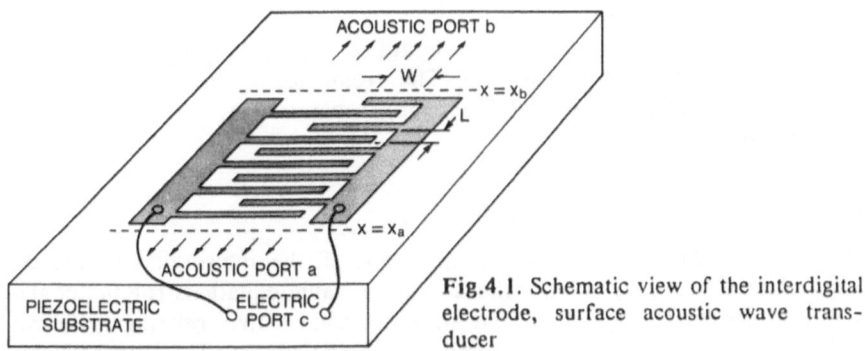

Fig. 4.1. Schematic view of the interdigital electrode, surface acoustic wave transducer

Springer Series in Electronics and Photonics, Vol. 23
Guided-Wave Acoustooptics   Editor: C. S. Tsai
© Springer-Verlag Berlin, Heidelberg 1990

and the frequency dependence of the acoustic output. A series of practical examples are then discussed to illustrate the design of a transducer configuration and electric circuit coupling networks which will provide efficient acoustic excitation over bandwidths ranging from a few percent to more than an octave. Since the major emphasis of this book is on the development of broad-bandwidth acousto-optic devices, as treated in Chap.5, we will highlight transducer design approaches which appear capable of providing 40% to 70% bandwidth at near optimum conversion efficiency [4.4, 10-18].

4.1 Fundamentals of Transducer Operation

We shall consider interdigital electrode transducers of the form shown in Fig.4.1. Such transducers consist of an array of thin-film metal electrodes which are deposited on a smooth, piezoelectric surface by use of photolithographic methods borrowed from integrated circuit manufacturing [4.19]; the electrode array is characterized by its electrode spacing L and adjacent electrode overlap W. When a voltage is applied to the electric terminals at Port C, a fringing electric field is excited between the electrodes, as illustrated in Fig.4.2a. As a consequence of the interdigital electrode connection, the polarity of the electric field alternates from one electrode to the next, giving rise to a periodic excitation of acoustic waves via the piezoelectric effect. At some convenient reference plane (for example, the plane defined by $x = x_a$ in Fig.4.1, the waves from all electrodes will add in phase if the electrode spacing L is approximately equal to one-half the acoustic wavelength Λ, defined by the wave velocity V_R, and the excitation frequency f, $L = \Lambda/2 = V_R/2f$. Acoustic power output versus frequency then has the form [4.21]

$$
P_a(f) \simeq \left[\frac{\sin[N\pi(f-f_0)/f_0]}{\sin\pi(f-f_0)/f_0} \right]^2
\tag{4.1}
$$

where N is the transducer acoustic length measured in interdigital periods (number of electrodes equals 2N). The frequency f_0 is the synchronism frequency at which the electrodes are separated by exactly one-half wavelength ($2L = \Lambda_0$).

4.1.1 Equivalent Circuit Model

In principle, the electro-acoustic operation of the interdigital transducer could be described by solving the boundary-value equations for the configuration shown in Fig.4.1 to find an admittance matrix relating to terminal quantities at the one electric and two acoustic ports. This approach has been used with great success for analogous volume acoustic transducers [4.22]. However, the Surface Acoustic Wave (SAW) case is much more

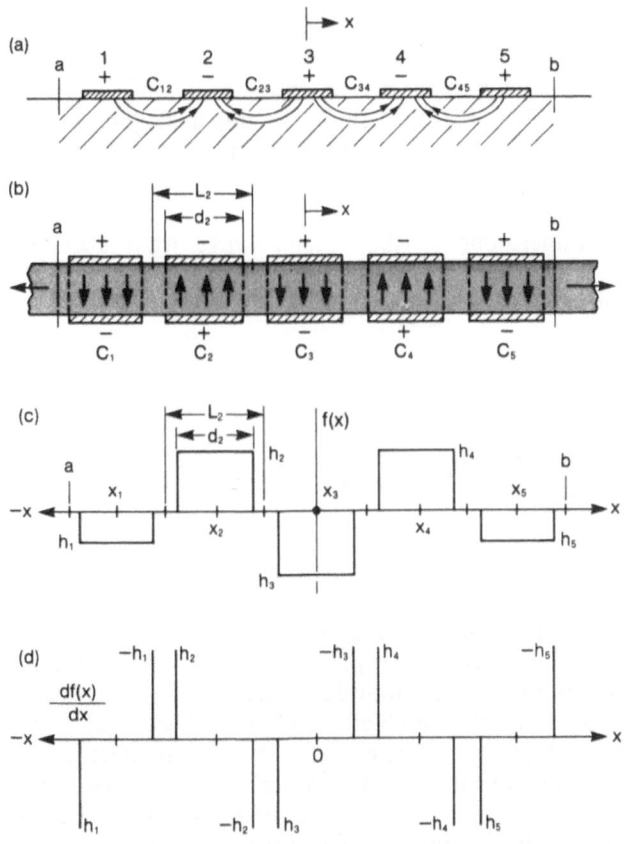

Fig.4.2a–d. Cross-sectional view of a SAW transducer illustrating the fringing electric field between electrodes and utilizing an analogous bulk acoustic wave transducer array to illustrate acoustic wave excitation [4.20]:
a) Cross-sectional view of the interdigital SAW transducer
b) Analogous bulk wave transducer array
c) Excitation function corresponding to bulk wave transducer array
d) Derivative of the excitation function

complex; the boundary value equations are difficult to solve because the configuration is two-dimensional and must utilize a piezoelectric aniso-tropic substrate. Instead, it has been expedient to analyse SAW transducers by use of equivalent circuits which are deduced by subdividing the inter-digital configuration into subsections that can be conveniently and accu-rately represented by analogous volume wave transducer models. A great deal of work has been done on such equivalent circuits [4.20,21,23–30]. Here we shall attempt to summarize, and where possible, to simplify, the equivalent circuit for use in modelling transducers for acousto-optic applications. In particular, we shall use a model developed by *Krimholz* [4.27], *Bahr* and *Lee* [4.26], and *Matthaei* et al. [4.20].

Since the acoustic fields associated with the interdigital transducer are largely bound to a region about one wavelength in depth below the elec-

trode array in Fig.4.2a, it makes sense that the transducer could be modelled by an array of "side-electroded" volume wave transducers which
have electrode periodicity and piezoelectric coupling that match those of
the desired SAW configuration. Figure 4.2b shows this analogous volume
wave array. The volume wave array can be thought of as being constructed on a thin piezoelectric bar, with acoustic propagation along the
length of the bar matching the velocity and power flow in the analogous
SAW transducer. The dimensions of the volume wave transducers are
chosen so that their capacitances C_m are equal to the fringing capacitance
between the corresponding SAW transducer electrodes,

$$C_1 = \frac{C_{12}}{2} \, ,$$

$$C_m \bigg|_{m=2 \text{ to } M-1} = \frac{1}{2} (C_{m-1,m} + C_{m,m+1}) \, , \quad C_m = \frac{1}{2} C_{M-1,M} \, . \quad (4.2)$$

Note that these capacitances will be proportional to the overlap length of
adjacent SAW electrodes (W in Fig.4.1).

Figure 4.2c shows a plot of an acoustic excitation function $f(x)$,
which is proportional to the effective piezoelectric field distribution
excited over the length of the volume transducer array. The piezoelectric
source term is proportional to the product $h(x) \cdot D(x)$, where $h(x)$ is the
appropriate piezoelectric coupling coefficient and $D(x)$ is the electric displacement [4.31]. If we were to attempt to rigorously solve the boundary
value equations for the array in Fig.4.2c, we should solve the inhomogeneous wave equations in particle displacement u and stress T, i.e.,

$$\frac{\partial^2 u}{\partial x^2} - \frac{1}{V_R^2} \frac{\partial^2 u}{\partial t^2} = \frac{1}{c} \frac{\partial hD}{\partial x} \, , \qquad (4.3)$$

$$\frac{\partial^2 T}{\partial x^2} - \frac{1}{V_R^2} \frac{\partial^2 T}{\partial t^2} = \frac{h}{V_R^2} \frac{\partial^2 D}{\partial t^2} \, , \qquad (4.4)$$

where c is the appropriate acoustic stiffness constant [4.32]. Equation (4.3)
implies that acoustic waves are excited only at points in the array where
the piezoelectric source term is changing [4.33]. Since the source term for
the volume transducer array changes very rapidly at the electrode edges,
we might expect that acoustic excitation would be proportional to a series
of delta function sources which alternate in sign, as illustrated in Fig.4.2d.
It can be shown that the excitation amplitudes h_m are related to the transducer physical parameters and to an effective piezoelectric coupling constant k_c by

$$h_m^2 = k_c^2 \, V_R \frac{C_m}{L_m} \left[\csc \frac{\pi}{2} \frac{d_m}{L_m} \right]^2 \, . \qquad (4.5)$$

As mentioned in early work on SAW transducers [4.21], the constant k_c can be thought of as the fractional acoustic velocity change that would occur if the piezoelectric field were shorted out, i.e.,

$$\frac{\Delta V_R}{V_R} = 1 - \sqrt{1 + k_c^2} \simeq -\frac{1}{2}k_c^2 \ . \tag{4.6}$$

For materials of interest, k_c^2 expressed in percentage varies from about 0.1 to 5% [4.34-36].

For the general case where the excitation function $f(x)$ may have arbitrary distribution, *Matthaei* et al. [4.20] shows that the total field seen at the acoustic port a is proportional to the Fourier transform

$$KF(K) = \int_{-\infty}^{+\infty} \frac{df(x)}{dx} \ e^{-jKx} \ dx \tag{4.7}$$

where $K = 2\pi/\Lambda = 2\pi f/V_R = \omega/V_R$ is the wave number for propagation in the x direction. In later discussions we shall refer to $KF(K)$ as the "frequency domain" excitation function.

Krimholtz [4.27] has shown that the volume transducer array can be exactly represented by the equivalent circuit in Fig.4.3. Here the voltages V_a and V_b represent the acoustic force at reference planes a and b in Fig.4.1; the acoustic transmission medium is modelled by an acoustic transmission line with characteristic impedance Z_0 and length L_T equal to the effective acoustic excitation length of the array. Piezoelectric coupling over the array is modelled by a transformer with frequency-dependent turns ratio ϕ and a special circuit element called a "J inverter" [4.20], the parameters of which are related to the even and odd parts of the integrated excitation function,

$$KF(K) = KF_e(K) + KF_0(K) \qquad \text{by} \tag{4.8}$$

$$\phi = -jKF_0(K)\sqrt{Z_0} \qquad \text{and} \tag{4.9}$$

$$J = |B_2| \ ; \quad B_2 = -KF_e(K)/\sqrt{Z_0} \ . \tag{4.10}$$

The capacitance C_T which appears across the electric-port terminals is the total electrostatic electrode capacitance, namely

$$C_T = \sum_{m=1}^{M} C_m. \tag{4.11}$$

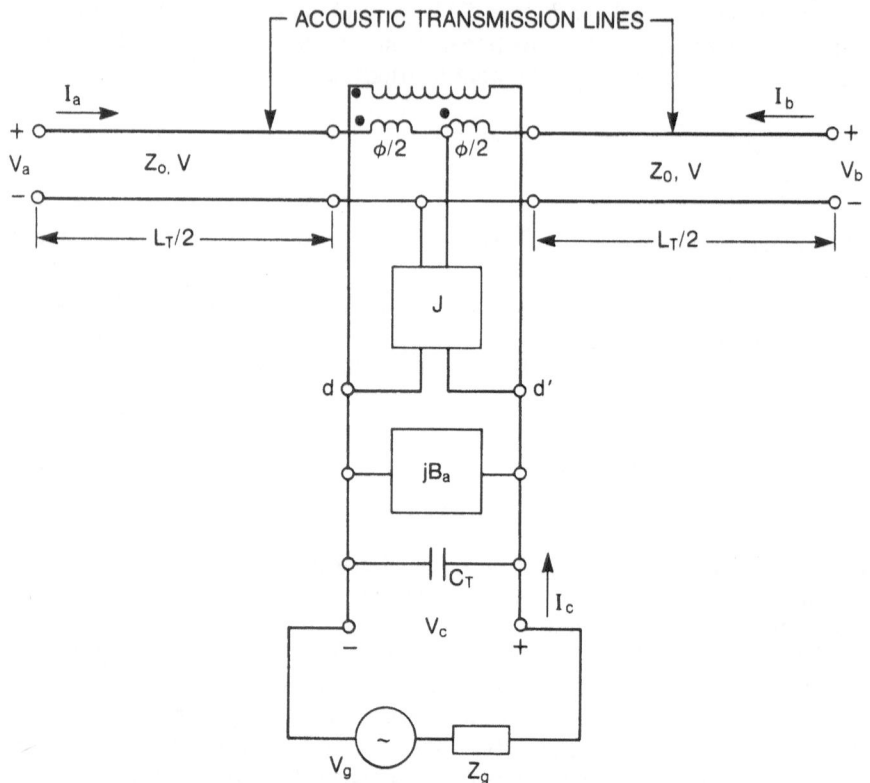

Fig.4.3. Equivalent circuit for the SAW transducer [4.20]

The last circuit parameter, jB_a, is an acousto-electrically excited susceptance which is obtained from a modified Hilbert transform of F(K). This
susceptance is usually much smaller than ωC_T, as it is proportional to
$k_c^2 \omega C_T$. We shall discuss its evaluation and importance later on.

　　To the extent that the volume wave analogy holds, the equivalent
circuit model in Fig.4.3 is a valid representation of the acousto-electric
operation of interdigital electrode SAW transducers over a very wide frequency range. This model and other approximately equivalent forms have
been used to represent the acousto-electric characteristics of SAW transducers to within a few dB over octave frequency bands centered at the
synchronism frequency f_0. It should also be pointed out that the above
model, due to its simplicity, neglects a number of second-order effects
which may be important in some applications. The primary sources of error here result from the use of an idealized excitation function in Fig.4.2
and the implicit assumption that the metal transducer electrodes do not
cause scattering of the excited SAW beam. Our simple model will tend to
err more in narrow-bandwidth applications (on the order of 10% or less
fractional bandwidth) where wave interactions over many electrodes add
in phase. *Smith* and co-workers [4.28-30] have shown that models similar
to ours, but with somewhat greater complexity, can be devised which ac-

curately account for arbitrary transducer excitation functions and electrode scattering. *Smith's* model is especially useful in analysing the effects of more complex electrode configurations, such as the "double electrode" configuration mentioned in Sect.4.2. and in predicting electro-acoustic efficiency at harmonics of the fundamental passband.

4.1.2 Transducer Admittance and Conversion Efficiency

We will now utilize the circuit in Fig.4.3 to calculate two quantities of central importance in SAW transducer operation: 1) electric port driving-point admittance and 2) acousto-electric conversion efficiency. In the applications considered here, we expect that the transducer acoustic transmission lines are well matched to the acoustic propagation medium; acousto-electric coupling at each electrode is small, being proportional to k_c^2. Thus, we may assume that the transmission lines in Fig.4.3 are terminated in Z_0 at the Ports a and b. Evaluation of the admittance Port c yields

$$Y_c = j\omega C_T + jB_a(K) + G_a(K) \qquad \text{where} \qquad (4.12)$$

$$G_a(K) = \frac{1}{2} (J^2 Z_0 - \phi^2/Z_0)$$

$$= \frac{1}{2} (K^2 F_e(K)^2 - K^2 F_0(K)^2) = \frac{1}{2} |K F(K)|^2 \qquad (4.13)$$

and $K = \omega/V_R$ as before. *Matthaei* et al. [4.20] further showed that the susceptance B_a is given by the Hilbert transform

$$B_a(K) = \frac{1}{\pi} \int_{-\infty}^{+\infty} \frac{G_a(K_0)}{K_0 - K} dK_0 . \qquad (4.14)$$

Thus, B_a is the minimum-susceptance function [4.37] that goes with the input conductance G_a. We note that the admittance $G_a + jB_a$ can be thought of as a "radiation admittance" which results from acousto-electric coupling. One may observe from the discussion associated with Fig.4.2 that the functions G_a and B_a are directly proportional to k_c^2; thus, these quantities reduce to zero with vanishing piezoelectric coupling. This will become even more clear in the example presented in the next section.

The acousto-electric conversion efficiency or power scattering coefficient [4.21] is given by

$$E(K) = P_a/P_{avail,c} \qquad (4.15)$$

where P_a is the acoustic power radiated away from Port a and $P_{avail,c}$ is the power available from the electric generator under matched conditions.

When expressed in terms of the model in Fig.4.3, the conversion efficiency becomes

$$E(K) = \frac{4Re\{Z_g\}}{Z_0} \left|\frac{V_a}{V_g}\right|^2 = \frac{(4/Z_0)Re\{Z_g\}}{1 + Z_g Y_c} \left|\frac{V_a}{V_c}\right|^2 \qquad (4.16)$$

where Z_g designates the output impedance of the electric generator, and Y_c is the transducer input admittance at Port c. Now (4.16) requires some interpretation because V_a has units of force, Z_0 is the *acoustic* transmission line impedance (i.e., Z_0 has dimensions kg/s), and the electrical circuit quantities have their usual volt-ohm definitions. However, *Krimholtz* [4.27] has shown that

$$\frac{V_a}{V_c} = j\frac{1}{2} \sqrt{Z_0} \, KF(K) \, \exp(jKL_T/2) ; \qquad (4.17)$$

where L_T is defined in Fig.4.3, and as we shall see in the next section, $KF(K)$ is proportional to and has the units of $(\omega C_T)^{1/2}$, the square root of the transducer array capacitive susceptance. Thus, the effect of Z_0 is cancelled out in the evaluation of (4.16).

Within the limits of the model, the above analysis applies to all interdigital electrode SAW transducers. The electrodes in Fig.4.1. may now be non-uniformly spaced and may have different overlap lengths (i.e., $W_m \neq W_{m+1}$). Although there are applications where the transducer may utilize electrodes of equal length and uniform spacing, in general, the electrode spacing and overlap are carefully chosen to synthesize a prescribed band-shape for the transducer efficiency versus frequency [4.24, 25].

In order to simplify the evaluation of transducer configurations for specific acousto-optic applications, we shall evaluate the frequency domain excitation function $KF(K)$ using the δ-function approximation described in Fig.4.2d. As shown in the figure, the piezoelectric excitation at each electrode is represented by a *pair* of δ-functions of value $\pm h_m$. Since piezoelectric excitation is expected to be uniform along the length of each electrode, the value of h_m will scale with electrode overlap length,

$$h_m = (W_m/\hat{W})\hat{h} . \qquad (4.18)$$

where $\hat{h}$ and $\hat{W}$ represent the magnitude of piezoelectric excitation at the largest value of electrode overlap. If we define x_m as the distance from the center of electrode m to reference plane a in Fig.4.2 and if d_m is the electrode width, then

$$KF(K) = \sum_{m=1}^{M} h_m \left(\exp[-jK(x_m + d_m/2)] - \exp[jK(x_m - d_m/2)] \right) \qquad (4.19)$$

where M is the toatal number of electrodes in the transducer. Note that because the field polarity alternates at adjacent electrodes, the δ-functions at adjacent electrode edges have the same sign. It is useful to further approximate the effective piezoelectric excitation at the mth electrode "gap" by

$$|h_{gm}| = |h_m| + |h_{m+1}| . \tag{4.20}$$

Thus, from (4.5) we have

$$h_{gm}^2 \simeq (2k_c)^2 \; (V_R/L_m) \; C_g W_m , \tag{4.21}$$

where $C_g = C_m/W_m$ is the capacitance per unit length of all adjacent e-lectrode pairs and the csc factor has been dropped since it is usually near unity. Moreover, the excitation at the outer most edges of the end electrodes will typically be small enough to be neglected. Consequently, it is a useful approximation to sum the piezoelectric contributions only at the e-lectrode fringing field gaps; i.e.,

$$KF(K) \simeq \sum_{m=1}^{M_g} h_{gm} \exp(-jKx_m) . \tag{4.22}$$

Here, M_g is the total number of gaps, not electrodes. The above approximation allows a straightforward evaluation of the transducer input admittance (4.13, 14) and conversion efficiency (4.16, 17).

4.1.3 An Important Example

The uniform SAW transducer, which has constant electrode spacing and length throughout the array, provides a useful benchmark for evaluating the characteristics of more complex configurations. For the array shown in Fig.4.1 the frequency domain excitation function becomes

$$KF(K) = h_g \sum_{m=1}^{M_g} \exp[-jK(x_0 + mL) + jm\pi]$$

$$= h_g \; \frac{\sin M_g \theta}{\sin \theta} \exp(-jKx_0 - jKL_T) , \tag{4.23}$$

where $h_g = (2k_c)[(V_R/L)C_g \hat{W}]^{1/2}$. We have defined x_0 as the distance from the nearest electrode to the reference plane defined by $x = x_a$, and

$$\theta = \frac{\pi}{2} \left[\frac{f-f_0}{f_0} \right] . \tag{4.24}$$

99

Here, $f_0 = \omega_0/2\pi = V_R/2L$ is the synchronism frequency introduced in (4.1). Noting that

$$h_g{}^2 = 4k_c{}^2 \left[\frac{V_R}{L}\right] C_g \hat{W} = \left(\frac{4k_c{}^2}{\pi}\right) \omega_0 \frac{C_T}{M_g} , \qquad (4.25)$$

where $C_T = M_g C_g \hat{W}$ is the total electrostatic capacitance defined earlier in (4.11). Substituting the above identity into (4.23) yields

$$|KF(K)|^2 = \frac{4k_c{}^2}{\pi} \omega_0 C_T M_g \left(\frac{\sin M_g \theta}{M_g \sin\theta}\right)^2 , \qquad (4.26)$$

where $\theta = (\pi/2)(f/f_0 - 1)$. We note that the factor in square brackets is unity at f_0 and odd harmonics of f_0, as was earlier predicted in (4.1). Using these definitions in (4.13-17), the acoustic radiation terms in the electric port admittance of the uniform array transducer are

$$G_a(f) = \hat{G}_a \left[\frac{\sin M_g \theta}{M_g \sin\theta}\right]^2 \qquad \text{and} \qquad (4.27)$$

$$B_a(f) = \hat{G}_a \frac{\sin 2M_g \theta - M_g \sin 2\theta}{2M_g(\sin\theta)^2} , \qquad \text{where} \qquad (4.28)$$

$$\hat{G}_a = (2k_c{}^2 M_g/\pi)\omega_0 C_T \qquad (4.29)$$

is the peak radiation conductance seen at $f = f_0$. These expressions are also seen to be identical to those used in the early description of SAW transducer models [4.24, 25]. If M_g, the number of electrode gaps, is set equal to N/2, these radiation terms can be expressed in terms of the "number of periodic sections, N", as was done in [4.21, 23].

The relatively simple form of radiation admittance derived in (4.27-29) allows a simple comparison of theory and experiment. Consider a uniform interdigital transducer with the following physical parameters:

$$f_0 = V_R/2L = 105 \text{ MHz},$$

$$M_g = 30 , \quad \hat{W} = 0.6 \text{ mm}.$$

It is known from a number of calculations on piezoelectric media [4.34] that for transducers fabricated on Y-cut lithium niobate (LiNbO$_3$) with electrodes oriented for acoustic propagation along the Z axis, $k_c{}^2 \simeq 0.05$. C_g is expected to be near $C_g \simeq \epsilon \simeq 5$pF/cm. Figure 4.4 shows a plot of transducer radiation admittance which compares experimentally measured

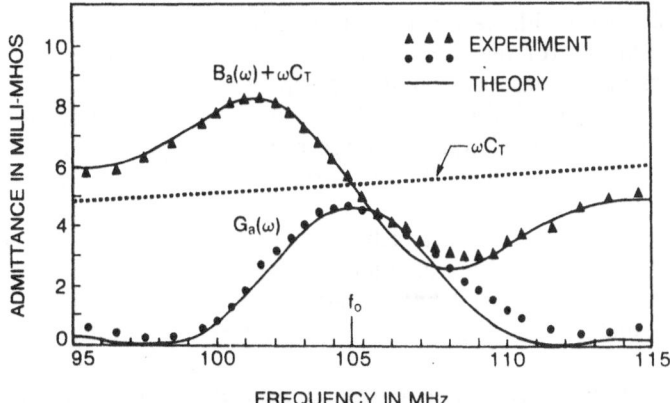

Fig.4.4. Theory-experiment comparison of Port c input admittance versus frequency for a uniformly weighted SAW transducer (f_0 = 104.5MHz, M_g = 30, $\hat{W}$ = 0.6mm) [4.21]

values with those calculated from (4.27-29). By comparing such data for transducers with different numbers of electrodes and a wide variety of piezoelectric substrate cuts and wave propagation directions, it is possible to assemble a very useful set of transducer design parameter data [4.21, 23,25,35,36]. Table 4.1 lists these data for piezoelectric media of current interest for acousto-optic applications.

We see also from the curves in Fig.4.4 that the radiation conductance for the uniform interdigital transducer has the $(\sin x)/x$ frequency response that is characteristic of spatially periodic, in-line wave emitters, such as the end-fire electromagnetic antenna array [4.3]. It is significant that the zeros of $G_a(f)$ are given by, see (4.27),

$$M_g \theta = M_g \frac{\pi}{2} \left[\frac{f_n - f_0}{f_0} \right] = n\pi \tag{4.30}$$

where n is an integer. Thus, $G_a(f)$ is zero at

$$\left| \frac{f_n}{f_0} \right| = 1 \pm \frac{2n}{M_g} \tag{4.31}$$

Table 4.1. SAW transducer parameters

Material	Cut	Propagation Direction	V_R [m/s]	k_c^2 [%]	C_g [pf/cm]
LiNbO$_3$	Y	Z	3488	4.5	4.6
LiTaO$_3$	Z	Y	3329	0.93	4.4
Quartz	Y	X	3159	0.23	0.50
GaAs	T	(110)	2860	0.14	1.1

and the fractional bandwidth between the first zeros above and below f_0 is $4/M_g$. One can verify for the data shown in Fig.4.4 that the bandwidth of these first zeros of response is accurately given by $4/M_g$.

We conclude that the admittance and frequency response characteristics of the uniform interdigital transducer are accurately specified by the parameters f_0, M_g, k_c^2, C_g, and $\hat{W}$. In particular, once a suitable frequency response is found through choice of f_0 and M_g, $\hat{W}$ can be adjusted within some limits to set a prescribed admittance level for the peak transducer admittance at f_0. For example, it is often desirable for efficient transducer operation to set $|Y_c(f_0)|$ near 20 m℧ (i.e., driving point impedance near 50 mΩ). This was done in the experiment illustrated in Fig.4.4.

In the more complex case, where the transducer electrode spacing and overlap length are variables, more elaborate synthesis methods are needed to design a specific transducer frequency response centered at f_0. We shall briefly address this problem in the next section, and the literature has many additional design examples available to the interested reader [4.38].

4.2 Design of Transducer and Coupling Networks for Broad-Band Operation

The above analysis has allowed us to describe the characteristics of acoustic radiation from the SAW interdigital transducer by the use of an electrical equivalent circuit. In this section we shall utilize the equivalent circuit in the design of transducer configurations and coupling circuits that enhance the broad band operation of the transducer for acousto-optic applications. We shall find that for transducer bandwidths of the order $f_0/4$ and larger, it is necessary to design both the transducer electrode *configuration* and the electric source to transducer coupling *network* for broad-band operation. Because broad-band SAW transducers have relatively weak acousto-electric coupling, it is possible to optimize the response of the transducer-circuit sub-system by studying separately the acoustic response (controlled by the transducer electrode configuration) and the electric response (controlled by the electric circuit) and then multiplying the two together to create the desired sub-system response.

4.2.1 The Transducer-Circuit Sub-System

In the general case, the interdigital transducer will be connected to the electric source through an electric coupling network, as shown in Fig.4.5. While the electric source is usually broad band and resistive ($Z_g = R_g + j0$), the effective source seen by the transducer will be frequency dependent $V_g = V_g'(f)$, $Z_g' = R_g'(f) + jX_g'(f)$, where the primes indicate source variables transformed by the coupling network. With reference to (4.12-17) and Fig.4.5 the conversion efficiency of the transducer sub-system is given by

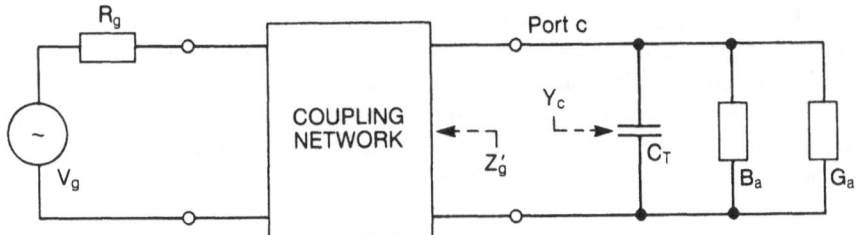

Fig.4.5. SAW transducer sub-system showing the use of a general coupling network between the electric source and the transducer elecrtic Port c

$$E(f) = \frac{4R_e|Z_g{}'(f)|}{|1 + Z_g{}'(f)Y_c(f)|^2} \frac{G_a(f)}{2} \qquad (4.32)$$

where $Y_c(f) = j\omega C_T + G_a(f) + jB_a(f)$ as before. Note that if we had looked only at Fig.4.5, we could have written down (4.32) by straightforward circuit analysis, except for the factor of two that divides $G_a(f)$. This factor comes from the bidirectional nature of the interdigital transducer [4.23]; half of the input power is radiated away from acoustic port b in Fig.4.1 and cannot be used. The simple form of (4.32) allows it to be expressed as the product of two separate efficiency factors, i.e.,

$$E(f) = E_c(f)E_a(f) \qquad \text{where} \qquad (4.33)$$

$$E_c(f) = \frac{2(R_e Z_g{}')\hat{G}_a}{|1 + Z_g{}'Y_c|^2} \qquad \text{and} \qquad (4.34)$$

$$E_a(f) = G_a/\hat{G}_a . \qquad (4.35)$$

With these definitions $E_a(f)$ becomes a dimensionless ratio which is near unity for frequencies near f_0 but decreases elsewhere. This factor is therefore a relative measure of the acoustic frequency response due to the transducer electrode configuration. On the other hand, since the acoustic radiation admittance is proportional to $k_c{}^2 M_g$, which will be small for broadband transducers,

$$G_a(f) \ll \omega C_T , \quad B_a(f) \ll \omega C_T , \qquad (4.36)$$

to first order. Thus, the electric circuit efficiency ratio is approximately equal to

$$E_c(f) \simeq \frac{2(R_e Z_g{}')\hat{G}_a}{|1 + j\omega C_T Z_g{}'|^2} \qquad (4.37)$$

which depends only on the frequency properties of the transducer capacitive susceptance and the electric coupling network.

The simple case where a uniform interdigital transducer is connected directly to an untuned, resistive source provides a good example for computing transducer efficiency versus operating bandwidth, one which will provide a useful comparison in later schemes for optimizing the acoustic and electric circuit bandwidth of the transducer sub-system. In this case, the acoustic efficiency ratio becomes

$$E_a(f) = \left[\frac{\sin M_g(\pi/2)(f-f_0)/f_0}{M_g \sin(\pi/2)(f-f_0)/f_0} \right]^2 \tag{4.38}$$

and the corresponding circuit efficiency ratio is

$$E_c(f) = \frac{2\hat{G}_a R_g}{(1+G_a R_g)^2 + (\omega C_T R_g)^2} . \tag{4.39}$$

Since the response of $E_a(f)$ peaks at f_0 and will typically vary much faster around f_0 than $E_c(f)$, it is a useful approximation to assume that $E_c(f)$ is constant at the value specified at f_0, so that the frequency response is entirely determined by $E_a(f)$. The 3 dB fractional bandwidth found from (4.38) is then well approximated by

$$BW = 2/M_g \tag{4.40}$$

and the transducer maximum efficiency derived from (4.39) is

$$E(f_0) = E_c(f_0) = \frac{\hat{G}_a R_g}{(1 + \hat{G}_a R_g)^2} \qquad \text{where} \tag{4.41}$$

$$\omega_0 C_T R_g = 1 + \hat{G}_a R_g \tag{4.42}$$

or using (4.29)

$$\hat{G}_a R_g = \frac{2k_c^2 M_g/\pi}{1 - 2k_c^2 M_g/\pi} . \tag{4.43}$$

Figure 4.6 shows a plot of (4.41) expressed in dB as a function of M_g and the fractional bandwidth for the transducer material configurations discussed earlier. Note that there is a smooth trade-off between efficiency and bandwidth; at small bandwidth (M_g large) the efficiency approaches -6 dB. At large bandwidth ($M_g < 10$) the efficiency approaches

$$E(f_0) \simeq 2k_c^2 M_g/\pi \tag{4.44}$$

104

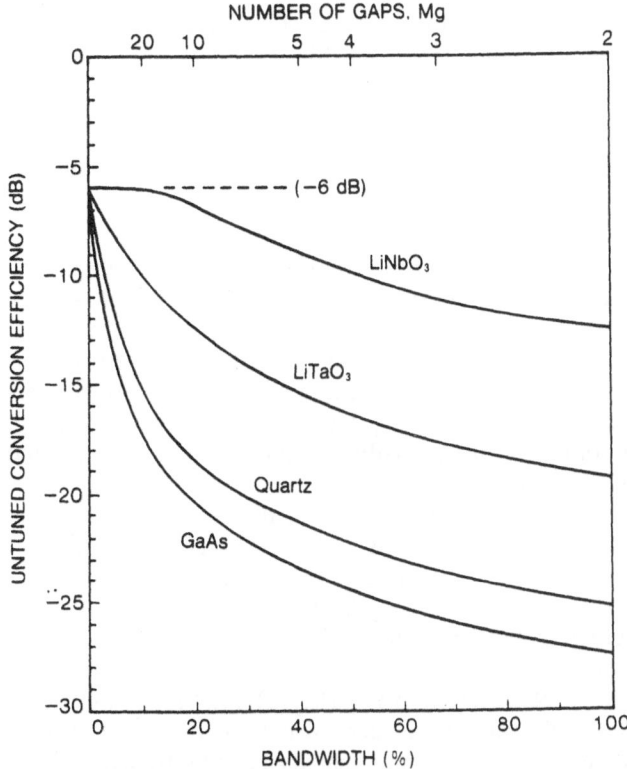

Fig.4.6. Calculated transducer sub-system efficiency versus fractional bandwidth for the case where a uniformly weighted SAW transformer is connected directly to a resistive source ($Z_g = R_g = 50\Omega$). The material constants used for each substrate are given in Table 4.1

which can be written in terms of the -3 dB fractional bandwidth BW as follows,

$$(\text{BW}) \; E(f_0) \propto 4k_c^2 \pi. \tag{4.45}$$

Thus, the product of bandwidth and efficiency tends to be a constant for broadband transducers, as was shown in early transducer studies [4.23, 25].

4.2.2 Broad-Band Acoustic Design

We saw earlier in (4.7, 13) that the acoustic radiation from an interdigital transducer is proportional to the spatial derivative of the piezoelectric excitation function ($f(x)$ in Fig.4.2). We also noted that the spatial derivative was well approximated by a series of δ-functions that represent acoustic excitation either at the electrode edges, or to a good approximation, at the center of "gaps" between electrodes of alternating potential. Moreover the δ-function distribution may be thought of as a sampled data representation of the transducer impulse response [4.24, 25]. Accordingly,

105

$G_a(f)$ may be viewed as the magnitude-squared Fourier transform of the transducer impulse response. Thus, from (4.7,13,22)

$$G_a(f) = \left| A \int_{-\infty}^{+\infty} h(t) \exp(-j2\pi ft) \, dt \right|^2 \qquad (4.46)$$

where A is a constant that depends on the transducer material parameters, and h(t) is the impulse response resulting from the transducer delta-function distribution. To synthesize a given conversion E(f), we must first find the corresponding $G_a(f)$ from (4.32) and calculate the inverse transform to specify the desired h(t). The transducer electrodes are then positioned (x_m) and their overlap length (W_m) proportioned or weighted to provide minimum sampling of the spatially distributed sine wave excited when an impulse is applied [4.25].

For example, in the uniform transducer case, the electrodes are equispaced at $L = V_R/2f_0$ and have constant weighting over the transducer length. When an electric impulse is applied to this transducer, a rectangular RF acoustic burst is excited with $M_g/2$ cycles. Actually, two bursts are excited, one traveling in the +x direction in Fig.4.2; the other traveling along -x. The wave form of this burst specifies h(t), here represented by a rectangular envelope of duration $\Delta T = (M_g/2)/f_0$. From (4.46) it is then easy to see that the uniform transducer will have the (sinx)/x frequency response predicted in (4.1,27) and that its bandwidth will be approximately $1/\Delta T = (2/M_g)f_0$.

The literature abounds with detailed descriptions of the design of SAW transducers to meet specific frequency response functions [4.38]. Although the (sinx)/x response provided by the SAW transducer is probably the easiest to achieve, much effort has been spent on the design of transducers which provide a steep-skirted rectangular shapes for frequency filtering applications. This type of response can be provided by either varying the electrode overlap (W_m) weighting to match a (sinx)/x impulse response [4.24,25] or the electrode spacing (x_m) may be varied in a square law pattern analogous to the phase variation of a radar "chirp" signal [4.5,6,28-30]. All of these approaches can be utilized to design transducers with fractional acoustic bandwidths on the order of 5 to 25%, but for applications requiring bandwidths in the range of 50 to 100%, the uniform and chirp weighted transducers are best.

Detailed calculations show that for any broad-band transducer, whatever the desired conversion efficiency, the maximum bandwidth-efficiency product is bounded by (4.45) if the transducer is connected directly to a resistive source (i.e., the transducer is "untuned"). Thus, an octave-bandwidth transducer fabricated on lithium niobate (BW = 67%, $k_c^2 = 0.05$) will have untuned efficiency near -10 dB at best. Typical experimental data, which include losses due to parasitic effects, have often yielded efficiencies near -14 dB for such broadband, untuned transducers [4.6].

4.2.3 Broad-Band Electric Circuit Design

In common with other electric circuit devices whose driving point impedance is predominantly reactive, the interdigital transducer poses a special problem for operation over very broad bandwidths. As shown by *Bode* [4.39] and *Fano* [4.40], even when an ideal coupling network is employed between a resistive source and reactive load, there are definite limits on the maximum bandwidth-efficiency product available for a given device Q. Here transducer Q may be defined as a "radiation" Q based on the susceptance to conductance ratio at f_0. For the uniform transducer the radiation Q is

$$Q_r = \omega_0 C_T / \hat{G}_a = \frac{\pi}{2k_c^2 M_g} \, . \tag{4.47}$$

Even for strongly piezoelectric materials like $LiNbO_3$, Q_r will typically be larger than $30/M_g$. Thus, for broad-band transducers where M_g is small, Q_r will be large.

In order to improve the electric circuit efficiency, it is expedient to tune out the transducer capacitive reactance with an inductor. While in some cases a shunt inductor has been used in parallel with the transducer to cancel the susceptance ωC_T at f_0, the residual transducer impedance $(1/\hat{G}_a)$ is often quite high for conventional transducer dimensions. Instead, it has been typical to put an inductor in series with the transducer, as shown in Fig.4.7. The apparent input impedance in this "series-tuned" case is

$$Z_c' = j\omega L_c + \frac{G_a - j(\omega C_T + B_a)}{G_a^2 + (\omega C_T + B_a)^2} \tag{4.48}$$

which for broad-band transducers is well approximated by

$$Z_c' \simeq G_a(\omega C_T)^{-2} + j(\omega L_c - 1/\omega C_T) \, . \tag{4.49}$$

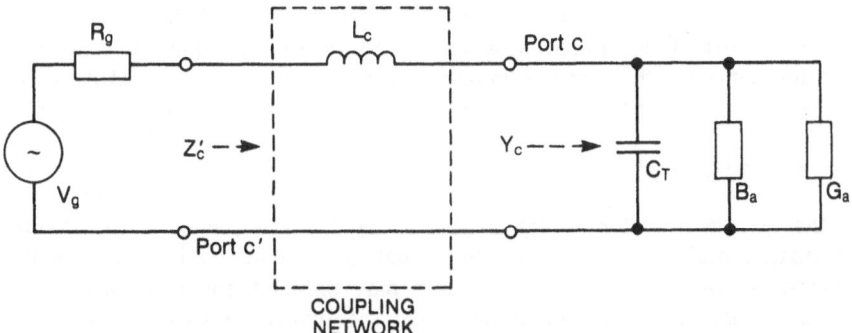

Fig.4.7. Simple SAW tranducer sub-system using a series inductor coupling network (i.e., the "series-tuned" transducer)

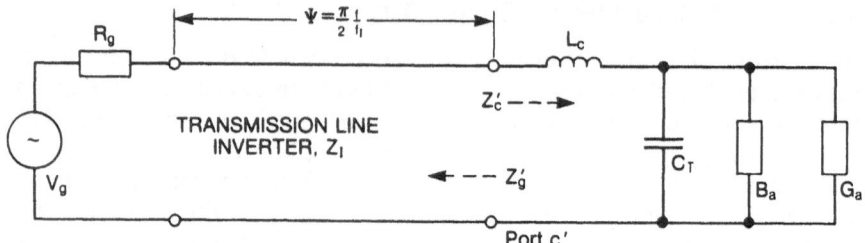

Fig.4.8. SAW transducer sub-system showing the use of a transmission-line inverter as the coupling network between a resistive source and the series-tuned transducer

In the most cases the transducer parameters can be chosen so that the real part of Z_c' is near $R_g = 50 \ \Omega$ at f_0. Because of the variation of G_a with frequency, the frequency response of the transducer sub-system will be dependent on both the electric circuit and acoustic parameters, as predicted earlier in (4.33-35). Studies on series-tuned transducers have shown that, for the uniform interdigital transducer, the maximum efficiency-bandwidth product occurs when

$$G_a'(\omega_0 C_T)^{-2} \simeq R_g \qquad \text{and} \tag{4.50}$$

$$M_g^2 \simeq \pi/k_c^2 \ . \tag{4.51}$$

For Y-cut lithium niobate, $k_c^2 \simeq 0.05$ from Table 4.1, yielding an optimum $M_g \simeq 9$, or from (4.40) an optimum -3 dB fractional bandwidth equal to 0.22. Diagnostic experiments on lithium niobate [4.23] have verified the above efficiency-bandwidth product criteria for the series-tuned transducer and have shown that transducers built on lithium niobate can achieve fractional bandwidths near 0.22 at center frequencies in the 100 to 500 MHz range with efficiencies in the range of -4 to -6 dB.

To achieve bandwidths larger than the series-tuned optimum one must trade reduced transducer efficiency for increased bandwidth as is specified by the Bode-Fano criteria [4.39,40]. One fairly simple way to do this is to couple the series-tuned transducer to the source through a transmission line "inverter" network as first described by *Reeder* and *Sperry* [4.14]. This approach is illustrated in Fig.4.8. The inverter is a quarter wavelength transmission-line transformer which transforms the resistive source impedance to the value

$$Z_g' \simeq (Z_I/R_g)^2 R_g \tag{4.52}$$

at the quarter wavelength frequency, f_I. Here Z_I is the transmission-line impedance and f_I is usually chosen equal to f_0. While it is clear that the transformation of R_g to a new value may enhance the transducer sub-system efficiency, it is not obvious that the inverter may enhance transducer bandwidth. However, if we evaluate the impedance Z_g' seen at the right-hand terminals of the inverter, we find for frequencies near f_I that $Z_g' = R_g'+jX_g'$, where

$$R_g' \simeq R_g (Z_I/R_g)^2 \frac{1}{\cos\Delta} , \qquad (4.53)$$

$$X_g' \simeq \frac{Z_I^2 - R_g^2}{R_g} \tan\Delta , \qquad (4.54)$$

and $\Delta = (\pi/2)(f-f_I)/f_I$. Note that for frequencies sufficiently near f_I

$$X_g' \simeq - \frac{Z_I^2 - R_g^2}{R_g} \frac{\pi}{2} (f/f_I - 1) . \qquad (4.55)$$

Similarly, if we look toward the series-tuned transducer in Fig.4.8 and evaluate the series reactance there near resonance of f_0, using (4.49), we find

$$X_c' \simeq \frac{f/f_0 - 1}{\omega_0 C_T} . \qquad (4.56)$$

We note that if $f_I = f_0$ and the transmission-line impedance Z_I is larger than the source resistance R_g, the reactance slope of X_g' will tend to cancel that of X_c'. Moreover, if we set

$$\frac{\pi}{2} \frac{Z_I^2 - R_g^2}{R_g} = \frac{1}{\omega_0 C_T} \qquad (4.57)$$

the series reactance at the inverter-transducer interface will be near zero over some bandwidth near $f_0 = f_I$. Using (4.29) we can put (4.57) in the form

$$Z_I^2 - R_g^2 = \frac{\hat{G}_a R_g}{(\omega_0 C_T)^2} \frac{1}{k_c^2 M_g} . \qquad (4.58)$$

If the transducer resistance at f_0 is near R_g, i.e.,

$$\hat{G}_a/(\omega_0 C_T)^2 \simeq R_g \qquad (4.59)$$

and if $k_c^2 M_g$ is of the order of 0.05 or larger, as it will be for lithium niobate, then the value of inverter impedance predicted by (4.58) will be less than $5R_g$. Thus, it is reasonable to expect that values of Z_I can be found which will permit the trade-off of efficiency for larger bandwidth. A trade-off study of this type was done numerically, assuming lithium-niobate transducer parameters [4.15]. Figure 4.9 shows plots of the transducer sub-system efficiency at f_0 and the -1.5 dB fractional bandwidth computed as a function of Z_I/R_g. This study considered a specific transducer configuration on lithium niobate (e.g., $k_c^2 = 0.05$, $M_g = 4$), and -1.5 dB bandwidth was computed because the transducers were used in

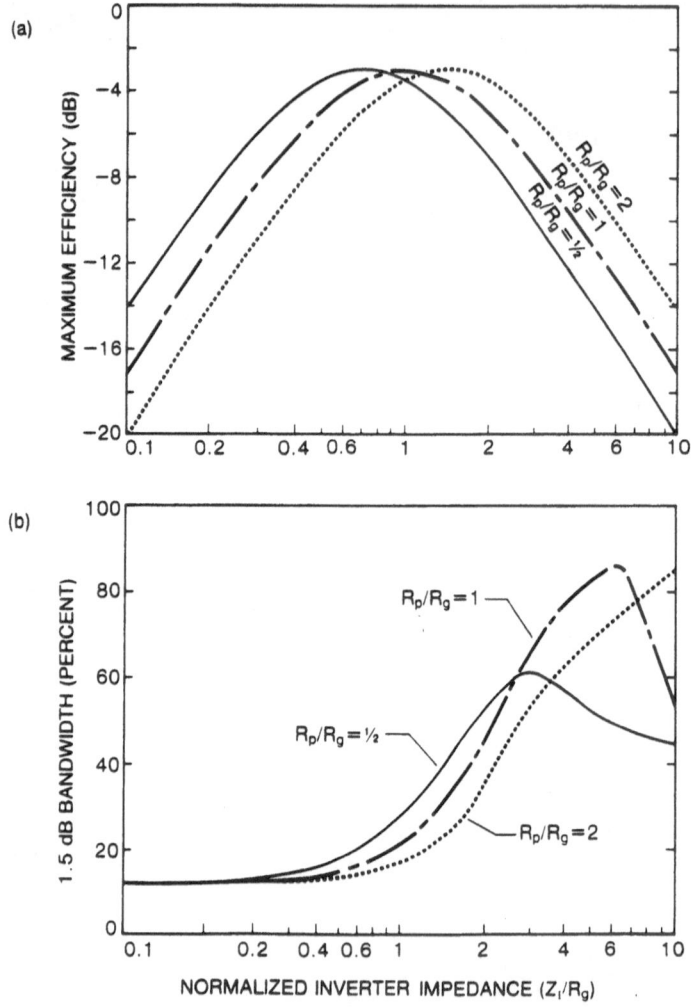

Fig.4.9 Calculated maximum transducer efficiency (a) and the -1.5 dB fractional bandwidth (b) for the inverter coupled, series-tuned SAW transducer sub-system (assume $k_c^2 = 0.05$, $M_g = 4$)

pairs to create a broadband, two-port delay line. The peak radiation resistance, $R_p = (\hat{G}_a/\omega_0 C_T)^2$, is used as a parameter in these calculations. Fractional bandwidths ranging from 40 to 70 are seen for Z_I/R_g in the range of 2 to 3, a range typically achievable with well-known transmission line fabrication methods.

The transmission-line inverter network may be realized in either distributed form or as a lumped element L-C approximation to the quarter-wavelength line [4.15]. In early studies carried out by *Reeder* and *Adams* [4.16], a lumped element line was utilized to provide broadband coupling to transducer pairs fabricated to form a delay line on Y-cut lithium niobate. Using the experimentally measured transducer parameters ($f_0 = 106$

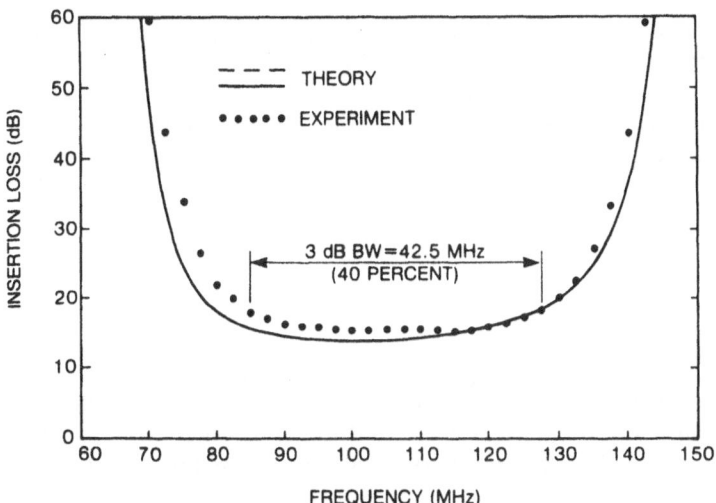

Fig.4.10. Insertion loss versus frequency for a broadband SAW delay line using a lumped elememt inverter network and uniformly weighted interdigital transducers [4.15]

MHz, $k_c^2 = 0.047$, $M_g = 5$, and $(\hat{G}_a/\omega_0 C_T)^2 = 85\Omega)$ and a lumped element inverter with $Z_I = 175 \, \Omega$, an almost maximally flat delay-line frequency response was obtained with efficiency and insertion loss very close to theoretical estimate (Fig.4.10). After subtracting out the estimated delay-line acoustic loss (2dB), the maximum transducer efficiency was estimated to be -8 dB, which is within 1 dB of the computer design.

Later experiments showed that the uniform transducer with lumped-element inverter approach could be extended toward octave bandwidth while retaining maximum conversion efficiency near the Bode-Fano optimum. One such experiment yielded approximately -11 dB conversion efficiency and 55% bandwidth centered at 235 MHz [4.16].

In other experiments *Reible* and co-worker [4.17,18] used 60 Ω coaxial transmission line inverters and 4:1 impedance matching transformers with uniform $M_g = 4$ transducers operating at 500 MHz on Y-cut lithium niobate to achieve a maximally flat response over a 44% bandwidth. In this case, the experimentally determined transducer efficiency was estimated to be -12 dB, a few dB higher than the computer estimate.

4.2.4 Parasitic and Second-Order Effects

The detailed design of transducer circuits and electrode configurations must account for a number of electric circuit parasitic and acoustic configuration second-order effects which tend, in most cases, to decrease bandwidth and efficiency and may add undesirable amplitude and phase ripple to the transducer sub-system frequency response. An excellent review of these effects has been given by *Smith* [4.30].

Perhaps the most serious of these effects is the undesirable excitation of inter-electrode acoustic reflections seen when the transducer electrodes are one quarter-wavelength wide (i.e., $d = \Lambda_0/4 = V_R/4f_0$). Small but finite wave reflections occur as a surface acoustic wave impinges on a metal transducer electrode in a piezoelectric medium. The presence of the electrode is, in effect, a perturbation of the acoustic transmission line which changes the effective wave velocity and impedance due to two effects: 1) shorting out the piezoelectric field and 2) mass loading of the wave surface. Piezoelectric shorting or "stiffening" alone can account for an effective change in impedance of order

$$\Delta Z/Z_0 \simeq (1/2)k_c^2 . \qquad (4.60)$$

If the transducer electrodes are $\Lambda_0/4$ wide and are separated by $\Lambda_0/4$, the reflections described by (4.60) will add in phase at f_0, creating very undesirable transducer response distortion for cases where more than a few electrodes are used (i.e., the (sinx)/x and chirp weighted transducer configurations).

Fortunately, this effect can be virtually eliminated [4.41] by constructing the transducer with "double electrodes" as in Fig.4.11. Here each electrode is $\Lambda_0/8$ wide so that the acoustic reflection cannot occur at fre-

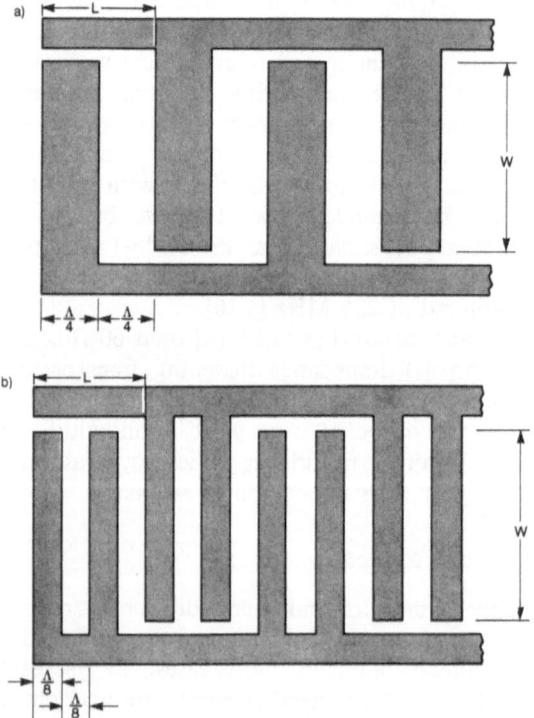

Fig.4.11. Schematic comparison of (a) interdigital SAW transducers utilizing quarter wavelength wide "single" electrodes spaced by one-half wavelength and (b) those using one-eighth wavelength wide "double" electrodes spaced by one-quarter wavelength

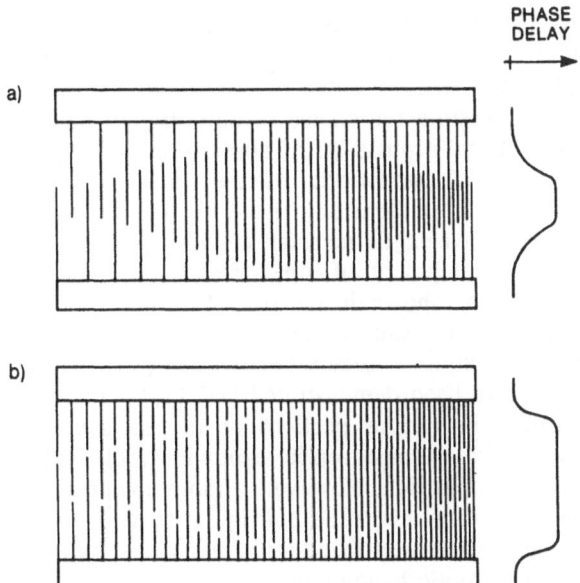

PHASE
DELAY

a)

b)

Fig.4.12a,b. Schematic view of an interdigital SAW transducer showing the use of "dummy electrodes" to prevent undesired wavefront distortion arising from wave propagation through a weighted transducer electrode configuration: (a) shows effect of wavefront distortion; (b) shows straight-crested wavefront created by use of "dummy" electrodes

quencies below $f = 2f_0$. Thus, low distortion, octave bandwidth transducers can be made using the "double electrode" approach. It should be pointed out that the effective k_c^2 is reduced in this case. However, accurate values have been calculated and experimentally verified for popular acoustic substrates and transducer configurations [4.28-30, 42].

Another important second-order effect is the acoustic wavefront distortion that can result from electrode overlap weighting variations along the acoustic length of the transducer [4.6, 43]. Since the wave velocity on a "free surface" or unmetallized portion of the acoustic substrate is higher than neighboring metallized regions due to piezoelectric stiffening, the absence of electrode material over some portions of the acoustic path will result in higher average wave velocity and reduced propagation delay in comparison to waves that pass through the maximum number of electrodes. This gives rise to a distortion in wavefront phase delay, as is illustrated schematically in Fig.4.12a. By adding "dummy electrodes", as shown in Fig.4.12b, this problem is effectively eliminated.

4.3 Summary

The analysis outlined in this chapter provides a useful first-order model of interdigital electrode SAW transducers and gives insight into methods of achieving broadband transducer sub-system operation through the use of coupling networks. As pointed out in Sect.4.1., the frequency range and accuracy of the transducer model can easily be extended by use of the somewhat more complex approach presented in [4.28-30].

We have left as an open question the choice of an optimum transducer configuration for broad band acousto-optic applications. For single transducer systems with fractional bandwidths of the order of 40% to 70%, it seems clear that either the uniform or chirp-weighted transducers can be used, and the quarter wavelength inverter network [4.14-16] can be applied to either in the development of sub-systems with optimum efficiency-bandwidth products. As shown in Chap.5, the choice becomes more complex when arrays of transducers are used to optimize the acousto-optic spatial phase match over broad bandwidths [4.1,4,11,12]. Moreover, chirp transducers have been demonstrated [4.4,11,12] that utilize a slight "tilt" of electrodes in various portions of the transducer to optimize the spatial phase match. Acousto-optic bandwidths of 50% to 60% have been achieved by this approach. However, a more complex analysis, considering both the transducer electrode configuration and electric circuit coupling network, will be required to design a transducer sub-system with optimally large acousto-optic bandwidth.

References

4.1   C.S. Tsai: IEEE Trans. CAS-64, 1072 (1979)
4.2   R.M. White, F.W. Voltmer: Appl. Phys. Lett. 7, 314 (1965)
4.3   R.M. White: IEEE Trans. ED-14, 181 (1967)
4.4   C.S. Tsai, M.A. Alhaider, L.T. Nguyen, B.Kim: Proc. IEEE 64, 316 (1976)
4.5   R.H. Tancrell, M.B. Schulz, H.H. Barrett, L. Davis, Jr., M.G. Holland: Proc. IEEE 57, 1211 (1969)
4.6   H.M. Gerard: Surface wave interdigital electrode chirp filters, in *Surface Wave Filters*, ed. by H. Matthews (Wiley, New York 1977) pp.347-380
4.7   R.M. Hays, C.S. Hartmann: Proc. IEEE 64, 652 (1976)
4.8   A.J. DeVries, R.Adler: Proc. IEEE 64, 671 (1976)
4.9   W.S. Drummond, S.A. Roth: Application of high-performance SAW transversal filters in a precision measurement station. Proc. IEEE Ultrasonics Symp. (1978) pp.494-499
4.10 R.C. Bray: Applications of SAW devices in high-performance instrumentation. Proc. IEEE Ultrasonics Symp. (1986) pp.299-307
4.11 C.C. Lee, K.Y. Liao, C.L. Chang, C.S. Tsai: IEEE J. QE-15, 1166 (1979)
4.12 K.Y. Liao, C.L. Chang, C.C. Lee, C.S. Tsai: Progress on wide-band guided wave acoustal optic Bragg deflector using a tilted finger chirp transducer. Proc. IEEE Ultrasonics Symp. (1979) pp.24-27
4.13 T.R. Joseph, B. Chen: Broad-band chirp transducers for integrated optics and spectrum analyzers. Proc. IEEE Ultrasonics Symp. (1979) pp.28-33
      T.R. Joseph, W.R. Smith: Wide-band SAW transducers for IOSA applications. 1981 Int'l Conf. Integrated Optics and Optical Fiber Communication (IEEE 81 CH1 649-3, 1981) pp.122-123
4.14 T.M. Reeder, W.R. Sperry: IEEE Trans. MTT-20, 453 (1972)
4.15 T.M. Reeder, W.R. Shreve, P.L. Adams: IEEE Trans. SU-19, 466 (1972)
4.16 T.M. Reeder, P.L. Adams: Electron Lett. 7, 570 (1971)

4.17   I. Yao, S.A. Reible: Wide-bandwidth acousto-electric convolvers. Proc. IEEE Ultrasonics Symposium (1979) pp.701-705

4.18   S.A. Reible: IEEE Trans. MTT-29, 463 (1981)

4.19   H.I. Smith: Surface Wave Device Fabrication, in *Surface Wave Filters*, ed. by H. Matthews (Wiley, New York 1977) pp.165-217

4.20   G.L. Matthaei, D.J. Wong, P.B.O'Shaughnessy: IEEE Trans. SU-22,105 (1975)

4.21   W.R. Smith, H.M. Gerard, J.H. Collins, T.M. Reeder, H.J. Shaw: IEEE Trans. MTT-17, 856 (1969)

4.22   D.A. Belincourt, D.R. Curran, H. Jaffe: *Physical Acoustics* 1A, 233-242 (Academic, New York 1964)

4.23   W.R. Smith, H.M. Gerard, J.H. Collins, T.M. Reeder, H.J. Shaw: IEEE Trans. MTT-17, 865 (1969)

4.24   R.H. Tancrell, M.G. Holland: Proc. IEEE 59, 393 (1971)

4.25   C.S. Hartmann, D.T. Bell, R.C. Rosenfeld: IEEE Trans. SU-20, 80 (1973)

4.26   A.J. Bahr, R.E. Lee: Electron. Lett. 9, 281 (1973)

4.27   R. Krimholz: IEEE Trans. SU-19, 427 (1972)

4.28   W.R. Smith: Wave Electron. 2, 25 (1976)

4.29   W.R. Smith, W.F. Pedler: IEEE Trans. MTT-23, 853 (1975)

4.30   W.R. Smith: Circuit model analysis and design of interdigital transducers for surface acoustic wave devices, *Physical Acoustics* 15, 99-189 (Academic, New York 1981)

4.31   R.F. Mitchell, M. Redwood: J. Acoust. Soc. Am. 47, 701 (1970)

4.32   B.A. Auld: *Acoustic Fields and Waves in Solids*, Vol. 1 (Wiley-Interscience, New York 1973) Chap.3

4.33   E.H. Jacobsen: J. Acoust. Soc. Am. 32, 949 (1960)

4.34   J.J. Campell, W.R. Jones: IEEE Trans. SU-15, 209 (1972)

4.35   M.B. Schulz, J.M. Matsinger: Appl. Phys. Lett. 20, 367 (1972)

4.36   A.J. Slobodnik: Proc. IEEE 64, 581 (1976)

4.37   L. Weinberg: *Network Analysis and Synthesis* (McGraw-Hill, New York 1962) p.282

4.38   See, for example, [4.20,21,23-30] and the proceedings of Ultrasonic Symposia (1975-1980)

4.39   H.W. Bode: *Network Analysis and Feedback Amplifier Design* (Van Nostrand, New York 1945) pp.360-371

4.40   R.M. Fano: J. Franklin Inst. 249, 57, 139 (1950)

4.41   T.W. Bristol et al.: Application of double electrodes in acoustic surface wave device design. Proc. IEEE Ultrasonic Symp. (1972) pp.343-345

4.42   T.W. Bristol: Analysis and design of surface acoustic wave transducers. IEEE Specialist Seminar on Component Performance and Systems Applications of Surface Acoustic Wave Devices, IEEE Publication No. 109 (1973) pp.115-129

4.43   R.H. Tancrell, R.C. Williamson: Appl. Phys. Lett. 19, 456 (1971)

5. Wideband Acousto-Optic Bragg Diffraction in LiNbO₃ Waveguide and Applications

Chen S. Tsai

With 54 Figures

Chapter 2 has presented the basic principles and analytical techniques for bulk-wave Acousto-Optic (AO) interactions. In this chapter, a detailed treatment of wide-band AO Bragg diffraction in a planar LiNbO₃ waveguide, the resulting devices, and some potential applications is given. The sequence of presentation for the content of the Chapter now follows. First, the basic configuration and mechanisms for planar guided-wave AO Bragg diffraction from a single Surface Acoustic Wave (SAW) in a LiNbO₃ waveguide and the resulting diffraction efficiency and frequency response are analyzed in detail using the coupled-mode technique. As a comparison some calculated and measured performances with the other two potential materials, namely GaAs and nonpiezoelectrics such as thermally-oxidized Si or As_2S_3, are also discussed. The key parameters of the resulting AO Bragg modulators and deflectors or cells and their inherent limitations are then identified and discussed. The coupled-mode technique is thereafter extended to analyze AO Bragg diffraction from multiple SAWs. The unified theory developed is applied to the two wide-band device configurations, namely, multiple tilted SAWs and phased SAWs. Subsequently, a number of SAW transducer configurations for realization of wide-band Bragg cells are described and compared. Design, fabrication, testing, and measured performance of wide-band AO Bragg cells in y-cut LiNbO₃ substrates are also presented. Finally, some potential applications of such wide-band AO Bragg cells in optical communications, Radio Frequency (RF) signal processing, and computing are described. As a continuation to this chapter, a progress report on realization of integrated AO device modules in planar and planar-channel waveguides as well as spherical waveguides in LiNbO₃ substrates, AO spectrum analyzer in silicon substrate, and Gigahertz AO Bragg cells and integrated AO spectrum analyzer modules in GaAs planar waveguide substrate is given in Chap.8.

5.1 Guided-Wave Acousto-Optic Bragg Interactions in Planar Waveguides

5.1.1 Basic Planar Acousto-Optic Bragg Interaction Configuration and Mechanisms

A basic planar guided-wave AO Bragg interaction configuration is shown in Fig.5.1 [5.1]. The optical-waveguide material should possess desirable optical, acoustical, and AO properties. The waveguide can be either a

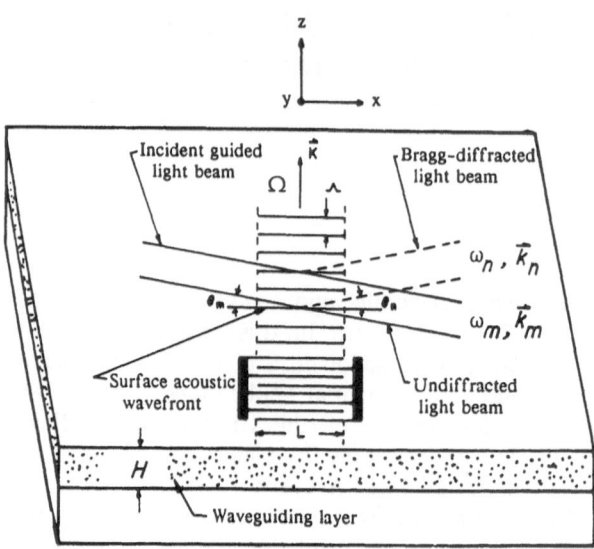

Fig.5.1. Guided-wave acousto-optic Bragg diffraction from a single surface acoustic wave in LiNbO$_3$ substrate

graded-index layer created within the substrate or a step-index layer deposited on the top of the substrate. Examples of the graded-index guides are the LiNbO$_3$ waveguides formed by out- and in-diffusion techniques [5.2,3], while examples of the step-index guides are the glass [5.4] or As$_2$S$_3$ film [5.5] waveguides formed by sputtering or deposition. In either case, the waveguide layer is assumed to have a thickness or an equivalent penetration depth H. Optical waveguide modes in such types of structure have been studied in detail (Chap.3) [5.6-10]. The SAW is commonly excited by an interdigital electrode transducer (IDT) (Chap.4). If the substrate material is sufficiently piezoelectric such as a Y-cut LiNbO$_3$, the IDT may be deposited directly on the optical waveguide (Chap.4). Otherwise, a piezoelectric film such as ZnO must be deposited upon the optical waveguide and the interdigital-electrode array be deposited either beneath or above it (Chap.6).

Propagation of the SAW creates both a moving optical grating in the optical waveguide and moving corrugations at both the air-waveguide and the waveguide-substrate interfaces. The moving grating and corrugations, in turn, cause diffraction of an incident guided-light wave [5.1,4,5,11-21]. Thus, except for the corrugations, the underlying mechanisms in coplanar guided-wave AO interaction are analogous to those of bulk-wave AO interaction [5.22-30], as presented in Chap.2. While the contribution due to the corrugations can be significant for certain ranges of waveguide thickness and acoustic wavelength [5.19], the optical grating is the dominant mechanism for diffraction in most practical cases.

As in the bulk-wave AO interaction, the diffraction may be of either the Raman-Nath type or the Bragg type. The diffraction is of the Raman-Nath type and consists of a number of side orders when the AO factor $Q \equiv 2\pi\lambda_0 L/n\Lambda^2$ is less than or equal to 0.3. The symbols λ_0 and Λ designate, respectively, the wavelengths of the guided optical wave (in free space) and the SAW, n is the effective refractive index of the medium, and L designates the aperture of the acoustic wave. In the other extreme, when the light wave is incident at the Bragg angle to be defined later, and Q is larger than 4π, the diffraction is of the Bragg type and consists mainly of one side order. We shall henceforth limit our discussion to the Bragg type diffraction because this type of diffraction has the desirable characteristics of higher acoustic center frequency, wider modulation bandwidth, higher diffraction efficiency, and larger dynamic range. Similarly, the diffraction may be either the isotropic type or the anisotropic type [5.20,29,31]. The diffraction is of the isotropic type when the polarization of the incident light is preserved in the diffracted light. In the anisotrpic type the polarization of the diffracted light differs by a 90-degree from that of the incident light. In other words, there exists a mode-conversion of TE-mode to TM-mode and vice versa in anisotropic diffraction. Finally, for guided-wave AO interaction in an anisotropic waveguide it is also possible for the diffracted light to propagate as leaky waves [5.32].

It is important, however, to note that while the basic interaction mechanisms for planar guided-wave acousto-optics are analogous to those of bulk-wave acousto-optics, the number of parameters involved in the guided-wave case is greater and the interrelation between them is much more complex [5.1]. For example, for a given optical wavelength the diffraction efficiency is a sensitive function of the spatial distributions of both optical and acoustic waves which in turn depend, respectively, on the guided optical modes and the acoustic frequency involved. In addition, in the case of piezoelectric and Electro-Optic (EO) substrate such as LiNbO$_3$ and ZnO the piezoelectric field accompanying the SAW can be so large that the induced index changes due to the EO effect become very significant [5.1,14-16]. In fact, such is the case for a Z-propagation SAW in a Y-cut LiNbO$_3$ substrate. Thus the contribution from the EO effect can either add to or subtract from that by the elasto-optic effect.

The relevant momentum (wave vector) and energy (frequency) conservation relations between the incident light waves, the diffracted light waves and the SAW are expressed by (5.1), where m and n designate the indices of the waveguide modes. The symbols $\mathbf{k}_n$, $\mathbf{k}_m$, and $\mathbf{K}$ are, respectively, the wave vectors of the diffracted light, undiffracted light, and the SAW, and ω_n, ω_m, and Ω are the corresponding radian frequencies. Figure 5.2a shows the wave-vector diagram for the general case in which the diffracted light propagates in a waveguide mode (nth mode) different from the incident light (mth mode). The diffracted light may have a polarization parallel or

$$\mathbf{k}_n = \mathbf{k}_m \pm \mathbf{K} , \tag{5.1a}$$

$$\omega_n = \omega_m \pm \Omega \tag{5.1b}$$

119

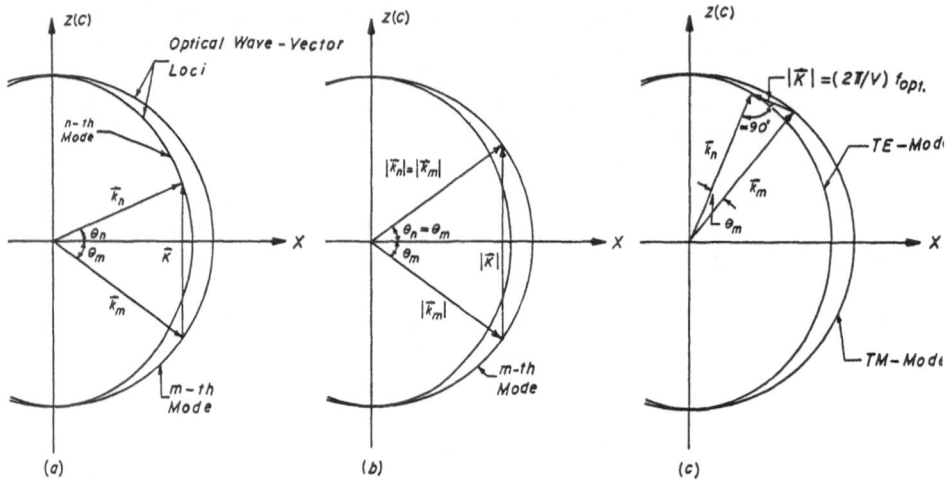

Fig.5.2a–c. Phase-matching diagrams indicating wave vectors of incident light, Bragg-diffracted light and surface acoustic wave in a y-cut LiNbO$_3$ waveguide: (a) general case, (b) special isotropic case, (c) optimized anisotropic case

orthogonal to that of the incident light. As mentioned above, these two classes of interaction are called isotropic and anisotropic AO Bragg diffraction, respectively. As a special isotropic case, the diffracted light may be in the same mode and thus have the same polarization as the incident light, as depicted in Fig.5.2b (n=m). Figure 5.2c depicts a particularly interesting case of the anisotropic Bragg diffraction in which the wave vector of the diffracted light is perpendicular or nearly perpendicular to that of the acoustic wave, and is commonly called optimized anisotropic diffraction [5.11, 12, 20]. Specific examples for both cases in a Y-cut LiNbO$_3$ substrate, as depicted in Fig.5.2b,c, are discussed in Sects.5.1.4 and 5.4.1, respectively.

The angles of incidence and diffraction measured with respect to the acoustic wavefront, $\theta_i = \theta_m$ and $\theta_d = \theta_n$, are

$$\sin\theta_m = \frac{\lambda_0}{2 n_m \Lambda}\left[1 + \frac{\Lambda^2}{\lambda_0^2}(n_m^2 - n_n^2)\right] , \qquad (5.2a)$$

$$\sin\theta_n = \frac{\lambda_0}{2 n_n \Lambda}\left[1 - \frac{\Lambda^2}{\lambda_0^2}(n_m^2 - n_n^2)\right] , \qquad (5.2b)$$

where n_m and n_n are the effective refractive indices of the undiffracted and diffracted light and, as defined earlier, λ_0 and Λ are the optical wavelength in free space and the wavelength of the SAW. Figure 5.3 shows plots for the angles of incidence and diffraction versus the SAW frequency for diffraction between distinct waveguide modes and that between the same mode. Clearly, when the undiffracted and diffracted light propagate in the same waveguide mode (Fig.5.2b), we have $n_n = n_m$. For this particular case, (5.2) reduces to the well-known Bragg condition

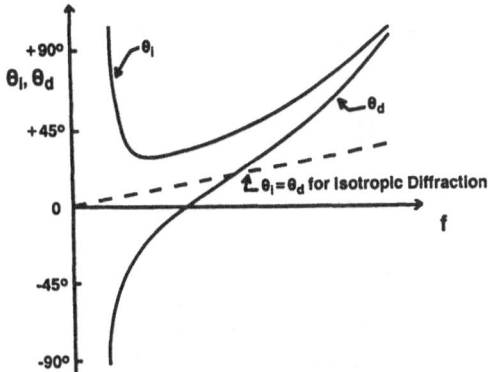

Fig.5.3. Angles of incidence and diffraction in planar guided-wave acousto-optic Bragg diffraction vs. SAW frequency: Diffraction between distinct modes (*solid curves*) and diffraction between the same mode (*dotted line*)

in isotropic diffraction, namely, $\sin\theta_n = \sin\theta_m = (\lambda_0/2n_m\Lambda) = (\lambda_0\Omega/4\pi n_m V_R)$, where $\theta_n = \theta_m$ is the so-called Bragg angle, and V_R designates the propagation velocity of the SAW. Thus, as shown in Fig.5.3 the diffraction angle is identical to the incidence angle, and both angles increase linearly with the acoustic frequency since in practice the optical wavelength in the medium (λ_0/n_m) is much smaller than the acoustic wavelenth Λ.

5.1.2 Coupled-Mode Analysis on Acousto-Optic Bragg Diffraction from a Single-Surface Acoustic Wave

The most common approach for treatment of acousto-optic interactions is the so-called coupled-mode technique [5.1,5,14,15]. This technique is now employed for the analysis of guided-wave AO Bragg diffraction from a single SAW in a LiNbO$_3$ planar waveguide. Such an analysis will reveal the physical parameters involved and the key device parameters as well as the performance limitations of the resulting devices. Both the analytical procedures and the methodology for numerical computation developed for this simple case can be conveniently extended to the case involving multiple SAWs. The general case involving N-SAW's and the specific case involving two tilted SAWs are treated in Sect.5.3.

Referring to the coordinate system of Fig.5.1 in which a guided-light wave propagating in the xz plane is Bragg diffracted by a z-propagating SAW. Assuming that the medium is lossless both optically and acoustically, the corresponding electric fields $\hat{E}_m$ and $\hat{E}_n$ of the undiffracted and diffracted light waves, and the strain field $\hat{S}$ of the SAW and its accompanying piezoelectric field $\hat{E}_p$ can be written as follows:

$$\hat{E}_m(x,y,z,t) = \tfrac{1}{2}E_m(x)U_m(y)\exp[j(\omega_m t - k_{mx}x - k_{mz}z)] + \text{c.c.} , \qquad (5.3a)$$

$$\hat{E}_n(x,y,z,t) = \tfrac{1}{2}E_n(x)U_n(y)\exp[j(\omega_n t - k_{nx}x - k_{nz}z)] + \text{c.c.} , \qquad (5.3a)$$

121

$$\hat{S}(y,z,t) = \frac{1}{2}S_I U_{aI}(y)\exp[j(\Omega t - Kz + \phi)] + \text{c.c.} , \quad I = 1,2,3,4,5,6 , \quad (5.3c)$$

$$\hat{E}_p(y,z,t) = \frac{1}{2}E_{pi} U_{pi}(y)\exp[j(\Omega t - Kz + \phi)] + \text{c.c.} , \quad i = 1,2,3 , \quad (5.3d)$$

where $E_m(x)$ and $E_n(x)$ are the as yet undetermined spatial distributions of the undiffracted and diffracted light waves; S_I and E_{pi} are the components of the strain field and its accompanying piezoelectric field, respectively. Also, $U_m(y)$, $U_n(y)$, $U_{aI}(y)$, and $U_{pi}(y)$ designate, respectively, the normalized field distributions (along the waveguide thickness) of the light waves, the acoustic wave, and the piezoelectric field. It is to be noted that for simplicity the subscripts I and i will be dropped henceforth. The frequencies of the light waves and the acoustic wave are designated as previously by ω_m, ω_n, and Ω, respectively. Similarly, k_m, k_n, and K designate the corresponding wavenumbers and the suffixes x and z represent the x and z components. Finally, ϕ designates the phase of the SAW.

When (5.3a-d) are substituted into the wave equation for the electric field of the optical waves, a set of coupled wave equations for $E_m(x)$ and $E_n(x)$ are obtained. These coupled wave equations are then readily reduced to the following decoupled form [5.14, 15]

$$\frac{\partial^2 E_m(x)}{\partial x^2} - jK\Delta\theta\frac{\partial E_m(x)}{\partial x} + ABE_m(x) = 0 , \qquad (5.4a)$$

$$\frac{\partial^2 E_n(x)}{\partial x^2} + jK\Delta\theta\frac{\partial E_n(x)}{\partial x} + ABE_n(x) = 0 . \qquad (5.4b)$$

From (5.4) the general solutions for $E_m(x)$ and $E_n(x)$ are easily found. By matching the boundary conditions $E_m(0) \equiv 1$ and $E_n(0) \equiv 0$ to these general solutions, the electric fields of the undiffracted and diffracted light waves at the output of the interaction region (x=L) are obtained as follows [5.14, 15]

$$E_m(L) = [\cos(qL) - j(K\Delta\theta/2q)\sin(qL)]\exp[j(K\Delta\theta L/2)] , \qquad (5.4c)$$

$$E_n(L) = j(B/q)\sin(qL)\exp[-j(K\Delta\theta L/2) - j\phi] , \qquad (5.4d)$$

where

$$q^2 = (K\Delta\theta/2)^2 + AB , \qquad (5.4e)$$

$$A = \frac{\omega_n{}^2 n_m{}^2 n_n{}^2}{4c^2 k_m \cos\theta_m}$$

$$\times \frac{\displaystyle\int_0^\infty U_m(y)U_n(y)[p\!:\!SU_a(y) + r\!:\!E_p^2 U_p(y)]dy}{\displaystyle\int_{-\infty}^{+\infty} U_m\,dy}, \tag{5.4f}$$

$$B = \frac{\omega_m^2 n_m^2 n_n^2}{4c^2 k_n \cos\theta_n}$$

$$\times \frac{\displaystyle\int_0^\infty U_m(y)U_n(y)[p\!:\!SU_a(y) + r\!:\!E_p^2 U_p(y)]dy}{\displaystyle\int_{-\infty}^{+\infty} U_n\,dy}. \tag{5.4g}$$

Here L is the aperture of the SAW, c the velocity of light in free space, and $\Delta\theta$ the deviation of the incidence angle of the light from the Bragg angle. The following physical constants with suppressed vector and tensor subscripts are employed; p: relevant photoelastic constant or constants and r: relevant electrooptic coefficient or coefficients. Finally, the symbol ":" designates multiplication of the two neighboring tensors, or the neighboring tensor and vector.

It is to be noted that for the case involving multiple SAWs [5.15] the input boundary values to an adjacent SAW are simply those given by (5.4 c,d) with appropriate phase factors added to account for the propagation delay between the adjacent SAWs.

5.1.3 Diffraction Efficiency and Frequency Response

From (5.4) the diffraction efficiency $\varsigma(f)$, which is defined as the ratio of the diffracted light power at the output (x=L) and the incident light power at the input (x=0) of the interaction region, is found [5.14]

$$\varsigma(f) = g^2(f)\left[\frac{\sin\sqrt{g^2(f) + (K\Delta\theta L/2)^2}}{\sqrt{g^2(f) + (K\Delta\theta L/2)^2}}\right]^2, \tag{5.5a}$$

where

$$g^2(f) \equiv \frac{\pi^2}{4\lambda_0^2}\left[n_m^3 n_n^3 |\Gamma_{mn}(f)|^2 \frac{L^2}{\cos\theta_m \cos\theta_n}\right], \tag{5.5b}$$

$$|\Gamma_{mn}(f)|^2 \equiv \frac{\left| \int_0^\infty U_m(y)U_n(y)\{p{:}SU_a(y) + r{:}E_p \, U_p(y)\}dy \right|^2}{\left| \int_{-\infty}^{+\infty} U_m^2(y)dy \right| \left| \int_{-\infty}^{+\infty} U_n^2(y)dy \right|} \tag{5.5c}$$

and $f = \Omega/2\pi$ = frequency [Hz] of the SAW.

The general expression given by (5.5a) can be used to calculate the diffraction efficiency and its frequency response (diffraction efficiency versus the RF or acoustic frequency f) as a function of both the polarization and the angle of incidence of the light [5.1]. Equation (5.5a) reduces to the following simple form when the Bragg condition is satisfied ($\Delta\theta \equiv 0$):

$$\zeta(f) = \sin^2 g(f) \; . \tag{5.6}$$

It is seen that in contrast to bulk-wave AO Bragg interaction the diffraction efficiency is a sensitive function of the spatial distributions of the diffracted and the undiffracted light waves and the SAW, as determined by the coupling function $|\Gamma_{mn}(f)|^2$. Since the confinement of the SAW is proportional to the acoustic wavelength, $|\Gamma_{mn}(f)|^2$ is also dependent on the thickness of the optical waveguide with respect to the acoustic wavelength. Consequently, $|\Gamma_{mn}(f)|^2$ varies strongly with the acoustic frequency. An efficient diffraction can occur only in the frequency range for which the confinement of the SAW matches that of the diffracted- and undiffracted-light waves. The dependence of the diffraction efficiency on acoustical, optical and AO parameters is further complicated by the accompanying EO effect. Thus, in general, the coupling function $|\Gamma_{mn}(f)|^2$ is more complicated than the so-called overlap integral [5.4, 5]. Only for the special case in which the accompanying EO contribution is either negligible (e.g., in a glass waveguide) or proportional to the elasto-optic contribution will the coupling function be proportional to the overlap integral.

For the same reason, it is impossible to define an AO figure of merit as simply as in the bulk-wave interaction [5.30, 31]. However, if we again consider the special case just mentioned and employ a model similar to that described in [5.13], we may define the total power flow of the SAW, P_a, as

$$P_a \equiv \frac{1}{2}\rho V_R^3 LS^2 \int_0^\infty U_a^2(y)dy \tag{5.7}$$

where ρ and V_R designate the density and the SAW propagation velocity of and in the interaction medium, S denotes the acoustic strain, and

$\int_0^\infty U_a{}^2(y)dy$ carries a dimension of length. Equations (5.5b,c and 7) are now combined to give the following expression for $g^2(f)$:

$$g^2(f) = \left[\frac{\pi^2}{2\lambda_0{}^2}\right] \left[\frac{n_m{}^3 n_n{}^3 p^2}{\rho V_R{}^3}\right] C_{mn}^2(f) \left[\frac{L}{\cos\theta_m \cos\theta_n}\right] P_a, \qquad (5.8a)$$

where the frequency-dependent coupling coefficient $C_{mn}^2(f)$ is defined as follows

$$C_{mn}{}^2(f) \equiv \frac{\left[\int_0^\infty U_m(y)U_n(y)U_a(y)dy\right]^2}{\left[\int_{-\infty}^\infty U_m{}^2(y)dy\right]\left[\int_{-\infty}^\infty U_n{}^2(y)dy\right]\left[\int_0^\infty U_a{}^2(y)dy\right]}. \qquad (5.8b)$$

Note that $C_{mn}{}^2(f)$ takes a form similar to the overlap integral with its value depending upon the optical and the acoustical modes of propagation and that it equals unity for bulk-wave AO interaction. Also, the factor $(n_m{}^3 n_n{}^3 p^2/\rho V_R{}^3)$ is similar to the bulk-wave AO figure of merit M_2 [5.30, 31], and henceforth designated by M_{2mn}.

From (5.6,7,8a) it is seen that for the case in which $g^2(f) \ll 1$ and $\Delta\theta \equiv 0$, the diffraction efficiency is approximately proportional to the total acoustic power P_a. The effect of this quasi-linear dependence on the dynamic range of the resulting AO Bragg modulators and deflectors is discussed in Sect.5.2. It is also seen that for a fixed total acoustic power and a relatively low diffraction efficiency (e.g., 50% or less) the diffraction efficiency is linearly proportional to the acoustic aperture.

We shall now examine the quantitative dependence of the diffraction efficiency on the acoustic frequency. From (5.5a and 6) it is first noted that in contrast to the bulk-wave AO interaction, even at $\Delta\theta \equiv 0$ the diffraction efficiency is a sensitive function of the acoustic center frequency f_0 because of the frequency dependence of the SAW confinement. Thus the bandwidth of a guided-wave AO Bragg modulator is mailnly determined by the frequency dependence of three factors: the transducer conversion efficiency, the Bragg condition (phase matching), and the SAW confinement. The individual and combined effects of these three factors on the frequency response can be determined using (5.5) and a digital computer. For example, in order to study the effect of the SAW confinement alone we may assume that the transducer bandwidth is so large that its frequency dependence can be ignored, and the acoustic aperture is so small that Bragg condition can be satisfied at all frequencies of interest. Similarly, by setting the right-hand side of (5.5a) equal to one half of its value at the acoustic center frequency f_0 and $\Delta\theta \equiv 0$, the -3 db modulator bandwidth $\Delta f_{-3dbBragg}$, as determined by the last two factors can be calculated [5.14]. The two types of calculations just described as applied to the Y-cut LiNbO$_3$ waveguides are presented in the following subsection.

For purpose of illustration we now calculate the diffraction efficiency at the acoustic center frequency f_0 for the special case in which the EO contribution is either negligible or proportional to the elasto-optic contribution. This calculation is applicable to the case involving the non-piezoelectric materials such as glasses, oxidized silicon, and As$_2$S$_3$. We further assume that the power of the SAW is uniformly distributed in a depth of one acoustic wavelength [5.15,16,17]. Thus as a special form to (5.7) we have

$$P_a = \frac{1}{2}\rho V_R{}^3 S^2 L \Lambda_0 = \frac{1}{2}\rho V_R{}^4 S^2 \frac{L}{f_0} \qquad \text{or} \qquad (5.9a)$$

$$S = \sqrt{\frac{2P_a}{\rho V_R{}^4}} \sqrt{\frac{f_0}{L}} \qquad (5.9b)$$

where Λ_0 designates the acoustic wavelength at the center frequency. Substituting (5.9b) into (5.5c) and combining (5.5b,c and 6), and restricting to the case of moderate diffraction, we obtain the following expression for the diffraction efficiency

$$\varsigma(f_0) \simeq g^2(f_0) = \frac{\pi^2}{2\lambda_0{}^2} M_{2mn} \left[\frac{f_0 L}{V_R \cos\theta_m \cos\theta_n} \right] P_a \, . \qquad (5.10)$$

This equation shows that for this special case the diffraction efficiency is proportional to the product of the center frequency and the aperture of the SAW as well as the total acoustic power.

We now turn to the AO Bragg bandwidth. In contrast to the bulk-wave interaction [5.27] the AO Bragg bandwidth cannot, in general, be expressed explicitly in terms of the center frequency and the aperture of the acoustic wave. Only for the special case in which $g^2(f) \ll (K\Delta\theta L/2)^2$ would (5.5a) lead to an expression similar to that of the bulk-wave case. For this special case the absolute AO Bragg bandwidth for isotropic interaction with TE$_0$ mode is given as follows:

$$\Delta f_{-3\,db\,Bragg} \simeq \frac{1.8 n_0 V_R{}^2 \cos\theta_0}{\lambda_0 f_0 L} \qquad \text{or} \qquad (5.11a)$$

$$\Delta f_{-3\,db\,Bragg}/f_0 \simeq \frac{1.8 n_0 V_R \cos\theta_0}{\lambda_0 f_0} \frac{\Lambda_0}{L} \qquad (5.11b)$$

where n_0 and θ_0 designate the effective refractive index and Bragg angle for the TE$_0$ mode, and Λ_0 again designates the wavelength of the SAW at the center frequency. Earlier measurements in the Y-cut LiNbO$_3$ wave-

126

guides have verified (5.11) [5.12-14]. The absolute Bragg bandwidth is inversely proportional to the product of the center frequency and the aperture of the SAW. Equations (5.10, 11) indicate clearly that the diffraction efficiency and the AO Bragg bandwidth impose conflicting requirements on the acoustic aperture. In fact, the diffraction efficiency-Bragg bandwidth product is a constant, independent of both the center frequency and the aperture of the SAW.

In summary, because of the complicated spatial distributions of the Guided-Optical Waves (GOW) and the SAW as well as the frequency dependence of the latter, numerical calculations using digital computers are required to determine the efficiency and the exact frequency response of a guided-wave AO Bragg modulator. The major procedures involved are first to obtain appropriate analytical expressions for the field distributions of the optical waves and the SAW based on their directions and modes of propagation; second, to include the frequency dependence of the amplitudes, phases and penetration depth of the SAW; third, to identify the relevant photo-elastic and EO constants; and finally, to calculate the diffraction efficiency versus the acoustic frequency, with the acoustic drive power as a parameter, using a digital computer by inserting the above information together with the remaining optical and acoustical parameters into (5.5a). Note that the frequency response of the SAW transducer can be incorporated in the second step of the aforementioned procedures.

5.1.4 Three Potential Acousto-Optic Substrate Materials

Among the many materials that have been explored for guided-wave AO interactions, Y-, X- or Z-cut $LiNbO_3$, nonpiezoelectrics such as thermally oxidized Si, and GaAs are the three that have demonstrated the highest potential. Excitation and propagation of the GOWs and the SAWs, and the AO interactions involved have been studied most thoroughly for the Y-cut plates of the first substrate material, and to a lesser extent for the second and third substrate materials. Some significant experimental results have been made recently for the third substrate material. This subsection will present the detailed calculated performances for the first substrate material and briefly decribe those obtainable with the second and the third substrate materials. Detailed treatments on the second and the third substrate materials are given in Chaps.6 and 8, respectively.

a) LiNbO₃ Substrate

Detailed numerical calculations and experiments have been carried out for both isotropic (Fig.5.2a, b) and anisotropic (Fig.5.2c) Bragg diffraction in Y-cut $LiNbO_3$ waveguides. While the latter is discussed in Sect.5.4.1, some results of the former are presented in this subsection. The isotropic Bragg diffraction involves a He-Ne laser light ($0.6328\,\mu m$) with a TE_0 mode propagating in the Y-cut $LiNbO_3$ out- and indiffused optical waveguides and a SAW propagating in either the Z(c) direction or $\pm 21.8°$ from the Z direction [5.14]. As indicated earlier, the contribution to Bragg diffraction by the EO effect is very important in a Y-cut $LiNbO_3$ substrate.

The spatial distributions of the optical fields in these waveguides [5.6-10] and those of the strain fields and the accompanying piezoelectric fields are available from the literature [5.33-37]. Relevant photo-elastic constants and EO coefficients have also been tabulated [5.37]. Specifically, for the case with a Z-propagation SAW the appropriate photo-elastic constants are $p_{31} = p_{32}$ and p_{33}. The only relevant EO coefficient is r_{33}. For the case with a 21.8°-propagation SAW matrix transformation must be carried out to determine the appropriate photo-elastic constants and EO coefficients. This transformation results in effective photo-elastic constants which are linear combinations of p_{11}, p_{12}, p_{13}, p_{14}, p_{31}, p_{33}, p_{41}, and p_{44}. Similarly, the resulting effective EO coefficients are linear combinations of r_{12}, r_{13}, r_{33} and r_{42}.

First, the frequency response, as determined by the confinements of the optical waves and the SAW, is calculated. This is equivalent to calculating the coupling coefficient $C_{mn}^2(f)$ as a function of the acoustic frequency under the assumptions that the transducer bandwidth is so large that it does not introduce any band-limiting effect and that the Bragg condition ($\Delta\theta \equiv 0$) is satisfied at all frequencies of interest. While the first assumption leads to a constant total acoustic power, the second assumption is equivalent to utilization of a sufficiently small acoustic aperture. Subsequently, from (5.5,6) the relative Bragg diffraction efficiency as a function of the acoustic frequency is calculated with the penetration depth of an optical waveguide mode as a parameter. Figure 5.4 shows the results for the two directions of SAW propagation involving a TE_0 mode for both the diffracted and undiffracted light waves. Note that in these calcu-

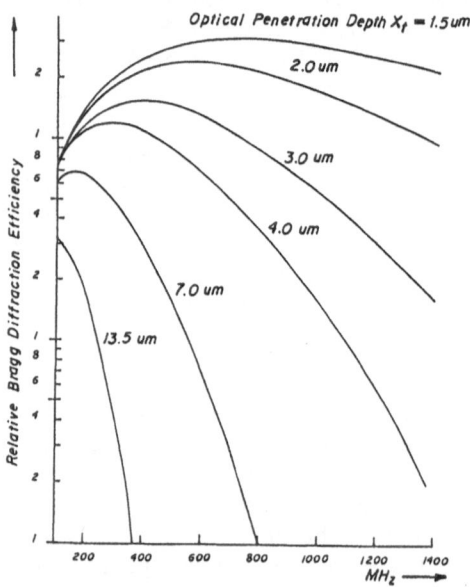

Fig.5.4a. Frequency response based on coupling coefficient alone with penetration depth of TE_0 mode in a y-cut $LiNbO_3$ waveguide as a parameter (for z-propagation SAW)

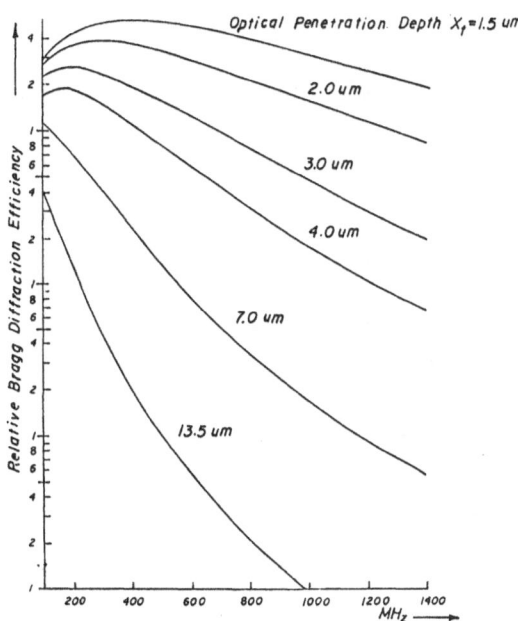

Fig.5.4b. Frequency response based on coupling coefficient with penetration depth of TE_0 mode in a y-cut $LiNbO_3$ waveguide as a parameter (for $\pm 21.8°$-propagation SAW)

lated plots the total acoustic power is assumed to be a constant, a consequence of the assumption that the transducer bandwidth is sufficiently large so that it does not introduce any band-limiting effect. These plots clearly show that the smaller the optical penetration depth, the higher will be the optimum acoustic frequency and the diffraction efficiency. It is seen that at the optical wavelength of 0.6328 μm the desirable range of optical penetration depth is from 1.0 to 2.0 μm. Furthermore, for the optical waveguide of smaller penetration depth the Z-propagation SAW provides the best diffraction efficiency at a higher acoustic frequency than the 21.8°-propagation SAW. Figure 5.4a specifically indicates that the optimum range of acoustic frequency should be centered around 700 MHz if the Z-propagation SAW is utilized in the waveguide with 1.0 to 2.0 μm penetration depth. Similarly, Fig.5.4b indicates that if the 21.8°-propagation SAW is utilized, the optimum range of acoustic frequency should be centered around 500 MHz. Clearly, as the optical wavelength increases from 0.6328 μm the corresponding optimum ranges of acoustic frequency will decrease from those mentioned above.

In the above discussion, the transducer bandwidth is assumed to be sufficiently large and the acoustic aperture sufficiently small so that no bandwidth limiting could occur. In practice, the transducer bandwidth as well as the finite acoustic aperture must be included in the calculation of the ultimate frequency response of the AO Bragg modulator. For example, a set of frequency responses has been obtained for an optical waveguide of 2.0 μm penetration depth with the fractional transducer bandwidth and

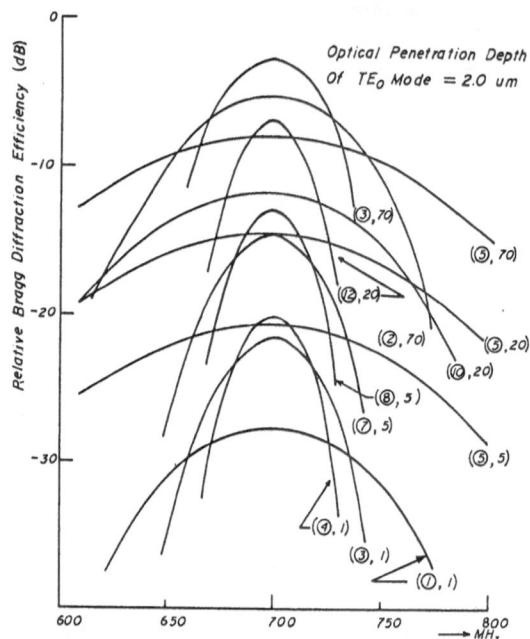

Fig.5.5a. Frequency response of a basic acousto-optic Bragg modulator using z-propagaion SAW in a y-cut LiNbO$_3$ waveguide with fractional transducer bandwidth and acoustic aperture as parameters

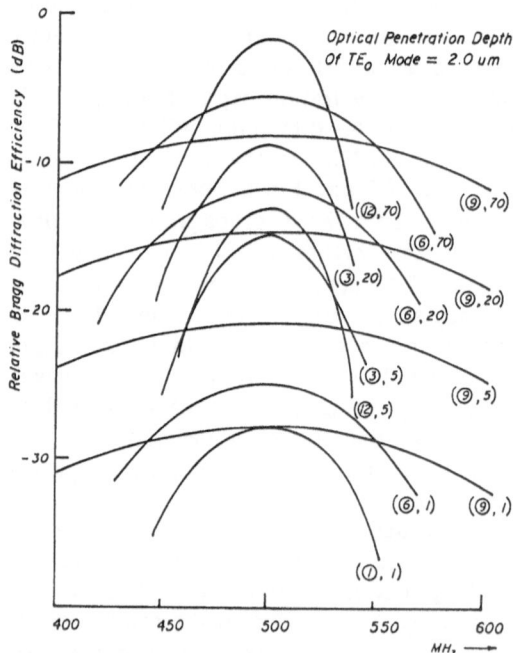

Fig.5.5b. Frequency response of a basic acousto-optic Bragg modulator using ±21.8°-propagation SAW in a y-cut LiNbO$_3$ waveguide with fractional transducer bandwidth and acoustic aperture as parameters

130

the acoustic aperture as parameters [5.1]. Some sample plots are shown in Fig.5.5 for the cases of z-propagation SAW and 21.8°-propagation SAW. The various combinations of fractional transducer bandwidth and acoustic aperture are designated by encircled numbers as listed in Table 5.1. The total acoustic drive power and the corresponding peak diffraction efficiency are listed in Table 5.2. The first number in parentheses which is attached to each sample plot designates the combination of transducer bandwidth and acoustic aperture, and is read from Table 5.1; the second number designates the diffraction efficiency (normalized with respect to a 3mm acoustic aperture), and the corresponding acoustic drive power required can be read from Table 5.2. The actual diffraction efficiency for the acoustic aperture of interest can also be read from Table 5.2. As an example, let us focus on the plot marked with $(1,1)$. The first number, namely 1, designates the combination of 15% transducer bandwidth and 0.5 mm acoustic aperture. The second number, also 1, designates a diffraction efficiency of 1% for 3 mm acoustic aperture, or more directly, a diffraction efficiency of 0.17% for the 0.5 mm acoustic aperture of interest. From the first column of Table 5.2 the required acoustic drive power

Table 5.1. Encircled numbers designating combinations of fractional transducer bandwidth and acoustic aperture. (Valid for both z-propagation and ±21.8°-propagation SAWs)

Fractional transducer bandwidth [%]	Acoustic aperture [mm]			
	0.5	1.0	2.0	3.0
15	1	2	3	4
25	5	6	7	8
50	9	10	11	12

Table 5.2. Bragg diffraction efficiency [%] versus total acoustic drive power and acoustic aperture. (Valid for z-propagation SAW. For a ±21.8°-propagation SAW the acoustic drive power required is twice larger for the same combinations of acoustic aperture and diffraction efficiency)

Acoustic Aperture [mm]	Acoustic Drive Power [mW]			
	0.075	0.38	1.61	7.5
0.5	0.17	0.85	3.5	15.5
1.0	0.33	1.69	7.0	29.3
2.0	0,67	3.35	13.7	52.4
3.0	1.0	5.0	20.0	70.0

is found to be 0.075 mW. Finally, the corresponding RF drive power is determined by the conversion efficiency of the transducer (Sect.5.2.3).

From the sample plots of Fig.5.5 a number of conclusions can be drawn. First, for a given acoustic drive power the diffraction efficiency is proportional to the acoustic aperture, and similarly, for a given acoustic aperture the diffraction efficiency is proportional to the acoustic drive power. As indicated earlier, this linear dependence is valid for diffraction efficiency of 50% or less. Second, the AO Bragg bandwidth decreases as the acoustic aperture increases, and may become smaller than the transducer bandwidth at a large acoustic aperture. Also, for a given acoustic aperture the absolute bandwidth of a modulator which utilizes a single SAW of low center frequency is limited mainly by the bandwidth of a periodic IDT. On the other hand, for the same acoustic aperture the absolute bandwidth of a modulator which employs a single SAW of high center frequency is limited mainly by the AO Bragg bandwidth. Figure 5.5 provides the basic data and guidelines required for the design of AO Bragg modulators and deflectors at 0.6328 μm optical wavelength using Y-cut LiNbO$_3$ waveguides.

In summary, the bandwidth of a guided-wave AO Bragg modulator or deflector which employs isotropic Bragg diffraction and a single periodic IDT of relatively large acoustic aperture is rather limited. For example, from the plots marked $(4,1)$, $(8,5)$, and $(12,20)$ in Fig.5.5a, a maximum -3 db modulator bandwidth of only 34 MHz is possible for a device using a Y-cut LiNbO$_3$ indiffused waveguide of 2 μm penetration depth at the optical wavelength of 0.6328 μm in which the aperture and the center frequency of the Z-propagation SAW are 3 mm and 700 MHz, respectively. This small modulator bandwidth (34MHz) is caused by the small AO Bragg bandwidth associated with the 3 mm acoustic aperture, practically irrespective of the fractional bandwidth of the transducer.

b) GaAs Substrate

Despite the various successes of the LiNbO$_3$-based wide-band planar AO Bragg devices referred to above and in Sect.5.4, the ultimate advantages of integrated optics cannot be fully realized in such substrate until a technology has been developed to incorporate suitable semiconductor materials for accomplishing total or monolithic integration. In the meantime, the GaAs substrate by itself can potentially achieve these ultimate advantages because both active and passive components such as the laser sources and the photodetectors can be integrated in the same substrate. Clearly, one of the key components in such future GaAs-based integrated optic circuits is an efficient wide-band AO Bragg modulator or deflector. Detailed theoretical treatment on planar guided-wave AO Bragg diffraction and some experimental study using both GaAs/GaAsP and GaAs/ GaAlAs optical waveguides have been carried out [5.38-40]. Successful realization of compact and miniaturized AO Bragg cells most recently using ZnO thin film overlay transducer [5.40] together with GaAs/GaAlAs waveguides is presented in Chap.8. In this subsection, only some examples of calculated results are given.

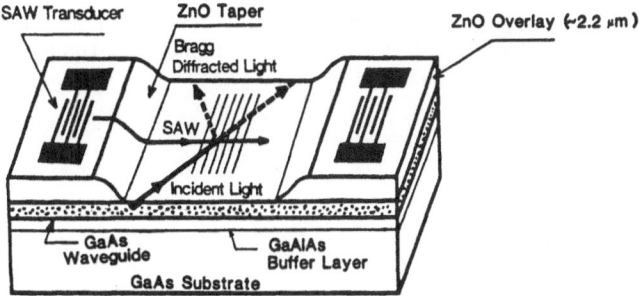

Fig.5.6a. Planar guided-wave acousto-optic Bragg diffraction in a GaAs-GaAlAs waveguide

For example, the Bragg diffraction efficiency and its frequency response, and the acoustic drive power requirements have been calculated for all directions of the SAW propagation in a Z-cut GaAs substrate, as shown in Fig.5.6a, again using the coupled-mode technique and the digital computer. It is to be noted that unlike the Y-cut LiNbO$_3$ substrate, the AO Bragg diffraction in the Z-cut GaAs substrate is dominated by the elasto-optic effect. Presented in Chap.8 are the plots showing the frequency dependences of the acoustic drive power for the $\langle 100 \rangle$-propagating SAWs in the (001)-substrate and the $\langle 121 \rangle$-propagating SAW in the $(1\bar{1}1)$-substrate with the thickness of the GaAs waveguide as a parameter. Note that these plots were generated for the case of 100% diffraction involving the interaction between a TE$_0$-mode incident (undiffracted) light and a diffracted light of the same mode. The plots clearly show that very efficient diffraction, namely a 100% diffraction efficiency at an acoustic power-beam width product (in mW mm) of 23.9 and 18.8 and very large AO Bragg bandwidth, namely 1.6 GHz and 1.4 GHz for $\langle 100 \rangle$- and $\langle 110 \rangle$-propagation SAWs, respectively, are theoretically possible using the 0.4 μm thick GaAs waveguides that support TE$_0$ mode at 1.15 μm wavelength. As will be shown in Chap.8, very efficient diffraction has been demonstrated experimentally [5.40] using the SAW with the frequency ranging from 190 to 1200 MHz.

c) SiO$_2$, As$_2$S$_3$ or SiO$_2$-Si Substrates

The third major substrate type concerns the nonpiezoelectric materials such as fused quartz (SiO$_2$) [5.4], arsenic trisulphide (As$_2$S$_3$) [5.5] and thermally oxidized silicon (SiO$_2$-Si). Interest in these substrate materials is based on the fact that the first is a common optical waveguide material, the second is an amorphous material of very high AO figure of merit, while the third may be used to capitalize on the existing silicon technology to further optical integration. While it is common to deposit a piezoelectric zinc oxide (ZnO) film on such substrate materials for the purpose of SAW generation similar to that depicted in Fig.5.6a, the ZnO-SiO$_2$ composite waveguide (Fig.5.6b) has also been utilized to facilitate guided-wave AO interaction because the ZnO film itself also possesses favorable optical waveguiding and acousto-optic properties (Chap.6) [5.41-46].

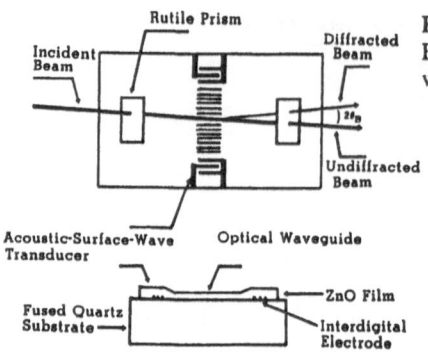

Fig.5.6b. Planar guided-wave acousto-optic Bragg diffraction in a ZnO-SiO$_2$ composite waveguide

Guided-wave AO Bragg diffraction in such substrate materials has been analyzed in detail using the coupled-mode technique [5.44] similar to that used for the LiNbO$_3$ substrate. The expression for the Bragg diffraction efficiency takes similar form as that for the LiNbO$_3$ substrate. However, it should be noted that for the case involving the ZnO-SiO$_2$ composite waveguide the distribution of the SAW fields exhibits a marked variation with the thickness of the ZnO film, while the distribution of the GOW fields is practically unaffected [5.42]. It was shown accordingly that the Bragg diffraction efficiency depends strongly on the thickness of the ZnO film and that for each optical waveguide mode there exists a film thickness for which the diffraction efficiency diminishes. A series of experiments performed at 0.632 μm optical wavelength using the dc sputtered ZnO-film waveguide on the fused-quartz substrate and the SAW centering at 130 MHz have verified the predictions. In this particular experimental study the thickness of the ZnO film was varied from 0.6 to 8.0 μm and the Bragg diffraction involved the interaction between the same GOW modes, namely TE$_0$ to TE$_0$ and TM$_0$ to TM$_0$. Subsequently, a strain-controlling film of fused quartz was added on top of the ZnO film to enhance the diffraction efficiency [5.43]. In a separate study, the ZnO film was used solely to generate the SAW for AO Bragg diffraction experiments in the SiO$_2$-Si substrate at a SAW frequency up to 700 MHz.

At the present time, the measured propagation losses of the SAW in the ZnO-SiO$_2$-Si composite substrate [5.45, 46] are much higher than those in the LiNbO$_3$ substrate and also considerably higher than those in the GaAs substrate. This contrast is accentuated as the acoustic frequency goes beyond 200 MHz. Thus it may be concluded that until a composite substrate of greatly reduced SAW propagation loss has been developed, the guided-wave AO Bragg devices using the foregoing nonpiezoelectric substrate materials have to be limited to a considerably lower RF center frequency and smaller bandwidth than those using the LiNbO$_3$ substrate. Finally, it should be mentioned that efficient diffraction was demonstrated recently at 2.0 GHz using a ZnO-sapphire substrate despite its high propagation loss for the SAW, namely, 30 dB/cm [5.47].

5.2 Key Performance Parameters of Basic Planar Acousto-Optic Bragg Modulators and Deflectors

The planar AO Bragg diffraction treated in Sect.5.1 can be utilized to modulate and/or deflect a light beam. The resulting light-beam modulators and deflectors can operate over a wide RF band by using a variety of SAW transducer configurations to be described in Sect.5.4, and thus constitute useful devices for integrated- and fiber-optic communication, computing, and signal processing systems. The key device parameters that determine the ultimate performance characteristics of the guided-wave AO modulators and deflectors are bandwidth, time-bandwidth product, acoustic and RF drive powers, nonlinearity, and dynamic range. We discuss each of these performance parameters in the following.

5.2.1 Bandwidth

As shown in Sect.5.1, the diffraction efficiency-bandwidth product of a planar AO modulator that employs isotropic Bragg diffraction and a single periodic SAW IDT [5.48,49] is rather limited. However, if the absolute modulator bandwidth is the sole concern a large bandwidth can be realized by using either a single periodic SAW IDT with a small acoustic aperture and a small number of finger pairs, or a single aperiodic SAW IDT with a small acoustic aperture and a large number of finger pairs (chirp transducers) [5.50-52]. It should be emphasized, however, that in either case the large bandwidth is obtained at the expense of a drastically reduced diffraction efficiency due to the very small acoustic aperture. Consequently, a higher diffraction efficiency will necessarily require a large acoustic and thus RF drive power. Unfortunately, a large acoustic drive power will not only increase the risk of transducer failure but also results in an excessive acoustic power density, especially at high acoustic frequencies, and thus increases the deleterious effects due to acoustic nonlinearity. A small transducer aperture will also result in a large acoustic radiation impedance [5.48,49] and thus greater complexity in the electrical matching circuit required for optimum electrical to acoustic transduction. Therefore we conclude that for applications requiring both wide bandwidth and high diffraction efficiency, more sophisticated device configurations must be employed. The unified analysis on AO Bragg diffraction from multiple SAWs, to be presented in Sect.5.3, suggests that this requirement can be met by employing multiple transducers of proper design and placement. Section 5.4.1 lists the five interaction and transducer combinations that have been explored for this purpose. It is now possible to realize high performance guided-wave AO modulators and deflectors with GHz center frequency and GHz bandwidth using these wide-band configurations. Design, fabrication, testing, and measured performance figures for a variety of wide-band devices are presented in Sect.5.4.2.

5.2.2 Time-Bandwidth Product

The time-bandwidth product of an AO modulator, TB, is defined as the product of the acoustic transit time across the incident light beam aperture and the modulator bandwidth [5.1]. It is readily shown that this time-bandwidth product is identical to the number of resolvable spot diameters of an AO deflector, N_R, which is defined as the total angular scan of the diffracted light divided by the angular spread of the incident light.

Thus the following well-known identities hold

$$TB = N_R \equiv \frac{\Delta f}{\delta f_R} = \frac{D}{V_R} \Delta f = \tau \Delta f , \tag{5.12a}$$

$$\delta f_R = V_R / D , \tag{5.12b}$$

$$\tau = D / V_R , \tag{5.12c}$$

where D designates the aperture of the incident light beam, V_R the velocity of the SAW, Δf the device bandwidth, τ the transit time of the SAW across the incident light beam aperture, and δf_R the incremental frequency change required for deflection of one Rayleigh spot diameter. The acoustic transit time τ may be considered as the minimum AO switching time if the switching time of the RF driver is sufficiently smaller than the acoustic transit time.

The desirable value for TB or N_R depends upon the individual application. For example, in light modulation and single-port switching applications, it is desirable to have TB as close to unity as possible so that the highest modulation or switching speed may be achieved. For this particular purpose, the incident light is focused to a small beam diameter at the interaction region so that the corresponding acoustic transit time is a minimum [5.24, 53]. Obviously, for multi-port switching applications a large value of N_R and thus a large number of switching ports could be accomplished using a large light beam aperture. However, this large multi-port capacity is obtained at the expense of switching speed. On the other hand, in signal processing applications it is desirable to have N_R as large as possible because this value is identical to the processing gain. Thus, for this particular area of applications it is also desirable to have a collimated incident light beam of large aperture. The light beam aperture is limited by the quality of the optical waveguide, the excitation mechanism, and the acoustic attenuation in the waveguide. For a Y-cut LiNbO$_3$ optical waveguide the acoustic attenuation is only about 1 db per cm at 1.0 GHz. Using a rutile prism coupler, a guided-light beam aperture as large as 1.5 cm and good uniformity in this type of waveguide was obtained previously at the author's laboratory. This light beam aperture would result in an acoustic transit time of about 4.4 μs for a Z-propagation SAW (V_R = 3.488·10^5 cm/s). It will be shown in Sect. 5.4 that a deflector bandwidth of up to 1.0 GHz can be realized using multiple SAW transducers, so that a time-bandwidth product as high as 4,400 is achievable. At present the acoustic attenuations in all other waveguide materials at 1.0 GHz are much higher, and thus considerably limit the maximum time-bandwidth product attainable.

5.2.3 Acoustic and RF Drive Powers

From (5.6 and 8) of Sect.5.1.3 we have

$$\zeta(f) = \sin^2\left\{\left(\frac{\pi}{\lambda_0\sqrt{2}}\right)\sqrt{M_{2mn}}\,C_{mn}(f)\sqrt{\frac{L}{\cos\theta_m\cos\theta_n}}P_a}\right\}. \qquad (5.13)$$

Note that M_{2mn} has been defined as $(n_m^3 n_n^3 p^2/\rho V_R^3)$ in Sect.5.1.3.

Following the common practice of specifying the drive power requirement for 50% diffraction, we simply set $\zeta(f)$ of (5.13) equal to 0.50 to arrive at the following expression for the required acoustic drive power at the center frequency f_0

$$P_a(50\% \text{ diffraction}) \simeq \left(\frac{\lambda_0^2\cos\theta_m\cos\theta_n}{8}\right)\left(\frac{1}{M_{2mn\,eff}}\right)L^{-1} \qquad (5.14a)$$

where

$$M_{2mn\,eff} \equiv C_{mn}^2(f_0)M_{2mn} . \qquad (5.14b)$$

Assuming that the transducer conversion efficiency is T_c at the center frequency, namely $P_a(f_0) = T_c P_e(f_0)$ in which $P_e(f_0)$ is the total RF drive power, we obtain the following expression for the total RF drive power at 50% diffraction:

$$P_e(50\% \text{ diffraction}) \simeq \left(\frac{1}{T_c}\right)\left(\frac{\lambda_0^2\cos\theta_m\cos\theta_n}{8}\right)\left(\frac{1}{M_{2mn\,eff}}\right)L^{-1} . \qquad (5.14c)$$

Clearly, both the total acoustic and RF drive powers required are inversely proportional to the acoustic aperture, as in the bulk-wave modulator. However, in contrast to the bulk-wave case, because of the coupling factor $C_{mn}^2(f)$ both drive powers are frequency dependent even if the conversion efficiency of the transducer is a constant over the frequency band of interest.

In order to determine explicitly the total RF drive power P_e we must first calculate the electrical to acoustic conversion efficiency T_c of the transducer used in the modulator or deflector. The frequency response and conversion efficiency of a periodic SAW IDT on LiNbO$_3$ substrates have been studied in detail in terms of one-dimensional equivalent circuits [5.48, 49]. A detailed treatment has been provided in Chap.4. The relevant quantities of the transducer as derived from the so-called "crossed-field" model equivalent circuit (Fig.5.7) are listed below

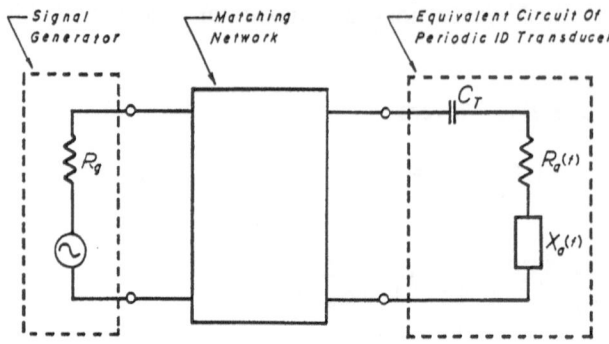

Fig.5.7. Schematic of an interdigital transducer driven by an RF generator through a matching network

$$\Delta f_{-3db(Acoustic)}/f_0 \simeq 1/N_e \ , \tag{5.15a}$$

$$R_a(f) \simeq R_a(f_0) \left[\frac{\sin x}{x} \right]^2 \ , \tag{5.15b}$$

$$R_a(f_0) = \frac{4}{\pi} \left[\frac{1}{2\pi f_0 C_s} \right] k_c^2 = \frac{2}{\pi^2} \left[\frac{1}{V_R(C_s/L)} \right] \frac{\Lambda_0}{L} k_c^2 \ , \tag{5.15c}$$

$$X_a(f) \simeq R_a(f_0) \left[\frac{\sin 2x - 2x}{2x^2} \right] \ , \tag{5.15d}$$

$$X_a(f_0) = 0 \ , \tag{5.15e}$$

$$C_T = N_e C_s \ , \tag{5.15f}$$

where

f_0, Λ_0: synchronous frequency and the corresponding acoustic wavelength,

$\Delta f_{-3db(Acoustic)}/f_0$: fractional acoustic bandwidth,

N_e: number of finger electrode pairs,

$R_a(f), X_a(f)$: radiation resistance and reactance, respectively, at the frequency f,

$R_a(f_0), X_a(f_0)$: radiation resistance and reactance, respectively, at the synchronous frequency f_0,

$x = N_e \pi (f-f_0)/f_0$,

k_c^2: electromechanical coupling coefficient,

C_s: static capacitance per finger pair of the electrode,

C_s/L: static capacitance per finger pair per unit acoustic aperture,

L: acoustic aperture of the transducer,

V_R: velocity of propagation of the SAW, and

C_T: total static capacitance of the IDT.

Note that the physical constants k_c^2, C_s and V_R for LiNbO$_3$ substrates have all been tabulated [5.48,49]. Equation (5.15a) indicates that the maximum possible fractional acoustic bandwidth is inversely proportional to the number of finger electrode pairs. Thus a small N_e must be chosen for a transducer to be used in wide-band AO interaction [5.1.14]. The ultimate transducer bandwidth is determined by the frequency dependence of the radiation impedance and that of the matching network [5.48,49]. Various types of wide-band matching networks have been developed [5.54-56], some of which have been discussed in Chap.4.

Finally, the peak electrical to acoustic conversion efficiency of the transducer and thus the total RF drive power can be calculated in terms of the synchronous radiation resistance as well as the impedance of the matching network and the RF generator. Equation (5.15c) shows that the synchronous radiation resistance is inversely proportional to the product of the synchronous frequency and the acoustic aperture, or inversely proportional to the acoustic aperture measured in terms of the synchronous acoustic wavelength. As will be discussed in Sect.5.4, this specific dependence imposes a practical limit in the realization of wide-band high-efficiency modulators and deflectors which utilize a single acoustic aperture to less than 1.0 GHz. It will be shown that using some of the wide-band transducer configurations to be discussed in Sect.5.4.1 the modulators or deflectors requiring only mW electric drive power per MHz bandwidth at 50% diffraction efficiency with a bandwidth approaching 1.0 GHz can be realized.

5.2.4 Nonlinearity and Dynamic Range

The ultimate dynamic range of an AO modulator-detector combination is determined by a number of factors including the nonlinearity inherent in the acoustic wave propagation and AO interaction process, nonlinearity in the RF driver circuit, the optical background noises resulting from scattering in the waveguide and in all passive and active components, and the dynamic range of the photodetector. In this subsection the discussion is limited to those inherent in the AO modulator itself, namely, the acoustic and AO interaction nonlinearities as they are independent of the applications involved. Other factors are discussed in Chap.7.

We again focus the attention at the center frequency f_0. Based on the discussion in the previous subsection the diffraction efficiency or the modulation index (5.13) can be further reduced to the following form:

$$\zeta(f_0) = \sin^2(A\sqrt{LP_e}) \tag{5.16a}$$

where

$$A \equiv \frac{\pi}{\lambda_0(2\cos\theta_m\cos\theta_n)^{1/2}}\sqrt{M_{2mn\,eff}}\sqrt{1/T_c} \;. \tag{5.16b}$$

139

For small signal cases in which $A(LP_e)^{1/2} \ll 1$, (5.16a) reduces to the following expression

$$\zeta(f_0) \simeq A^2 LP_e \left[1 - \frac{2}{3} A^2 LP_e \right] . \tag{5.16c}$$

It is seen that as the diffraction efficiency or the modulation index is increased, the accompanying distortion due to nonlinearity is also increased. However, for a very small modulation index the intensity of the diffracted light is linearly proportional to the power of the electric drive signal. This approximately linear relationship is the basis of many modulation and signal processing applications such as the spectral analysis of RF signals to be treated in Sect.5.5 and Chap.7. Clearly, as the RF drive power increases to the extent that the inequality $A(LP_e)^{1/2} \ll 1$ no longer holds, distortion and nonlinearity become significant and should be included.

In applications such as the spectral analysis that involves multiple RF signals the inter- and cross-modulations due to multiple AO diffraction also constitute the sources of nonlinearity. Detailed measurements on such modulations in connection with bulk-wave AO spectrum analyzers [5.57] have been reported. Some measurements on guided-wave counterparts have also been made [5.58]. Note that since the proportional constant A^2L is frequency dependent, the accompanying nonlinearity is also frequency dependent and thus may be important in applications involving very wide-band RF signals. The AO nonlinearity described above sets an upper bound for the strength of the RF signal to be processed and thus limits the dynamic range for large signals. For small signals the dynamic range is limited by the scattered light and the noise current of the photodetector.

Finally, in some cases the effect of acoustic nonlinearity [5.59] on dynamic range must also be taken into account. Acoustic nonlinearity is particularly important at the higher acoustic frequencies because the absolute penetration depth of the SAW decreases, as the frequency increases. As a result, the acoustic power density becomes so large that a significant portion of the fundamental frequency acoustic power is converted into harmonics. In addition, this depletion of fundamental frequency acoustic power increases with the propagation distance. The end results are that an incident light beam of large aperture will incur significant nonuniformity in the intensity of the diffracted light as well as spurious diffracted light. These in turn degrade the linearity of modulation as well as the spatial resolution in deflection and spectral analysis applications. In fact, this deleterious effect has been observed in the experiment in the GHz frequency range in a Y-cut $LiNbO_3$ waveguide [5.60].

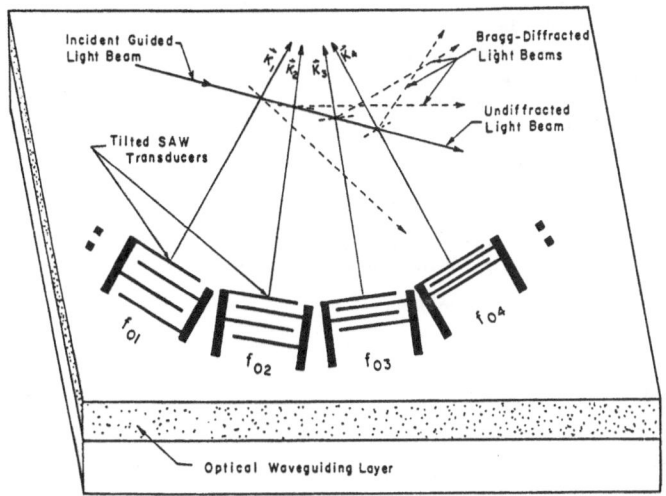

Fig.5.8. Guided-wave acousto-optic Bragg diffraction from multiple-tilted SAWs

5.3 Guided-Wave Acousto-Optic Bragg Diffraction from Multiple Surface Acoustic Waves

The coupled-mode analysis presented in Sect.5.1 has shown that the diffraction efficiency-bandwidth product of a planar guided-wave AO device which utilizes a single SAW is a constant of rather limited value. It is clear intuitively, however, that a larger composite bandwidth and thus a larger diffraction efficiency-bandwidth product can be achieved by employing a multiple of SAWs that are properly tailored and configured [5.12]. For example, as will be shown in Sect.5.4, multiple SAWs of staggered center frequency and tilted propagation direction [5.14] (Fig.5.8) as well as phased multiple SAWs of identical center frequency and propagation direction [5.61-64] (Fig.5.9) can be used to achieve this objective. This section presents a unified treatment on AO Bragg diffraction from multiple SAWs [5.15]. This general approach can be employed to analyze the AO interactions involving multiple tilted and phased SAWs or a combination of both. As an illustration, the case with two tilted SAWs is analyzed first [5.14]. An extension to the general case involving N-SAWs then follows. The analytical results obtained are then applied to the two specific cases involving multiple tilted SAWs and phased array SAWs.

5.3.1 Acousto-Optic Bragg Diffraction from Two Tilted SAWs

It is clear from Fig.5.10 that the input boundary values (the diffracted and undiffracted light fields) for the second interaction region are obtained by adding an optical phase factor, which accounts for the optical propagation delay and the relative phase between the two SAWs, to the output boundary values of the first interaction region as given in (5.4c

141

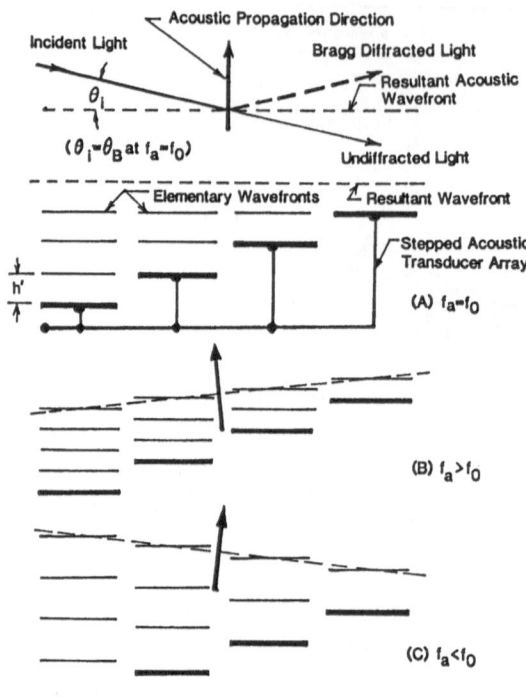

Fig.5.9. Steering of resultant acoustic wavefront from phased array and thus tracking of Bragg condition as the acoustic frequency f_a is varied from the center frequency f_0

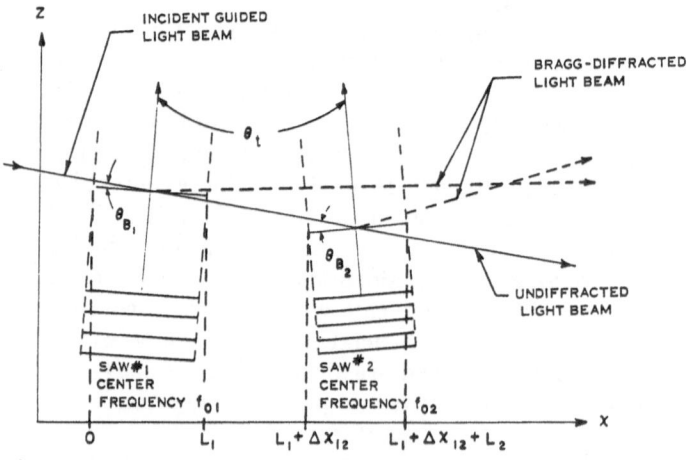

Fig.5.10. Guided-wave acousto-optic Bragg diffraction from two tilted SAWs

and d). Thus the methodology for numerical computation which has been developed for the case involving a single SAW can be conveniently extended to the present case [5.14]. Note that in Fig.5.10 the tilt angle θ_t between the two SAWs is set equal to the difference in the Bragg angles at the two acoustic center frequencies.

For convenience we shall henceforth add the second subscripts 1 and 2 to all relevant quantities to designate the two SAWs and express the boundary values at the input boundary of the second SAW as follows

$$E_{m2}(L_1 + \Delta x_{12}) = \alpha ,$$ (5.17a)

$$E_{n2}(L_1 + \Delta x_{12}) = \beta \exp(j\psi) ,$$ (5.17b)

where Δx_{12} designates the distance between the output boundary of the first SAW and the input boundary of the second SAW, and from (5.4,5) we have

$$\alpha = \sqrt{1 - \zeta_1} ,$$ (5.17c)

$$\beta = \sqrt{\zeta_1} ,$$ (5.17d)

$$\psi = \pi/2 + (K_1 \Delta\theta_1 \Delta x_{12})/2$$

$$+ \tan^{-1}[(K_1 \Delta\theta_1)/2q_1 \cdot \sin(q_1 L_1)/\cos(q_1 L_1)] + \Phi_1 ,$$ (5.17e)

where ψ designates the relative phase between the undiffracted and the diffracted light waves plus the phase shift Φ_1 of the first SAW. After substituting (5.17) into the general solutions $E_{m2}(x)$ and $E_{n2}(x)$ for the second interaction region which take the same form as $E_{m1}(x)$ and $E_{n1}(x)$, given in Sect.5.1, we obtain the electric field of the diffracted light wave at the output boundary of the second SAW as follows

$$E_{n2}(L_1 + \Delta x_{12} + L_2) = \{\beta \cos(q_2 L_2) + j(1/q_2)$$

$$\times [(K_2 \Delta\theta_2/2)\beta + \alpha B_2 \exp j(\Phi_2 - \psi)]\sin(q_2 L_2)\}$$

$$\times \exp\{j[-(K_2 \Delta\theta_2 L_2/2 + \psi]\} ,$$ (5.18)

where K_2, Φ_2, $\Delta\theta_2$, L_2, q_2, and B_2 are defined in the same manner as those in the first SAW.

The resultant overall diffraction efficiency ζ_T is simply

$$|E_{n2}(L_1 + \Delta x_{12} + L_2)|^2 .$$

Since the diffraction efficiencies ζ_1 and ζ_2 due to the first and the second SAW alone are $|\beta \exp(j\psi)|^2$, as given by (5.5a), and $(B_2^2/q_2^2)\sin^2(q_2 L_2)$ because $A_2 = B_2$, the overall diffraction efficiency is written in the following convenient form

$$\zeta_T = \zeta_1(1-\zeta_2) + \zeta_2(1-\zeta_1) + 2\{\zeta_1(1-\zeta_1)\zeta_2\}^{1/2}$$

$$\times [(K_2 \Delta\theta_2/2q_2)\sin(q_2 L_2)\cos(\Phi_2-\psi) - \cos(q_2 L_2)\sin(\Phi_2-\psi)] .$$ (5.19)

Since each individual SAW alone will provide a rather limited AO Bragg bandwidth, we shall focus our attention to the case in which the acoustic center frequency of neither SAW falls within the AO Bragg bandwidth of the other. Thus, we have $\zeta_1 \gg \zeta_2$ for the acoustic frequency range $f \leq f_{01}$. Equation (5.19) predicts that $\zeta_T \simeq \zeta_1$ for this lower frequency range. Similarly, we have $\zeta_2 \gg \zeta_1$ for the acoustic frequency range $f \geq f_{02}$, and thus $\zeta_T \simeq \zeta_2$ for this upper frequency range. On the contrary, for the frequency range $f_{01} \leq f \leq f_{02}$ both SAWs contribute to the diffraction, and the resultant diffraction efficiency is given by the sum of three terms. The effect of interference between the two SAWs on the resultant diffraction efficiency is represented by the cross term. It is clear that enhancement as well as reduction in the diffraction efficiency occurs as the phase shift $(\Phi_2 - \psi)$ varies.

Using (5.19), a family of plots have been generated for the resultant diffraction efficiency versus the acoustic frequency, with the phase shift between the two SAWs $\Delta\Phi_{12}$ as a parameter [5.14]. Some sample plots for a He-Ne laser light at 0.6328 μm wavelength are shown in Fig.5.11. In order to compare these calculated plots with the experimental data to be given in Sect.5.4, the acoustic and acousto-optic parameters for these two sets of plots are chosen to be identical, namely f_{01} = 170 MHz, f_{02} = 225 MHz, L_1 = 2.5 mm, and L_2 = 1.66 mm, and the tilt angle is identical to the difference in the corresponding Bragg angles at the center frequency for the individual SAWs, namely 3.4 mrad. From these plots it is seen that the phase shift may cause destructive as well as constructive interference in the resultant diffracted light power. Consequently, the composite bandwidth can be significantly larger than the sum of the two individual bandwidths when a proper phase shift is incorporated. Thus the phase shift as well as the tilt angle between adjacent SAWs are the impor-

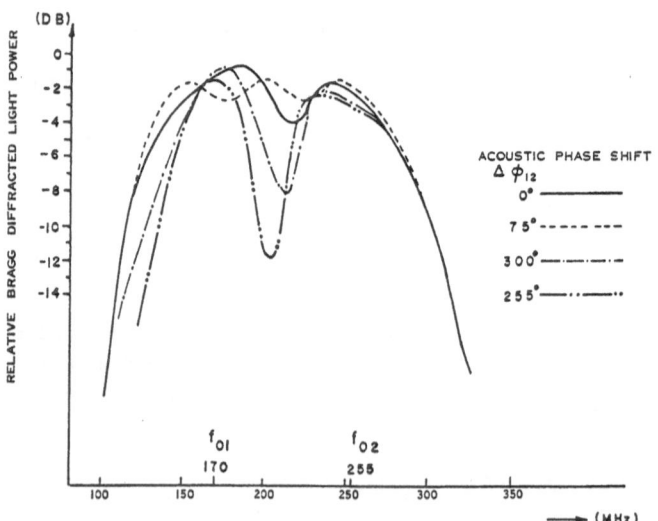

Fig.5.11. Effect of phase shift between two tilted SAWs on resultant frequency response (theoretical)

144

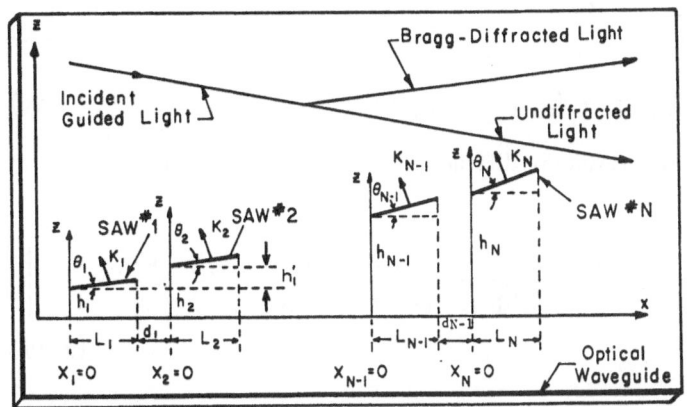

Fig.5.12. General interaction configuration for Bragg diffraction from multiple sur-face acoustic waves

tant parameters, in addition to the center frequency and the acoustic aperture of the individual SAWs, for the design of wide-band high-dif-fraction-efficiency guided-wave AO Bragg devices.

5.3.2 Acousto-Optic Bragg Diffraction from N-SAWs

We shall now extend the analysis to the general case, as shown in Fig. 5.12, in which an array of N-SAWs is involved [5.15]. The advantage of using a separate coordinate system for each transducer (i.e., $x_i = 0$ at the input of each transducer) will become clear later. It is to be noted that the tilt angles between each pair of adjacent transducers in the array are assumed to be sufficiently small so that the separation between adjacent SAWs in the interaction region may be approximated by the correspond-ing transducer separation.

Again, using (5.4c,d) the electric fields of the undiffracted and dif-fracted light at the output boundary of the (i-1)th SAW, namely $x_{i-1} = (L_{i-1}+d_{i-1})$ can be obtained. These output fields are just the input fields to the ith SAW, and thus may, in turn, be used to determine the output light fields from the ith SAW. Therefore, using the special coordinate system of Fig.5.12, (5.4c,d) can be applied successively to any number of SAWs. As in the example treated in the preceding subsection, the input boundary values for the ith SAW are written in the following form

$$E_{m,i}(x_i=0) = \alpha_i , \qquad (5.20a)$$

$$E_{n,i}(x_i=0) = \beta_i \exp(j\psi_i) , \qquad (5.20b)$$

where ψ_i is the relative phase between $E_{m,i}(0)$ and $E_{n,i}(0)$. By generaliz-ing (5.4c,d) and after some algebraic manipulation, α_i, β_i, and ψ_i are obtained as follows [5.15]

$$\alpha_i = \sqrt{1 - \zeta_{T,i-1}} , \qquad (5.20c)$$

145

$$\beta_i = \sqrt{\varsigma_{T,i-1}} \, , \tag{5.20d}$$

$$\psi_i = K_{i-1}\Delta\theta_{i-1}d_{i-1} + \psi_{n,i-1} - \psi_{m,i-1} \, , \tag{5.20e}$$

where $\varsigma_{T,i-1}$ is the resultant diffraction efficiency due to the first (i-1) SAWs, and

$$\psi_{m,i-1} = \tan^{-1}\left[\frac{\beta_{i-1}\cos(\phi_{i-1}-\psi_{i-1})\sqrt{\varsigma_{i-1}} - \alpha_{i-1}(K_{i-1}\Delta\theta_{i-1}/2q_{i-1})\sin(q_{i-1}L_{i-1})}{\alpha_{i-1}\cos(q_{i-1}L_{i-1}) + \beta_{i-1}\sin(\phi_{i-1}-\psi_{i-1})\sqrt{\varsigma_{i-1}}}\right] \tag{5.20f}$$

$$\psi_{n,i-1} = \psi_{i-1}$$

$$+ \tan^{-1}\left[\frac{\beta_{i-1}(K_{i-1}\Delta\theta_{i-1}/2q_{i-1})\sin(q_{i-1}L_{i-1}) + \alpha_{i-1}\cos(\phi_{i-1}-\psi_{i-1})\sqrt{\varsigma_{i-1}}}{\beta_{i-1}\cos(q_{i-1}L_{i-1}) - \alpha_{i-1}\sin(\phi_{i-1}-\psi_{i-1})\sqrt{\varsigma_{i-1}}}\right] \tag{5.20g}$$

where ς_{i-1} is the diffraction effciency due to the (i-1)th SAW with input electric fields $E_{m,i-1}(0) = 1$ and $E_{n,i-1}(0) = 0$, and ϕ_{i-1} is the total phase of the (i-1)th SAW which includes the transducer electric phase and the acoustic phase due to the step height $h'_{i-1} = (h_i - h_{i-1})$. Note that as depicted in Fig.5.12, the step height denotes the differential of the ordinates of adjacent SAWs. By deduction the electric fields of the light at the output boundary of the Nth SAW are obtained as follows [5.15]

$$E_{m,N}(x_N=L_N) = \Big[\alpha_N\cos(q_N L_N) + (j/q_N)[\beta_N A_N\exp[-j(\phi_N-\psi_N)]$$

$$- (K_N\Delta\theta_N L_N/2)\sin(q_N L_N)]\Big]\exp(jK_N\Delta\theta_N L_N/2) \, , \tag{5.21a}$$

$$E_{n,N}(x_N=L_N) = \Big[\beta_N\cos(q_N L_N)e^{j\psi_N} + (j/q_N)[\alpha_N B_N e^{j\phi_N}$$

$$+ (K_N\Delta\theta_N L_N/2)\beta_N e^{j\psi_N}]\Big]\exp(-jK_N\Delta\theta_N L_N/2) \, . \tag{5.21b}$$

Finally, the resultant diffraction efficiency is

$$\varsigma_{T,N} \equiv |E_{n,N}(L_N)|^2$$

$$= \varsigma_N(1 - \varsigma_{T,N-1}) + (1 - \varsigma_N)(\varsigma_{T,N-1})$$

$$+ 2\{\varsigma_{T,N-1}(1 - \varsigma_{T,N-1})\varsigma_N\}^{1/2} \tag{5.21c}$$

$$\times [(K_N\Delta\theta_N/2q_N)\sin(q_N L_N)\cos\Phi_N - \cos(q_N L_N)\sin\Phi_N] \, ,$$

where

146

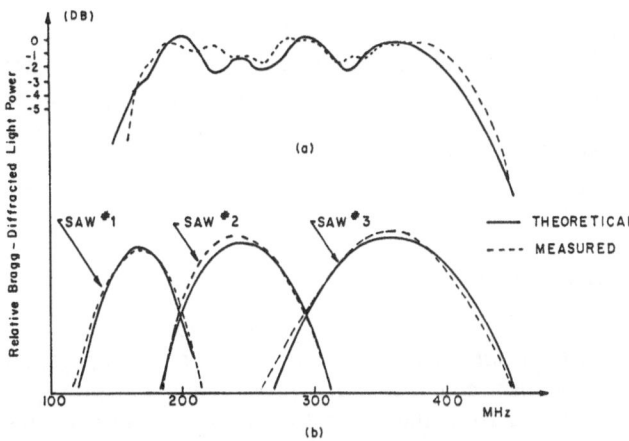

Fig.5.13. Individual and composite frequency responses of the Bragg-diffracted light power from three tilted SAWs, (a) composite response, and (b) individual responses

$$\Phi_N = \phi_N - \psi_N = \phi_N - K_{N-1}\Delta\theta_{N-1}d_{N-1} - \psi_{n,N-1} + \psi_{m,N-1} . \qquad (5.21d)$$

Equations (5.21c and d) clearly show the manner in which the resultant diffraction efficiency depends upon those of the first (N-1) SAW and the last (Nth) SAW, as well as the phase shifts of the light and the acoustic waves. It is to be noted that for the first SAW, we have $\varsigma_{T,0} = 0$, $\alpha_1 = 1$, $\beta_1 = 0$, and $\psi_1 = \phi_1$. Thus, by utilizing these input boundary values for the first SAW and successively applying the general expressions given by (5.21a,b), the resultant diffraction efficiency for any number of SAWs may be calculated numerically. The methodology for this numerical calculation has been established for the analysis and design of wide-band devices using either one or a combination of the two wide-band configurations to be discussed in the following subsection.

5.3.3 Two Specific Wide-Band Configurations

Application of the unified theory just presented to the two wide-band device configurations that have been successfully realized is now discussed.

a) Tilted-SAWs Configuration

The configuration involving N tilted SAWs that are staggered in center frequency and tilted in propagation direction, can be deduced from the general configuration of Fig.5.12. This is done by letting all step heights h_i equal zero and each tilt angle θ_i equal the difference between the Bragg angle of the ith SAW and the first SAW, both at their respective center frequencies. Consequently, the resultant frequency response for the diffraction efficiency can be conveniently obtained from (5.20,21). For example, Fig.5.13 shows the calculated frequency responses of the Bragg-diffracted light power from three tilted SAWs that cover the frequency range of 150 to 400 MHz together with the measured frequency responses [5.14, 15]. It is important to emphasize that the phase shift as well as the tilt angle between adjacent SAWs, in addition to the center frequency and the acoustic aperture of the individual SAWs, are the important parameters for the design of a Bragg device with a large composite bandwidth [5.14] which employs multiple tilted SAWs.

b) Phased-SAWs Configuration

Similarly, the configuration for a phased-array of N identical SAW element transducers that are of identical center frequency and propagation direction, but successively displaced from each other along the propagation direction can be deduced from the general configuration [5.15]. This is obtained by letting $L_1 = L_2 = ..., L_N$, $d_1 = d_2 = ..., d_N$, all tilt angles $\theta_i = 0$, and the center frequencies of all element transducers and all step heights between adjacent element transducers be identical. In addition, the separation and the step height between adjacent element transducers, d_i and $h_i' = (h_{i+1} - h_i)$, and the phase difference $\Delta\phi$ between adjacent SAWs are given by (5.22) for the case of first-order acoustic beam steering [5.64]

$$L_i + d_i = p(n_m \Lambda_0^2 / \lambda_0) , \tag{5.22a}$$

$$h_i' = h_{i+1} - h_i = p(\Lambda_0 / 2) , \tag{5.22b}$$

$$\Delta\phi = [\pi + p(f/f_0)\pi] \quad \text{for} \quad p \text{ being odd} , \tag{5.22c}$$

and

$$\Delta\phi = p(f/f_0)\pi \quad \text{for} \quad p \text{ being even} , \tag{5.22d}$$

where Λ_0 and λ_0 are the acoustic wavelength at the center frequency f_0 and the free-space light wavelength, respectively, and p designates the step height between adjacent element transducers in terms of the number of half-acoustic wavelengths at the center frequency. Also, n_m designates the refractive index of the m-th waveguide mode (Sect.5.1). Neglecting the overlap integral, it can be shown that as a result of the first-order acoustic beam steering, the Bragg bandwidth Δf_B is approximately N times larger than that of a single SAW with an aperture identical to the total aperture of the phased array [5.64]. Note that p is subject to conflicting requirements; that is, high diffraction efficiency requires a large p while a large bandwidth demands a small p. Also, the results given by (5.22a-c) are similar to those using bulk acoustic phased-array transducers [5.26, 65, 66]. Figure 5.14 shows a six-element stepped SAW transducer array and the corresponding planar transducer array together with the RF driver circuits. Both the calculated and the measured frequency responses of the Bragg-diffracted light power from such a phased SAW array and that from a single-element SAW alone are shown in Fig.5.15a and b, respectively. Note that both the calculated and the measured AO frequency

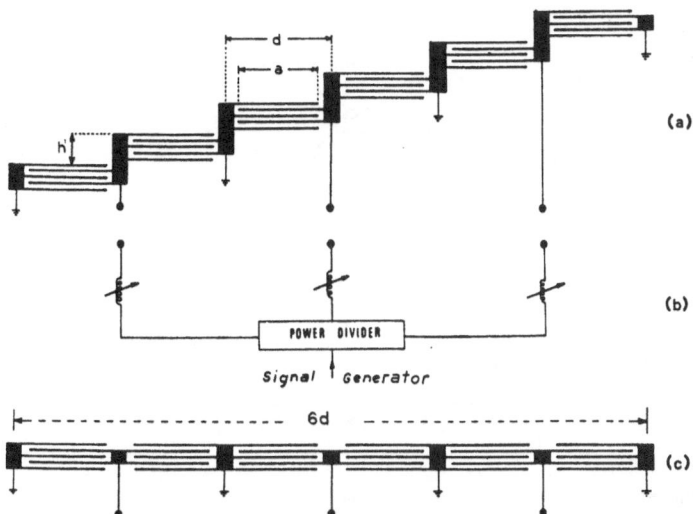

Fig.5.14. A six-element stepped-SAW transducer array (a), RF driver circuits (b), and the corresponding planar transducer array (c)

responses for the stepped array and its equivalent planar array shown in Fig.5.15a were obtained using the same total acoustic drive power. However, for comparison an acoustic drive power approximately six times larger was used in generating the calculated and the measured AO frequency responses for the single-element SAW shown in Fig.5.15b. Note that for comparison the absolute peak diffraction efficiencies in Fig.5.15 were set identical.

In summary, the analysis and the methodology for numerical calculation presented thus far may be extended to the case involving N SAWs and it can be concluded that by using multiple SAWs with proper arrangements in center frequency, propagation direction, and placement, guided-wave AO Bragg devices with large composite bandwidth and thus large diffraction efficiency-bandwidth product can be realized.

5.4 Realization of Wide-Band Planar Acousto-Optic Bragg Modulators and Deflectors

As indicated previously, the planar AO Bragg modulator and deflector of Fig.5.1 constitutes a highly useful component for multichannel integrated and fiber-optic systems. Consequently, it is desirable to realize such modulators and deflectors with as large a bandwidth and as high a diffraction efficiency as possible. Five wide-band device configurations have been identified and explored for this purpose. These wide-band configurations utilize the following interaction and SAW transducer combinations: (i) isotropic diffraction with multiple tilted transducers of staggered center (synchronous) frequency [5.14], (ii) isotropic diffraction with phased-array

149

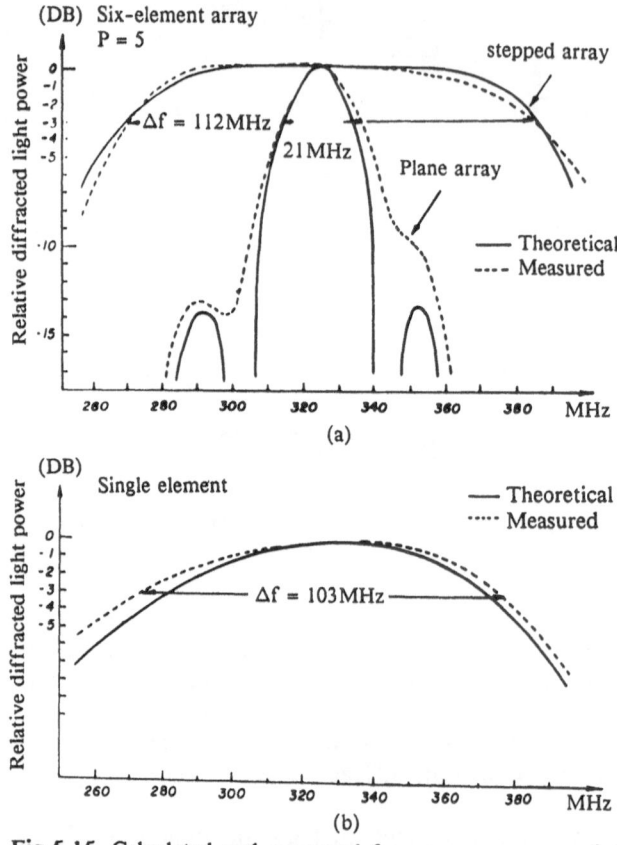

Fig.5.15. Calculated and measured frequency responses of the Bragg-diffracted light power for a six-element phased-SAW array and its corresponding planar array (a) and for a single-element SAW (b)

transducer [5.64], (iii) isotropic diffraction with a transducer array which combines that of (i) and (ii) [5.1], (iv) isotropic diffraction with a single tilted-finger chirp transducer [5.67] or an array of such transducers, and (v) optimized anisotropic diffraction with multiple transducers of staggered center frequency or a parallel-finger chirp transducer [5.20]. In the earlier work, all of the above transducer types and arrangements were fabricated in the Y-cut LiNbO$_3$ waveguides using the well-established photolithographic techniques [5.68]. In the most recent work, electron beam lithographic technique has been used to fabricate the transducers of GHz center frequency [5.69, 70, 71]. In this section, the principle of operation and key design parameters and procedure for each wide-band device configuration are first discussed. Fabrication, testing, and measured results for the specific devices in Y-cut LiNbO$_3$ are then described and an update is given of the performance figures most recently obtained. Some comments are added regarding the relative merits of a single transducer versus multiple transducers for the realization of wide-band AO Bragg devices.

150

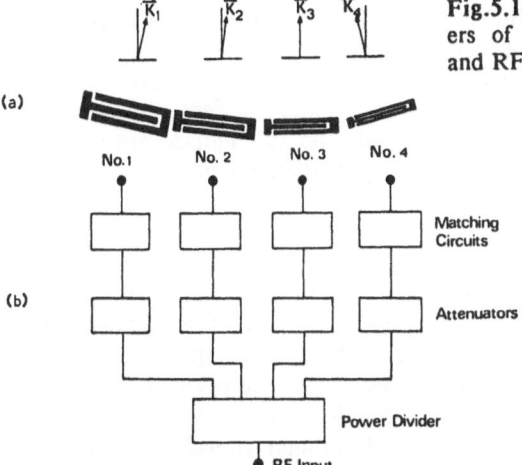

Fig.5.16. Multiple tilted SAW transducers of staggered center frequency (a) and RF driver circuits (b)

5.4.1 Principle and Design Procedure of Wide-Band Device Configurations

a) Isotropic Diffraction with Multiple Tilted Tansducers of Staggered Center Frequency

The transducer arrangement for this wide-band configuration is illustrated in Fig.5.16a. The individual periodic ID Transducers (IDTs) are staggered in center (synchronous) frequency and tilted in acoustic propagation direction [5.14]. The tilt angle between each pair of adjacent transducers is set equal to the difference of the two Bragg angles at the two corresponding center frequencies. As indicited in Fig.5.16b, each element transducer is incorporated with a matching network and the transducers are electrically driven in parallel through a power divider. Individual attenuators may also be incorporated between the outputs of the power divider and the inputs of the matching networks [5.1] to tailor the peak diffraction efficiency at each center frequency. It is clear that the multiple tilted SAWs generated by such a composite transducer satisfy the Bragg condition in each frequency band and thus enable a broad composite frequency response to be realized. It should be noted that a curved transducer as depicted in Fig.5.17 will evolve if the finger electrodes of a large array of small-aperture, tilted transducers are joined side by side. Like the tilted-finger chirp transducer to be described in the immediate

Fig.5.17. A curved transducer evolved from a large number of tilted transducers of staggered center frequency

following, this curved transducer lacks the flexibility for compensation and adjustments after it has been fabricated.

A rigorous analysis for this modulator/deflector configuration using the unified approach [5.15] has been presented in Sect.5.3. Numerical computation of the frequency response based on this unified approach has also been carried out [5.15]. The key design parameters and procedure that have been identified are now discussed.

The key design parameters for the multiple tilted transducers are: center frequency, bandwidth, and aperture of individual element transducers, tilt angle between adjacent element transducers, and relative positions of adjacent element transducers. For each pair of adjacent element transducers the two center frequencies are chosen such that the individual frequency responses intersect at -6 db down from the peak diffraction efficiency. The number of finger electrode pairs in each element transducer must be sufficiently small to ensure a fractional transducer bandwidth consistent with the required fractional AO Bragg bandwidth. Furthermore, the aperture of the element transducers must be sufficiently small to provide the fractional AO Bragg bandwidth required but sufficiently large to ensure efficient diffraction. In the meantime they must present suitable acoustic radiation impedances.

We now turn our attention to the relative positions of the element transducers. As a result of the difference in the phases of the SAWs generated by the adjacent transducers and in the acoustic propagation path, as measured from the front edge of the transducers to the interaction region, the individual components of diffracted light from the adjacent SAWs may differ in phase for the crossover frequencies. Adjustable electronic phase shifters were incorporated in the earlier devices to compensate for this phase difference and to ensure that the individual, diffracted light components add in phase. For example, in the earlier work two phase shifters were used to achieve bandwidths of 358 and 500 MHz [5.14]. However, it was subsequently discovered that the phase shifters may be eliminated by properly configuring the element transducers through proper choice of both the horizontal separation D_s and the vertical step height h′ between adjacent element transducers [5.1]. For the example involving transducers #1 and #2, as shown in Fig.5.18, the horizontal separation D_s and vertical step height h′ (between the center coordinates of the transducers) are given as follows [5.1]

$$D_s = M2\Lambda_i^2/\lambda ,$$ (5.23a)

$$h' = \left[\frac{\lambda}{\Lambda_i} - \frac{\lambda}{2\Lambda_{01}}\right] D_s \quad \text{where}$$ (5.23b)

λ: wavelength of the light wave in the waveguide,
Λ_{01}, Λ_{02}: wavelength of the SAW at the center frequency f_{01} for transducer #1 and the center frequency f_{02} for transducer #2, respectively,

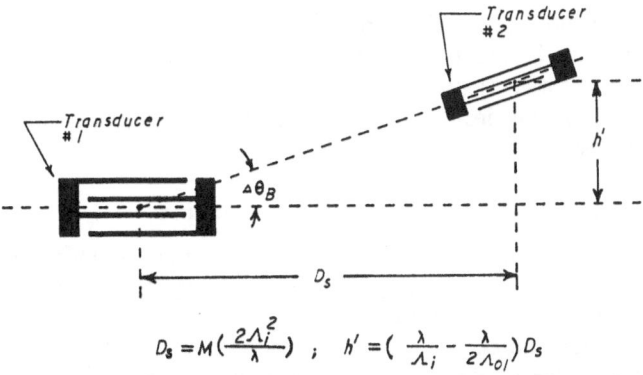

$$D_s = M\left(\frac{2\Lambda_i^2}{\lambda}\right) \quad ; \quad h' = \left(\frac{\lambda}{\Lambda_i} - \frac{\lambda}{2\Lambda_{0l}}\right)D_s$$

Fig.5.18. Geometrical configuration of element transducers in a frequency-staggered tilted transducer

Λ_i: wavelength of the SAW at the crossover frequency $(f_{01}+f_{02})/2$,
M: integer, and
$\Delta\theta_B = \lambda/2\cdot(1/\Lambda_{02}-1/\Lambda_{01})$ is the difference of the two Bragg angles at f_{01} and f_{02}.

Note that a proper choice of M is dictated by the apertures of the adjacent element transducers. The above design formulas can be successively applied to determine the relative locations of all element transducers in the array.

Based on the aforementioned design guidelines, useful design procedures may be established. For simplicity, it is assumed that the fractional transducer bandwidths and the corresponding fractional AO Bragg bandwidths are identical for all element transducers but with the latter being slightly smaller than the former. Note that in a more refined design the fractional transducer bandwidth and/or the fractional AO Bragg bandwidth may be varied among the element transducers. Thus, given the upper and lower -3 db points of the desired frequency response, the number of tilted transducers and their center frequencies can be determined, once the fractional AO Bragg bandwidth is chosen. Subsequently, the aperture of element transducers can be determined by substituting the center frequency and the fractional AO Bragg bandwidth into (5.11). By choosing a fractional transducer bandwidth slightly larger than the fractional AO Bragg bandwidth the number of finger electrode pairs can then be determined. The positions of the element transducers are then determined using (5.23). Using the center frequency and the aperture of the individual transducers and the number of finger pairs just determined, the electrical parameters such as radiation impedance, static capacitance and conductive resistance for all individual transducers can be calculated. The matching network [5.55,56] and the attenuator for each element transducer can be designed accordingly. Finally, based on the specification for the upper and lower -3 db frequencies the penetration depth (or the thickness) of the optical waveguide can be determined using the frequency response plots (Figs.5.4,5) given in Sect.5.1. In some cases it may be more expedient to follow the design procedures presented above in reverse order. As discussed in the following subsection, a deflector of 680 MHz bandwidth [5.67] was realized in a Y-cut titanium-indiffused LiNbO$_3$ waveguide using the design procedures and the improved transducer geometry just described.

b) Isotropic Diffraction with Phased-Array Transducers

This wide-band configuration is characterized by multiple, periodic ID SAW transducers of identical center frequency and propagation direction, but arranged in a stepped configuration, as shown in Fig.5.14 [5.64]. As a result of the finite step height, a variable phase shift is introduced between the adjacent SAWs as the acoustic frequency is varied, resulting in steering of the combined acoustic wavefront. Steering of the wavefront enables the acoustic beam of the entire aperture to track the varying Bragg angle and, therefore, produces efficient diffraction over a relatively wide band.

Like the configuration of Fig.5.16, the individual groups of elementary transducers are tuned separately and are driven electrically in parallel through a power divider. This wide-band configuration can also be analyzed as a special case of the unified approach presented in Sect.5.3. However, it has been found that an approximate approach similar to that adopted in bulk-wave acousto-optics [5.26,65,66] not only brings out the key design parameters but also permits simpler design procedures. Specifically, this approximate approach was used to derive the design formulas and procedures for the first-order beam steering [5.64] presented here.

The key design parameters include the number of finger electrode pairs N$_e$ and the aperture of the elementary transducer a, the separation d, and the step height h′ between adjacent transducers, and the number of elementary transducers N in the array. Note that in contrast to Fig. 5.12, d here is used to designate the total separation between adjacent elementary transducers. In other words, d is identical to L$_i$ + d$_i$ of Fig.5.12 for all i′s. Also, a is smaller than d by the width of the terminal electrode. The relationships among these parameters are given as follows [5.64]:

$$h' = p\Lambda_0/2 , \tag{5.24a}$$

$$p = 2n - \phi/\pi , \quad n = 1,2,3... , \tag{5.24b}$$

$$d = p\Lambda_0^2/\lambda , \tag{5.24c}$$

$$\Delta f_{-3db\,Bragg} \simeq (1.8/p)(d/a)f_0 , \quad \text{for} \quad (N/p)^{1/2} \ll 1.5a/d , \tag{5.24d}$$

$$\Delta f_{-3db\,Bragg} \simeq (1.8/Np)^{1/2} f_0 , \quad \text{for} \quad (N/p)^{1/2} \gg 1.5a/d , \tag{5.24e}$$

where f$_0$, Λ_0, and ϕ designate, respectively, the center frequency, the acoustic wavelength at center frequency, and the electrical phase shift between adjacent element transducers; p takes even and odd integers, respectively, when $\phi = 0$ and π. Also, $\lambda = \lambda_0/n_m$ is the wavelength of the guided light. Finally, numerical calcualtion is required to determine the fractional AO Bragg bandwidth when $(N/p)^{1/2}$ lies between the two limiting cases given in (5.24d, e).

A small N_e must be chosen to ensure a sufficiently large transducer bandwidth so that the large AO Bragg bandwidth made possible by the acoustic beam steering may be fully utilized. A higher diffraction efficiency can be obtained by utilizing a wider total acoustic aperture. This, in turn, may be obtained by employing either a small N but a large h' or a large N but small h' and, thus, large values of a and d in one case and small ones in the other case. The first choice is accompanied by a faster decrease in the AO Bragg bandwidth, while the second choice is accompanied by a more stringent requirement in lithography and fabrication as well as a greater complexity in the RF driver circuits. Design and construction of wide-band deflectors using six-element phased-array transducers in Y-cut LiNbO$_3$ outdiffused waveguides will be described in the following subsection.

In summary, a phased-array transducer incorporating first-order beam steering can be utilized to achieve simultaneously very high diffraction efficiency and moderate bandwidth in Y-cut LiNbO$_3$ waveguides. In comparison to frequency-staggered tilted-transducer arrays, the phased-array transducer is capable of higher diffraction efficiency because all elementary transuncers are excited simultaneously. It provides a much greater interaction length. However, the bandwidth of the resulting device is limited by the transducer bandwidth and is therefore considerably smaller than that of the device which utilizes the tilted transducer array. In addition, the design and fabrication involved are more complex and the flexibility for adjustments and compensations after the device has been fabricated is considerably reduced.

c) Isotropic Diffraction
with Multiple Tilted Phased-Array Transducers

In this wide-band configuration, the elements of the tilted transducer array are stepped-array transducers rather than single-aperture transducers, as shown in Fig.5.19 [5.1]. Implementation of this wide-band configuration has been reported [5.69, 70]. As in the first two wide-band configurations, the element transducers are incorporated with separate matching circuits and attenuators, and are driven electrically in parallel through a

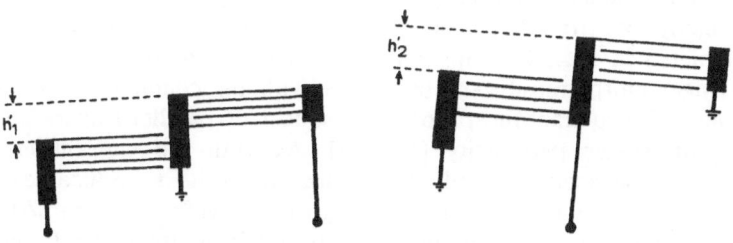

Fig.5.19. Multiple tilted phased-array transducers

power divider. Since the tilted transducer array is capable of providing very large device bandwidth at moderate diffraction efficiency while the phased-array transducer is capable of providing very high diffraction efficiency at moderate device bandwidth, such tilted phased-array transducer should simultaneously provide a very large device bandwidth and a very high diffraction efficiency. The design parameters and the general design guidelines and procedures established in the first two configurations are also applicable here. Clearly, this wide-band configuration involves a higher degree of complexity in transducer design, fabrication, and RF driver circuits.

As indicated in Sect.5.1, for a given fractional AO Bragg bandwidth the maximum allowable acoustic (transducer) aperture of a single-aperture transducer is inversely proportional to the square of the acoustic center frequency. As a result, the transducer aperture and thus the diffraction efficiency decrease drastically as the center frequency increases. This undesirable degradation in performance, which becomes increasingly severe at the gigahertz frequency range and above, can be overcome by using the phased-array transducer. However, for the lower frequency range (up to say 700 MHz) this undesirable performance degradation is not significant and the single-aperture transducers should be adequate. Thus, in order to reduce the complexity in design and fabrication, one may employ a simplified version consisting of single-aperture transducers for the lower frequency elements and phased-array transducers for the higher frequency elements.

Based on the projected improvement in the conversion efficiency of GHz transducers, planar AO Bragg modulators and deflectors with a 1.0 GHz bandwidth at an electrical drive power of 1 mW per megahertz bandwidth with 50% diffraction efficiency should be achievable using the multiple-tilted phased-array transducer.

d) Isotropic Diffraction with a Single Tilted-Finger, Chirp Transducer or an Array of Such Transducers

We now return to Fig.5.16a and consider the situation involving a large number of element transducers. Specifically, the element transducers are assumed to have closely spaced synchronous frequencies and each has a single pair of finger electrodes. Clearly, in this situation the appropriate tilt angle between adjacent element transducers will be very small. If we now place all such element transducers together, one behind the other in a straight line, and connect them electrically in parallel, we will have constructed a composite transducer of varying finger periodicity and tilt angle, as depicted in Fig.5.20. This resulting composite transducer is called "tilted-finger chirp transducer", consistent with the common usage of the terminology "chirp transducer" for a transducer of parallel fingers but linearly-varying finger periodicity [5.50,51]. As with the conventional chirp transducer the acoustic bandwith of this composite transducer can be very large. Furthermore, with proper design the wavefront of the SAW generated by this composite transducer can be made to track the Bragg

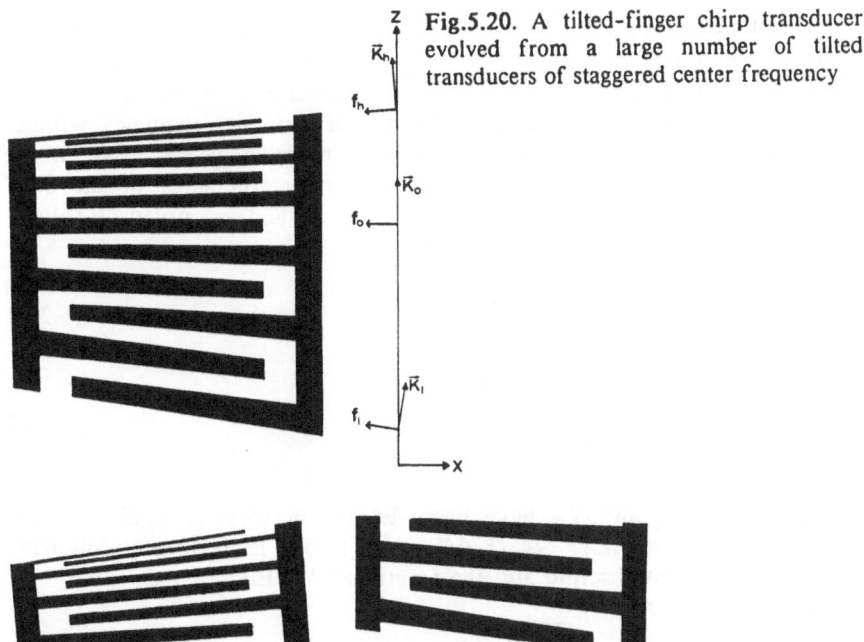

Fig.5.20. A tilted-finger chirp transducer evolved from a large number of tilted transducers of staggered center frequency

Fig.5.21. Multiple tilted-finger chirp transducer array

condition for the entire frequency band, thus providing a large AO Bragg bandwidth. Hence, to the extent that the bandwidth is not limited by the coupling coefficient $C_{mn}^2(f)$, this composite transducer should also be capable of providing a large device bandwidth. Note that a conventional chirp transducer of parallel fingers and small aperture can also be employed to obtain a large device bandwidth [5.72]. However, in this case the large bandwidth is obtained at a drastically reduced diffraction efficiency. Verification of the concept described above was first carried out using a simple tilted-finger chirp transducer fabricated in a titanium-indiffused Y-cut LiNbO$_3$ waveguide [5.67]. Transducers of slightly different finger electrode arrangements were also realized subsequently [5.73,74]. Obviously, an array of such tilted-finger chirp transducers, as shown in Fig.5.21, all tilted properly such that individual transducers satisfy the Bragg angles over the corresponding ranges of acoustic frequency, can be employed to realize a very large composite AO Bragg bandwidth.

e) Optimized Anisotropic Diffraction with Multiple Transducers of Staggered Center Frequency or a Parallel-Finger Chirp Transducer

As in bulk-wave interaction, for a given acoustic aperture a considerably larger AO Bragg bandwidth can be realized by utilizing anisotropic Bragg diffraction. Anisotropic Bragg diffraction also provides the inherent advantage of reducing that portion of the background noise which is caused by the undiffracted light. In the *optimized* anisotropic Bragg dif-

157

fraction between TM_0 and TE_0 modes, the wave vector of the diffracted light is orthogonal or nearly orthogonal to that of the acoustic wave of relatively large aperture (Fig.5.2c). Thus, optimum phase-matching condition can be maintained over a wide range of acoustic frequency without requiring acoustic beam steering. From the wave vector diagram of Fig.5.2c we can readily show that the optimum frequency of the SAW (f_{opt}) and the incident angle of the light wave (θ_m) are given as follows [5.1, 13]

$$f_{opt} = \frac{V_R}{\lambda_0}(n_m + n_n)^{1/2}(n_m - n_n)^{1/2} , \tag{5.25a}$$

$$\theta_m = \sin^{-1}\left[\frac{\lambda_0 f_{opt}}{n_m V_R}\right] , \tag{5.25b}$$

where λ_0, n_m, n_n and V_R have been defined previously. Again, for the special case in which $g^2 \ll (K\Delta\theta L/2)^2$, (5.25a) leads to the following expressions for the absolute and the fractional AO Bragg bandwidths:

$$\Delta f_{-3db\,Bragg} \simeq 2V_R \sqrt{\frac{2n_m}{\lambda_0 L}} \quad \text{(optimum anisotropic) ,} \tag{5.25c}$$

$$\frac{\Delta f_{-3db\,Bragg}}{f_0} \simeq 2\frac{V_R}{f_0}\sqrt{\frac{2n_m}{\lambda_0 L}} \quad \text{(optimum anisotropic) .} \tag{5.25d}$$

Comparing (5.25c,d) with (5.11a,b) for isotropic Bragg diffraction one finds the following relationship between the two fractional AO Bragg bandwidths

$$\left.\frac{\Delta f_{-3db\,Bragg}}{f_0}\right|_{\text{optm. anisotropic}} \simeq \sqrt{2\left.\frac{\Delta f_{-3db\,Bragg}}{f_0}\right|_{\text{isotropic}}} . \tag{5.25e}$$

It is seen from (5.25c) that the AO Bragg bandwidth for optimized anisotropic diffraction is inversely proportional to the square root of the acoustic aperture and may therefore be considerably larger than that for isotropic diffraction. Consequently, it is possible to achieve simultaneously high diffraction efficiency and large AO Bragg bandwidth. Also, (5.25e) shows that the fractional AO Bragg bandwidth with optimum anisotropic diffraction can be much larger than that with isotropic diffraction, as the latter is, in general, less than unity. Experimental demonstration of this unique feature using the optical modes of orthogonal polarizations (TM_0 and TE_0) in a Y-cut Ti-diffused $LiNbO_3$ waveguide [5.20] will be described in the following subsection.

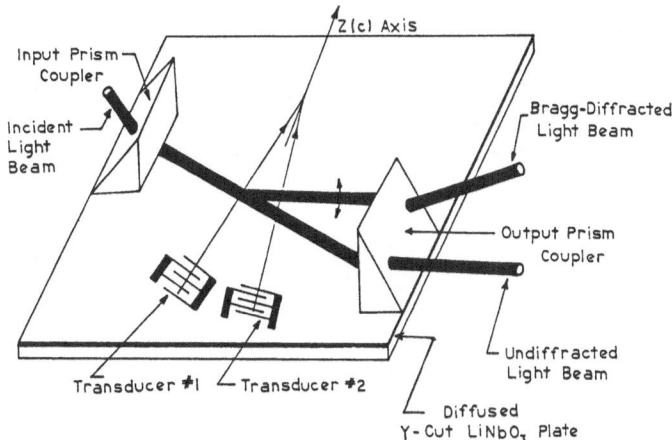

Fig.5.22. Measurement of planar guided-wave acousto-optic Bragg diffraction using a pair of prism couplers

5.4.2 Design, Fabrication, Testing and Measured Device Performance

Using the design procedures and guidelines presented in Sect.5.4.1, a variety of wide-band AO Bragg modulators and deflectors were designed and fabricated, mostly on Y-cut $LiNbO_3$ substrates. The single-mode optical waveguides involved were grown using either the outdiffusion [5.2] or the indiffusion [5.3] technique, and the interdigital SAW transducers were fabricated using the well-established lift-off method [5.68].

Testing and performance measurement of the devices were carried out using mostly the He-Ne laser light at 0.6328 μm. Excitation of the optical guided waves at TE_0 and TM_0 mode was accomplished using a rutile prism [5.75], and a second rutile prism was used to couple out both the diffracted and the undiffracted light beams for detailed measurement (Fig.5.22). Typically, the aperture of the guided light beam could be varied from 1 to 15 mm, with some degradation in the uniformity of the light beam for the widest aperture. The best throughput coupling efficiency was as high as 35%. A photograph of one of the complete devices is shown in Fig.5.23. Note the prism-couplers $LiNbO_3$-plate combination in the middle of the brass platform. The input and output RF connectors and the matching circuits for the transducers were located at the right and the left of the combination. The complete device was attached to a precision holder (micro-manipulator) to facilitate precision optical alignment and adjustment. The wide-band device configurations described in Sect. 5.4.1, each demonstrating certain distinct features, were tested and the device performances measured. Detailed performances data together with the design specifications are given in the following.

Fig.5.23. Photograph of a guided-wave acousto-optic Bragg deflector using multiple tilted SAWs

a) Isotropic Device with Multiple Tilted Transducers of Staggered Center Frequency

The first device to be discussed is the one that was used to study the composite frequency response, as well as the effect of phase shift on the composite frequency response, and the electric drive power requirement. The device employed a three-element tilted transducer [5.14] with the center frequencies at 170, 255, and 382 MHz. The corresponding acoustic apertures were 2.50, 1.66, and 1.11 mm, respectively, and the tilt angles between adjacent transducers were 3.4 and 5.2 mrad, corresponding to the difference in the Bragg angles at the center frequency of the adjacent transducers. In order to obtain as wide an acoustic bandwidth as possible, the number of finger-electrode pairs for each transducer was chosen to be as small as two and a half. Measured acoustic bandwidths of approximately 30% of the center frequencies, namely 50, 68, and 115 MHz, were obtained by inserting a single inductance of proper value in series with each transducer. The acoustic wavelengths at the center frequency were 20.5, 13.7, and 9.3 μm, respectively.

The individual transducers were excited in parallel using power dividers, and the adjustment of phase shifts between adjacent SAWs was implemented by using sections of coaxial cables. Figure 5.13 shows a recorder plot of the resultant frequency response from the three-element tilted SAWs together with those from the individual SAWs. It is seen that the device has a -3 db resultant bandwidth of approximately 265 MHz while the bandwidths using the individual SAWs are 45, 65, and 82 MHz, respectively. Clearly, the resultant bandwidth was considerably larger than the sum of the three individual bandwidths. The allowable variation of the incident angle of the light beam (outside the crystal) to maintain the same bandwidth was approximately 0.5°. Typical diffraction efficiency versus the total electric drive power at the center frequency is shown in

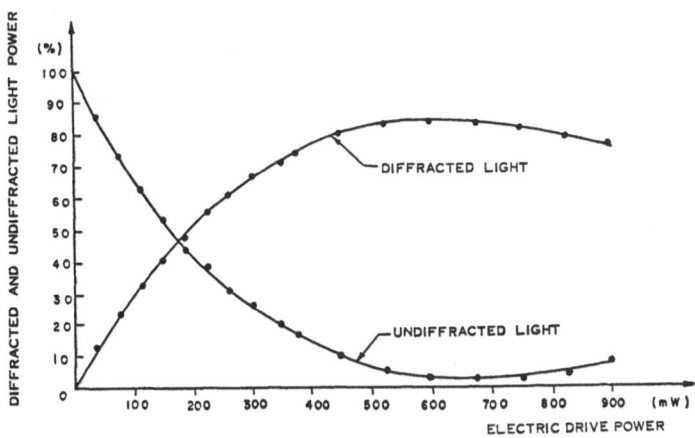

Fig.5.24. Percentage diffracted and undiffracted light power versus total electric drive power

Fig.5.24. It is seen that a total electric drive power of 220 mW was required to diffract 50% of the incident light. The corresponding total acoustic power was estimated to be at most 15 mW because the best conversion efficiency of the transducers was measured to be -13 db. Based on the above performance figures, the diffraction efficiency-bandwidth product of this particular device is substantially larger than all previous devices that employed a single SAW.

As discussed in Sect.5.3, interference between adjacent SAWs becomes important in the range of frequencies at which both adjacent transducers excite SAWs efficiently. This interference will, in turn, affect the diffraction efficiency, and thus the resultant frequency response. Figure 5.25 illustrates this interference effect in the frequency bands around 210 and 310 MHz for the three-element unit [5.14]. It is apparent that as the electrical phase shift $\Delta\phi_{12}$ was varied from -22° (Fig.5.25a) to +177° (Fig.5.25b), the diffraction efficiency varied by as much as -11 db. Similarly, when the electrical phase shift $\Delta\phi_{23}$ was varied from +84° (Fig.5.25c) to -121° (Fig.5.25d), the diffraction efficiency varied by -14 db. Although it is difficult to compare the theoretical results with the experimental results based on the absolute phase shift (since in the experiments the electrical phase was varied externally by known increments, while in the analysis the phase shift is assumed to be that between the adjacent SAWs), the effect of the phase shift between the adjacent SAWs upon the resultant frequency response has been clearly shown in the experiments. Figure 5.25 has clearly demonstrated that the resultant frequency response of the devices which employ multiple tilted SAWs can be made flat by inserting appropriate phase shifters between adjacent transducers and attenuators. Also, in doing so, the resultant device bandwidth may be made significantly larger than the sum of the individual bandwidths [5.14].

We now turn our attention to the number of resolvable spot diameters and the beam quality of the deflected light. Figure 5.26 shows the

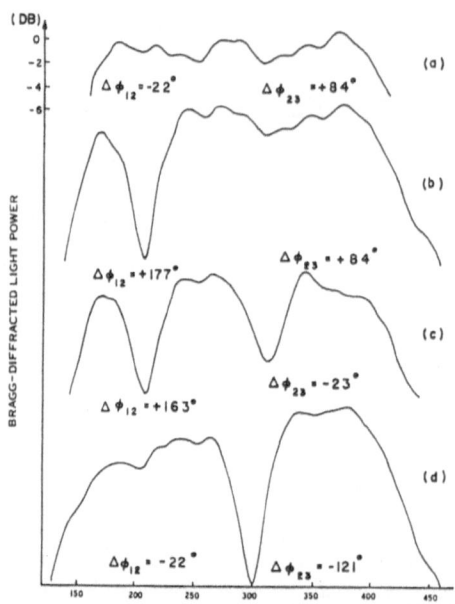

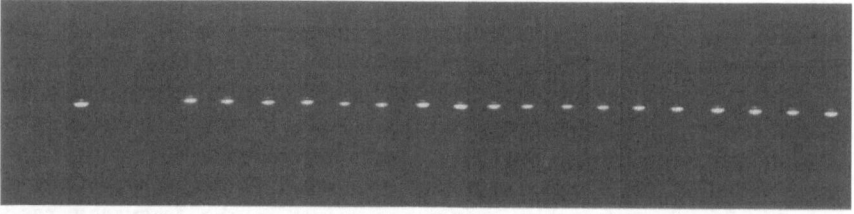

Fig.5.26. Undeflected light spot (*left*) and deflected light spots (*right*) obtained using tilted-SAW array as the SAW frequency was varied from 155 to 410 MHz at 15 MHz per step

composite photograph of the undeflected light spot (when no RF power was applied to the device) and the deflected light spots, both at the far field, as the frequency of the driving signal was varied from 155 to 410 MHz [5.14]. It is to be noted that the appropriate size of this composite photograph was obtained by shortening the spacing between the undeflected light spot and the first deflected light spot (produced at 155MHz). The aperture of the incident light beam employed for this particular experiment was approximately 0.1 cm. The fine quality of the undeflected light beam (RF power off) was preserved in the deflected light beam. The measured values for the number of resolvable spot diameters N_R and the incremental frequency change δf_R required for deflection of one Rayleigh spot diameter were found to be in good agreement with those predicted using (5.12).

Finally, a device bandwidth of 358 MHz was obtained with a second unit that utilized a four-element tilted SAW transducer [5.14] (Fig.5.27). The beam profile of the deflected light for an incident light beam aper-

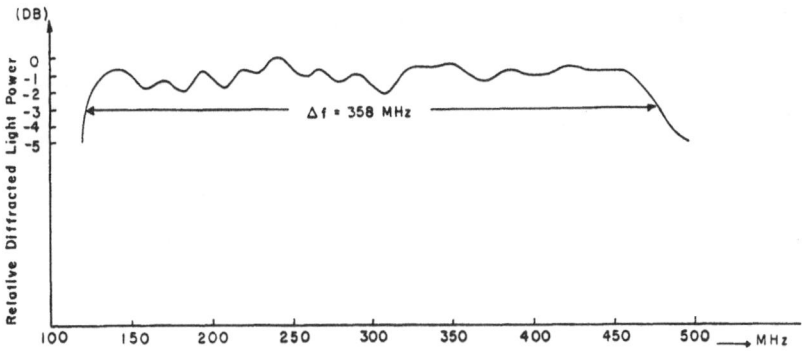

Fig.5.27. Resultant frequency response of a guided-wave acousto-optic Bragg deflector using four tilted transducers

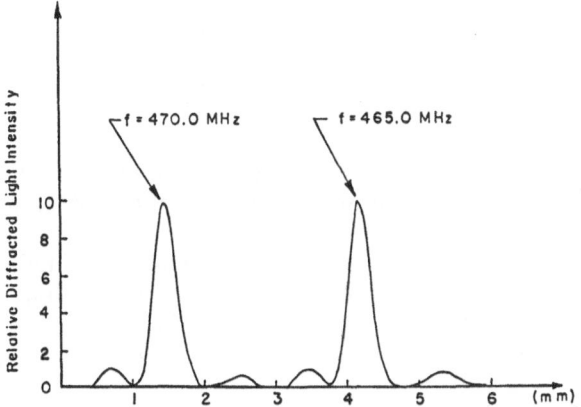

Fig.5.28. Beam profile of deflected light at two acoustic frequencies

ture of approximately 0.45 cm, as recorded by a fiber-optic probe, is shown in Fig.5.28. From this plot, the measured frequency change required for one Rayleigh spot diameter was determined to be approximately 0.8 MHz in comparison with the calculated value of 0.78 MHz. This close match between calculated and measured values indicates that the quality of the deflected light beam was not degraded appreciably by the diffraction process in agreement with the fine beam quality illustrated in Fig.5.26. Since the total bandwidth of this device, Δf, was 358 MHz, the device could deflect a light beam of 0.45 cm aperture into 450 Rayleigh spot diameters at a random-access switching time of 1.28 μs. If the aperture of the light beam were enlarged to 1 cm, the light beam could be deflected by the same device into 1000 Rayleigh spot diameters at a random-access switching time of 2.85 μs. Similarly, the same device was capable of switching a guided light beam of 100 μm aperture into 10 beam positions (channels) at a switching time of 29 ns. In this connection it should be mentioned that a light beam aperture as small as 60 μm has been obtained recently using a channel-planar composite waveguide to-

Table 5.3. Design parameters of the acoustooptic Bragg defelector using a four-tilted transducer array

Element transducer	#1	#2	#3	#4
Center frequency, f_{0i} [MHz]	379	524	724	1000
Cross-over frequency, f_{0ij} [MHz]		400	608	840
Wavelenght, Λ_i [μm]	9.20	6.66	4.82	3.49
Electrode width, $\Lambda_i/4$ [μm]	2.30	1.67	1.21	0.87
Aperture, L_i [mm]	0.644	0.466	0.337	0.244
Acoustooptic Q-factor, Q_{Bi}	13	19	26	36
Static capacitance, C_{Ti} (pF)	1.19	0.86	0.62	0.45
Capacitive reactance, $X_{C_{Ti}}$ [Ω]	352	352	352	352
Radiation impedance, R_{ai} [Ω]	75	75	75	75
Radiation Q, Q_{ri}	4.7	4.7	4.7	4.7
Conductive resistance, R_{ci} [Ω]	16	16	16	16
Loaded Q, Q_{LD}	2.5	2.5	2.5	2.5
Tilt angle, $\Delta\theta_{Bij}$ [mrad]		5.95	8.21	11.33
Center-to-center separation, D_{sij} [mm]		0.8785	0.9205	0.4812
Step height, h'_{ij} [μm]		18.05	26.13	19.81

gether with a TIPE linear microlens array [5.76]. A discussion on this new and novel microlens fabrication technique and some applications of such microlenses will be given in Chap.8.

Subsequently, tilted SAW transducer arrays fabricated using photolithography and e-beam generated masks were utilized to realize AO Bragg cells at center frequencies of 500 MHz [5.77,78] and above, namely in the band from 770 to 1470 MHz [5.69-71].

b) Isotropic Device with Multiple, Tilted Transducers of Improved Geometry

Note that adjustable electronic phase shifters were incorporated in all of the deflectors described in the last subsection. Using the improved transducer geometry in which the relative positions of the element transducers were determined by (5.23a, b) and related design procedures presented in Sect.5.4.1, a deflector of 680 MHz composite bandwidth was realized in a Y-cut Ti-diffused LiNbO$_3$ waveguide without any electronic phase shifter [5.67]. The deflector utilized four tilted transducers with measured center frequencies of 380, 520, 703, and 950 MHz. Other design parameters of this deflector is detailed in Table 5.3 [5.67]. Note that the measured center frequencies of the element transducers differ slightly from the design values. Figure 5.29 shows a photograph of this deflector. A deflected light beam of high quality was again observed. The measured conversion efficiency of the four-element transducers were -7.5, -7.0, -10, and -15 db, respectively. The measured frequency response of this deflector is shown in Fig.5.30. The corresponding diffraction efficiency was 8% at a total RF drive power of 1 W for the entire 680 MHz bandwidth. Since the measured diffraction efficiency of the deflector with only the first three trans-

Fig.5.29. 680 MHz bandwidth deflector with four tilted transducers of improved geometry: (*left*) complete deflector assembly without prism couplers, (*right*) complete deflector assembly with prism couplers

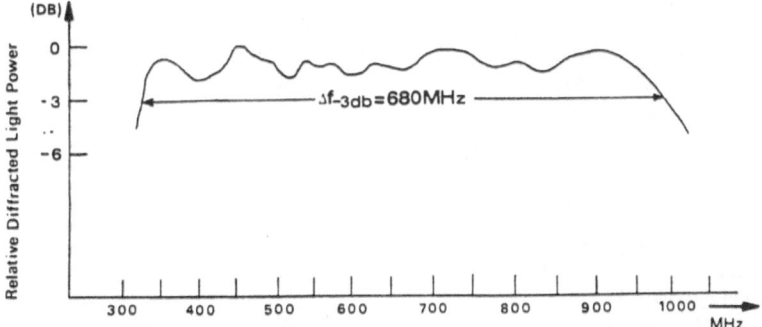

Fig.5.30. Measured frequency response of a wideband deflector employing four tilted transducers of improved geometry in a y-cut LiNbO$_3$ waveguide

ducers activated was nearly five times higher, a considerably better diffraction efficiency could be expected if the conversion efficiency of the fourth transducer (the one with highest center frequency) were improved to that of the first three. Also, once a better transducer conversion efficiency is achieved, a deflector with GHz bandwidth can be realized by slightly increasing the center frequencies of the element transducers. Based on this projection a performance figure of approximately 1 mW electric drive power per MHz bandwidth with 50% diffraction efficiency and 1 GHz bandwidth should be realizable. Consequently, one remaining task with this improved wide-band deflector configuration lies in improvement of the conversion efficiency of transducers in the GHz frequency range.

c) Isotropic Device with Phased-Array Transducers

A number of deflectors were designed and fabricated using phased (stepped)-SAWs of 325 MHz center frequency and first-order beam steering in y-cut LiNbO$_3$ outdiffused waveguides of 7 μm penetration depth [5.64]. The details of design and measured parameters are given in Table 5.4. In one of the deflectors employing a six-element array (Fig.5.14) with a total acoustic aperture of 10.44 mm, only 68 mW of electrical drive power or

165

Table 5.4. Design and measured parameters of the acoustooptic Bragg deflectors using phased-SAW array

	p = 1	p = 5
Center frequency, f_0 [MHz]	325	325
Aperture, a [mm]	0.373	1.74
Separation, d [mm]	0.464	1.93
Step height, $h' = p(\Lambda_0/2)$ [μm]	5.27	26.3
Fresnel distance at center frequency [cm]		
Single element: a^2/Λ_0	1.3	28.7
Six-element stepped array: $\simeq(4d+2a)^2/\Lambda_0$	64	1,189
Acoustooptic quality factor Q	37.7	188.4
Acoustic scanning bandwidth, Δf_{Am} [MHz]		
Single element aperture	606	130
Six-element plane array aperture	82	19
Six-element stepped array aperture	728	130
Bragg tracking bandwidth, Δf_B [MHz]	356	160
Number of finger electrode pairs, N_e	2.5	2.0
Capacitance of each transducer pair [pF]	1.7	6.4
Tuning inductance for each transducer pair [μH]	0.139	0.037
Total input impedance of each transducer pair at the center frequency [Ω]	63	33
Acoustic bandwidth $\Delta f_a = 1/N_e$ [MHz]	130	162
Electrical bandwidth based on the input impedance [MHz]	200	150
Transducer bandwidth [MHz]	113	120

3.5 mW of acoustic drive power was required to diffract 50% of the light over a bandwidth of 112 MHz. As shown in Fig.5.15, this bandwidth was a nearly sixfold increase over that of the deflector employing a single transducer of equivalent aperture or an unphased (unstepped or linear) transducer of identical dimensions, in good agreement with the theoretical prediction. Also, at the same absolute diffraction efficiency, the acoustic drive power required in the single-element SAW was found to be approximately six times larger than the phased-SAW. It should be noted that the 112 MHz deflector bandwidth was limited by the transducer bandwidth as the measured AO Bragg bandwidth was at least twice as large. Again, as shown in Fig.5.31, the quality of both deflected and undeflected light beams in the deflectors just described was high [5.64]. Based on the design plots shown in Figs.5.4,5, even higher diffraction efficiency would result if the penetration depth of the LiNbO$_3$ optical waveguide were smaller than 7 μm. Thus, assuming that the bandwidth of the transducer is greatly increased by incorporating a suitable wide-band matching network, further improvement in the performance figures should be achievable by employing an indiffused LiNbO$_3$ waveguide [5.3] of smaller penetration depth. In addition, as previously discussed, very large bandwidth and very high diffraction efficiency can be achieved simultaneously by employing a combination of tilted and phased transducers. In fact, this possiblity has been verified in an experiment that utilized As$_2$S$_3$ wave-

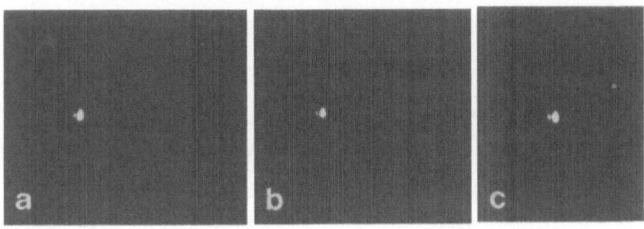

Fig.5.31a-c. Far-field undeflected and deflected light spots using phased-SAW array: (a) undeflected light with no RF power, (b) undeflected light with RF power, (c) deflected light

guide [5.78]. Finally, we note that a wideband phased-array SAW transducer which utilizes metallization of the piezoelectric LiNbO$_3$ substrate to provide the required frequency-dependent phase shift has also been suggested [5.79].

d) Isotropic Device with a Tilted-Finger, Chirp Transducer

The first tilted-finger chirp transducer of the simplest form, as shown in Fig.5.20, was designed and fabricated in a Y-cut LiNbO$_3$ waveguide, and Bragg diffraction experiments were carried out using a 0.6328 μm He-Ne laser light propagating at TE$_0$-mode [5.67]. The synchronous frequency of the finger electrodes was designed to vary linearly from 320 MHz (f$_\ell$) at one end to 630 MHz (f$_h$) at the other. The corresponding finger electrode width at the center of the finger aperture varies from 2.7 to 1.4 μm. The transducer contains a total of 51 finger electrodes, each with an aperture of 0.55 mm. A photograph showing portion of the transducer and the details of the tilted-finger electrodes is given in Fig.5.32. The transducer was driven directly with an RF signal generator of 50 Ω source impe-

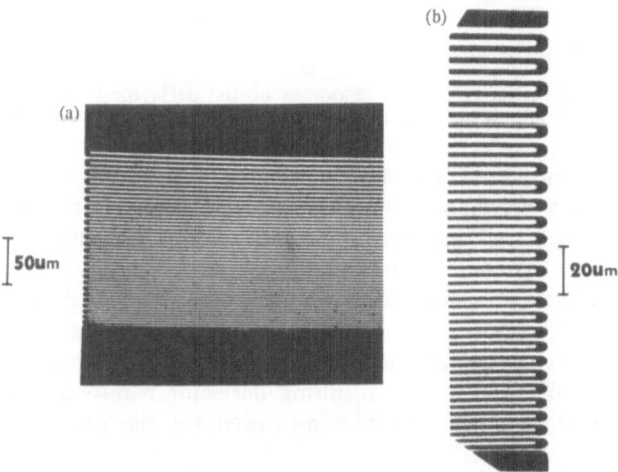

Fig.5.32. (a) Photograph showing portion of a tilted-finger chirp transducer (500 MHz center frequency) in a y-cut LiNbO$_3$ waveguide. (b) Photograph showing details of the tilted-finger electrodes

167

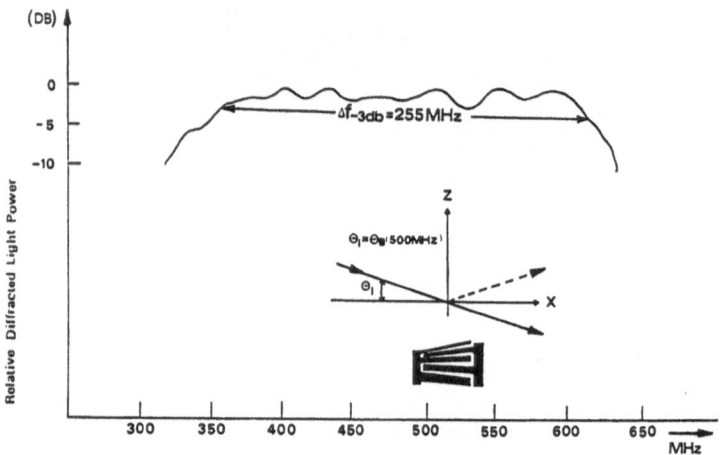

Fig.5.33a. Measured frequency response of the deflector with tilted finger chirp transducer at an optimum incident light angle (Bragg angle)

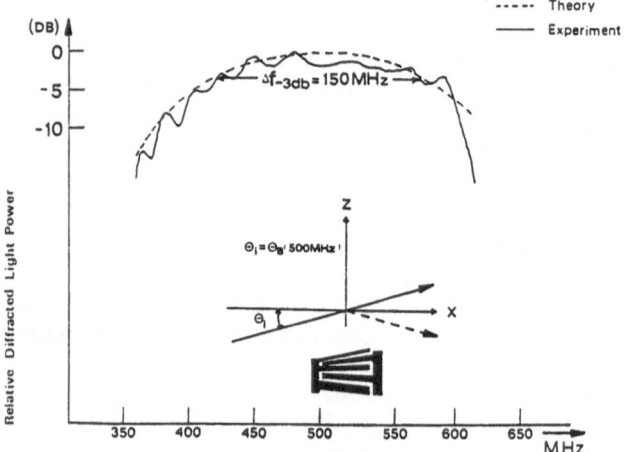

Fig.5.33b. Measured and calculated frequency responses of the deflector with tilted finger chirp transducer at an incident light angle other than optimum

dance. The measured -3 db transducer bandwidth was 255 MHz. Figure 5.33a shows the measured frequency response for the Bragg diffracted light power under the preferred incident angle indicated in the inset. A performance figure of 2.4 mW electrical drive power per MHz bandwidth with 50% diffraction efficiency and 255 MHz deflector bandwidth was demonstrated. Note that Fig.5.33b shows the AO frequency response measured at the other incident angle indicated in the inset. As expected, due to lack of Bragg condition tracking, the resulting deflector bandwidth of 150 MHz was significantly smaller than that measured for the preferred incident angle.

Subsequently, a tilted-finger chirp transducer of improved design using "dog-leg" configuration, as shown in Fig.5.34a [5.73], was used to realize a deflector bandwidth as large as 470 MHz (Fig.5.34b). Note that

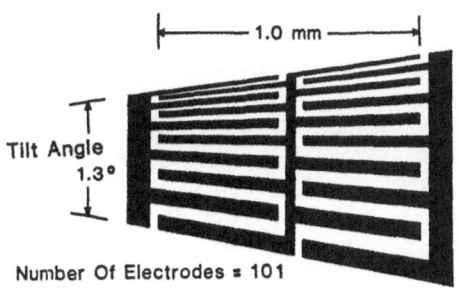

Fig.5.34a. Sketch of a series-connected (dog-leg) tilted-finger chirp transducer

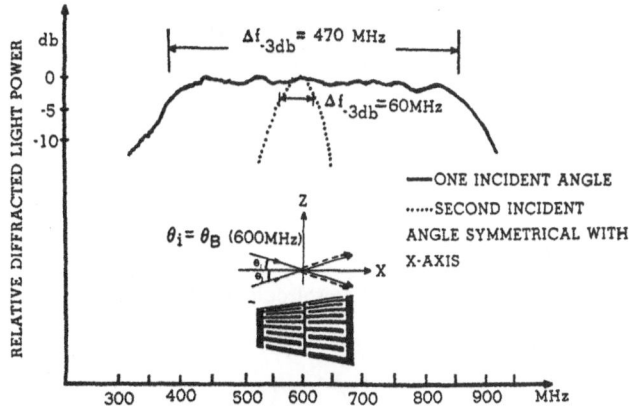

Fig.5.34b. Measured frequency responses of the AO deflector at two symmetrical incident light angles (Bragg angle at 600 MHz: 1.41°)

for this deflector the difference between the bandwidth measured under the preferred incident angle and that measured under the other incident angle was even more dramatic than the first deflector just described. The measured diffraction efficiency was 16% at 200 mW RF drive power. This improved deflector was subjected to 1 W of cw RF drive power without failure. A wide-band transducer of slightly different electrode arrangement was also designed and fabricated subsequently [5.74]. Most recently, such tilted-finger chirp transducers were employed in further realization of wide-band AO Bragg cells for RF spectrum analysis [5.79].

e) Optimized Anisotropic Device with a Single Transducer of Large Aperture

An optimized anisotropic Bragg deflector that utilizes the fundamental modes of orthogonal polarizations (TE_0 and TM_0 modes) was designed and fabricated in a Y-cut Ti-diffused $LiNbO_3$ waveguide [5.20]. The directions of propagation for both the incident and the diffracted lights and for the SAW are shown in Fig.5.35a. The aperture and the center frequency of the SAW transducer were 0.7 cm and 400 MHz, respectively. This center frequency was chosen to coincide with the calculated optimum frequency f_{opt} as determined by (5.25a). The device configuration and the arrangement for diffraction measurement are shown in Fig.5.35b. Effi-

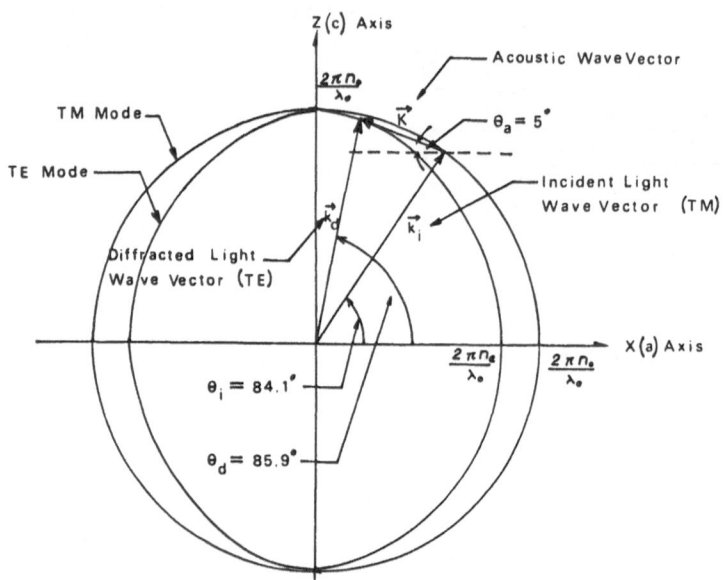

Fig.5.35a. Phase-matching diagram for optimized anisotropic AO Bragg diffraction in a y-cut LiNbO$_3$ substrate

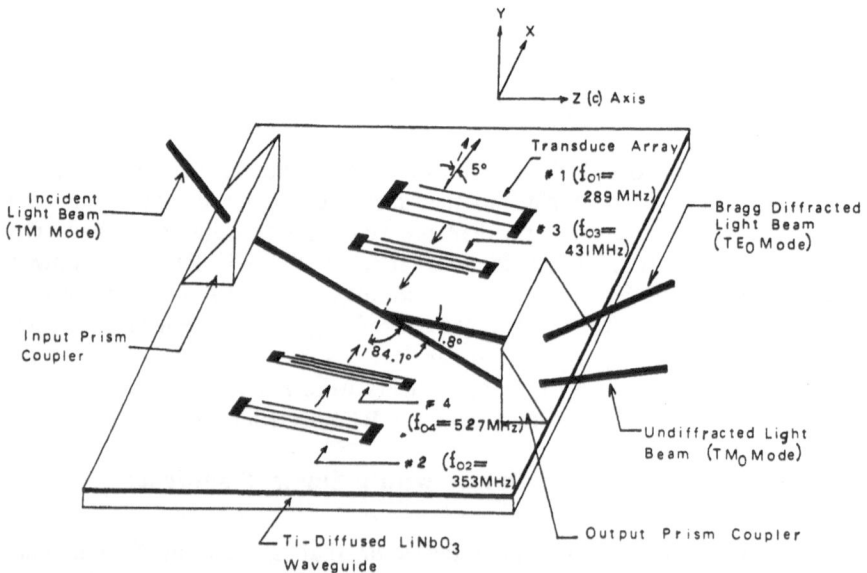

Fig.5.35b. Wideband guided-wave optimized anisotropic AO Bragg diffraction

cient and wide-band Bragg diffraction was obtained at 0.6328 μm optical wavelength. Figure 5.36 clearly shows that a measured deflector bandwidth of 222 MHz was obtained. This bandwidth was a sevenfold increase over the one based on isotropic Bragg diffraction from identical acoustic aperture. The only unfavorable finding encountered in this particular deflector was the requirement of a larger acoustic drive power,

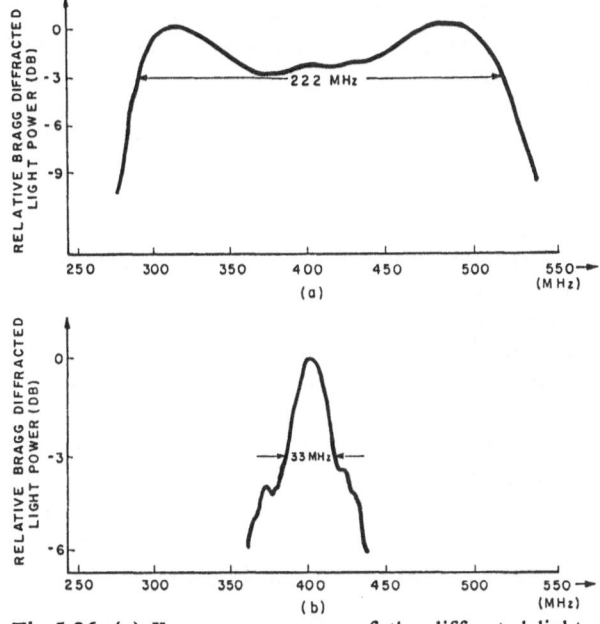

Fig.5.36. (a) Frequency response of the diffracted light power for optimized aniso-tropic Bragg diffraction (TM_0 mode diffracted into TE_0 mode). **(b)** Frequency response of the diffracted light power for isotropic Bragg diffraction (TE_0 mode diffracted into TE_0 mode)

namely about five times that for isotropic diffraction from the Z-propagation SAW of identical acoustic aperture. As shown in Fig.5.35b multiple parallel-finger periodic transducers of staggered center frequency but with identical acoustic propagation direction were used to obtain the large transducer bandwidth required in this experiment. A single transducer of parallel finger but varying finger periodicity (chirp transducer) with large acoustic aperture would be a better choice for this purpose.

It should be noted that in contrast to the bulk-wave anisotropic Bragg diffraction which can occur only in anisotropic materials, the guided-wave anisotropic Bragg diffraction can occur in either isotropic or anisotropic waveguide materials. Furthermore, as shown in Chap.3, through the choice of the waveguide thickness the effective refractive indices of the optical modes involved can be controlled. Thus the resulting waveguide dispersion or birefringence between modes can be made sufficiently small to enable f_{opt} to be placed at a conveniently low-frequency range. This is normally impossible for the bulk-wave case because the natural birefringence of the crystal is so large that f_{opt} is typically in the GHz frequency range.

5.4.3 Relative Merits of Single Transducer Versus Multiple Transducers

As discussed in the previous subsections, both single transducers and multiple transducers (or array transducers) have been utilized successfully to

171

construct wide-band planar AO Bragg cells. It is thus appropriate to comment on the relative merits of a single transducer versus multiple transducers. A Bragg cell that employs a single tilted-finger chirp transducer will require no matching circuits and takes less space than the deflector that employs multiple transducers. However, for the deflector that employs multiple transducers, the availability of multiple and independent RF inputs provides the flexibility for compensation and adjustments after the deflector has been fabricated. Such compensation and adjustments may be required because of the fabrication errors as well as variations in the physical properties of the AO material and in the quality of the waveguide and the transducers. By means of simple electrical attenuators and filter circuits such compensation and adjustments can be easily made. Thus it is reasonable to conclude that each wide-band deflector configuration has its relative merits, and selection of a particular one among the five just described would largely depend upon the flexibility and the performance required in a specific application.

5.5 Applications in Optical Communications, Signal Processing, and Computing

As described in Sects.5.4.1,2, it is now possible to realize high-performance guided-wave AO Bragg modulators and deflectors of GHz center frequency and large bandwidth in $LiNbO_3$ and GaAs waveguides. Together with the fabrication of miniature laser sources [5.80,81], waveguide lenses [5.82], and photodetector arrays [5.83], integration of all passive and active components on a single substrate or a small number of substrates has become a reality [5.84]. The resulting integrated AO device modules or subsystems possess a number of attractive features such as low electrical drive power, small size, light weight, less susceptibility to environmental effects, and potentially low cost due to batch fabrication. A number of such integrated AO device modules that are being developed are detailed in Chaps.7 and 8. Clearly these integrated AO device modules or subsystems will find a number of unique applications in wide-band multichannel optical communications, signal processing [5.1,85], and computing [5.85]. In this section some of the potential applications of the guided wave AO Bragg modulators and deflectors together with the performance figures that have been measured, are described.

5.5.1 Optical Communications

High-speed light-beam modulation, deflection and switching, optical wavelength multiplexing and demultiplexing (filtering), and optical frequency shifting are among the important functions in future wide-band multichannel integrated and fiber-optic communication systems [5.1]. With regard to modulation, the existing guided-wave AO Bragg modulators may provide a maximum modulation bandwidth of around 1 GHz using focused SAWs [5.86], as set by the acoustic transit time across the focal region (beam waist) of the light beam. This maximum AO modulation bandwidth can be realized by using the recently developed TIPE lenses [5.75] which can readily produce the very small beam waist required. In the meantime, however, as discussed in Chap.3, a variety of electro-optic (EO) modulators of multigigahertz bandwidth requiring very low drive power have been developed (Chap.3 and [5.87]). Consequently, the existing AO Bragg modulators are in general not as competitive as the EO modulators for applications requiring subnanosecond modulation speed.

The situation is quite different with regard to multiport light beam deflection and switching. The AO deflectors to be discussed herein are capable of deflecting and switching a guided-light beam into a large number of ports at moderate speed (microseconds and submicroseconds) and with low electrical drive power per port. In contrast, the existing EO switches [5.87,88] and deflectors [5.89-93], although capable of a faster switching speed (nanoseconds or subnanoseconds), can only provide either two ports or a relatively small number of ports per device. Furthermore, the electrical drive power per port required for the latter is considerable higher than their AO counterpart. Consequently, the AO deflectors are unique and superior for applications in multi-port light beam deflection and switching. Hence, only the technical aspects and the projected performance figures relating to this area of applications are discussed here.

With regard to optical wavelength filtering, both passive [5.94] and active guided-wave devices [5.95-99] have been studied. Some of these devices have been described in Chap.3. In this section only the active AO filter which utilizes anisotropic and noncollinear guided-wave Bragg diffraction [5.99] is described. Subsequently, a brief discussion on the specific device configurations for AO frequency shifting is given. One of these device configurations which utilizes a $LiNbO_3$ spherical waveguide will be described in detail in Chap.8. Finally, it is to be noted that guided-wave AO diffraction in $LiNbO_3$ was recently utilized to construct an integrated optical bistable device [5.100].

a) Digital Deflection and Switching

Like the bulk-wave AO deflectors, guided-wave AO deflectors can function under either digital (random access) or analog (sequential) mode of operation [5.101]. In the digital mode of operation, the frequency of the RF signal applied to the deflector is varied in discrete steps to switch the light beam. The number of resolvable beam spots N_R, the frequency step required for deflecting one resolvable spot position δf_R based on Rayleigh criteria, and the minimum switching time τ have been given in Sect.5.2.2.

Equation (5.12c) indicates that the instantaneous scan rate R, that is, the number of resolvable beam positions or ports scanned per second, is V_R/D. To appreciate the orders of magnitude involved, we consider two specific examples using the $LiNbO_3$ deflector described in Sects.5.4.1,2. We assume that a deflector bandwidth of 500 MHz is used in both examples. In the first example, a light beam aperture of 3.45 mm is used and we have $N_R = 500$, $\delta f_R = 1$ MHz, and $\tau = 10^{-6}$ s. The corresponding in-

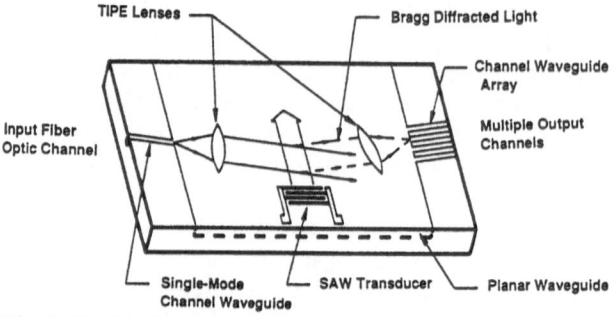

Fig.5.37. An integrated acousto-optic space-division demultiplexer switch module using channel-planar composite waveguide and TIPE lenses in LiNbO₃

stantaneous scan rate is 10^6 spots/s. In the second example, the same deflector bandwidth is used but a much smaller light beam aperture of 0.134 mm is assumed. We then have $N_R = 20$, $\delta f_R = 25$ MHz, and $\tau = 40 \cdot 10^{-9}$ s. The corresponding instantaneous scan rate is $25 \cdot 10^6$ spots/s. Thus, for a given deflector bandwidth the number of resolvable ports and the scan rate impose a conflicting requirement on the light beam aperture in a digital AO deflector. Nevertheless, the digital deflectors possess the unique capability for medium-speed random-access deflection and switching. One specific application involves a fiber-optic system with a large number of fan-out ports. A single-mode optical fiber which carries optical signals is coupled into a channel-planar composite waveguide together with a TIPE microlens [5.75]. The optical signals are then routed and coupled to any of an array of single-mode optical fibers that serve as the output terminals by means of a guided-wave digital AO deflector (Fig.5.37). A linear array of TIPE microlens channel waveguide combination may also be incorporated in the planar waveguide to facilitate efficient coupling to the output fiber array [5.75,85].

b) Analog Deflection and Switching

A guided-light beam may also be scanned and switched using an analog (linear F-M) mode of operation in which the frequency of the electrical drive signal varies linearly with time (linear F-M or chirp signal) [5.102]. Using this mode of operation it is possible to achieve a scan rate (number of scan spots per second) much greater than the scan rate which can be achieved using the digital mode of operation [5.101]. This higher scan rate results from the fact that the optical grating created by the linear FM acoustic wave acts as a moving Fresnel zone lens to an incident collimated light beam. Thus, the diffracted light is focused to a small size and simultaneously scanned at the acoustic wave velocity to result in a very high scan rate. The relevant parameters include the focal length, the focused spot size, number of scan spots, and scan rate. Note that some of these parameters and the notations involved are depicted in the inset of Fig. 5.38. A comparison between the analog deflector and the digital deflector in terms of the number of scan spots and scan rate is given in Table 5.5.

174

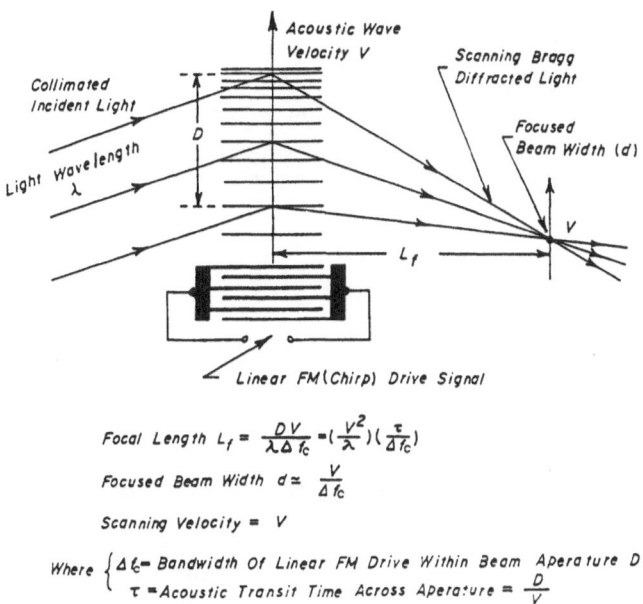

Focal Length $L_f = \frac{DV}{\lambda \Delta f_c} = (\frac{V^2}{\lambda})(\frac{\tau}{\Delta f_c})$

Focused Beam Width $d \approx \frac{V}{\Delta f_c}$

Scanning Velocity $= V$

Where $\begin{cases} \Delta f_c = \text{Bandwidth Of Linear FM Drive Within Beam Aperature D} \\ \tau = \text{Acoustic Transit Time Across Aperature} = \frac{D}{V} \end{cases}$

Fig.5.38. Guided-wave acousto-optic light beam deflection using analog mode of operation

Table 5.5. Performance comparison between digital and analog acousto-optic light-beam scanning (scanner bandwidth: Δf, acoustic transit time: τ) [5.1]

Mode of operation	Digital (using discrete frequency steps)	Analog (using linear FM of time aperture T)
Number of re-solvable spots, N_R	$\tau \cdot \Delta f$	$(T-\tau)/T \cdot \tau \Delta f$
Inst. scanning rate, R [spots/s]	$1/\tau$	$(\tau/T) \cdot \Delta f$
Numerical example $\tau = 1\mu s$, $T = 2\tau$ $\Delta f = 500\text{MHz}$	$N_R = 500$ $R = 1 \cdot 10^6$	$N_R = 250$ $R = 250 \cdot 10^6$

For the example given in this table, the scan rate for the analog deflector is larger than that of the digital deflector by two orders of magnitude, namely $250 \cdot 10^6$ spots/s versus $1 \cdot 10^6$ spots/s.

A typical arrangement for the analog deflection experiment using one of the wide-band deflectors described in Sect.5.4 is shown in Fig.5.39 [5.1, 101]. The deflected light spots thus produced were evolved (merged) into a straight line. A series of photomasks consisting of multiple slits of various apertures and spacings were inserted in front of the photomultiplier in order to accurately measure the scan rate of the light beam. Typical waveforms from the output of the photomultiplier, as displayed on a

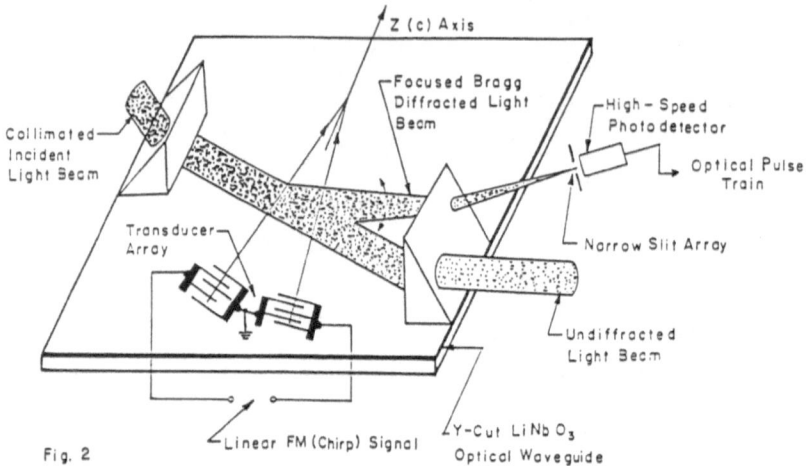

Fig. 2

Fig.5.39. Experimental arrangement for guided-wave analog acousto-optic light beam scanning

wide-band oscilloscope, are shown in Fig.5.40 for a photomask of 6 μm aperture and 508 μm periodicity. The center frequency and the total bandwidth of the chirp pulse employed are 430 and 180 MHz, respectively. The time for each line of scan is identical to the total sweep time for the chirp pulse, namely, 2.4 μs. It is seen that the periodicity of the optical pulse train is about 150 ns, in agreement with the fact that the focused light spot sweeps at the SAW velocity of $3.48 \cdot 10^5$ cm/s in a Y-cut Z-propagation LiNbO$_3$ substrate. Also, since the bandwidth of the chirp pulse within the light beam aperture is one half of the total bandwidth, namely, 90 Mhz the calculated focused beam width of the deflected light based on the formula in the inset of Fig.5.38 is 38.6 μm. Thus, the corresponding scan time per spot is $38.6 \cdot 10^{-4}$ cm divided by $3.48 \cdot 10^5$ cm/s, namely, 11.1 ns. The light beam aperture was 4 mm, which corresponds to 1.2 μs acoustic transit time. Thus the light beam was scanned at a speed 108 times faster than that obtainable using the digital mode of operation. Again, a considerably higher scan rate and a larger number of beam positions can be obtained with this wide-band deflector when a chirp generator of higher center frequency and wider bandwidth is used.

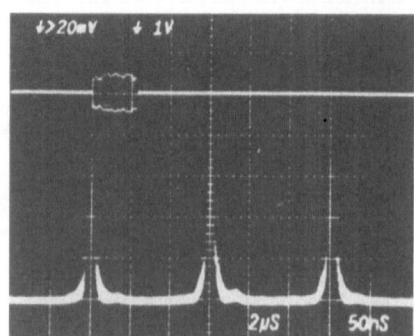

Fig.5.40. Very high-speed guided-light beam scanning. RF chirp waveform (*upper trace*): 2 μs per division, scanned light beam waveform (*lower trace*): 50 ns per division (periodicity of photomask: 508 μm, width of transparent region: 6μm)

176

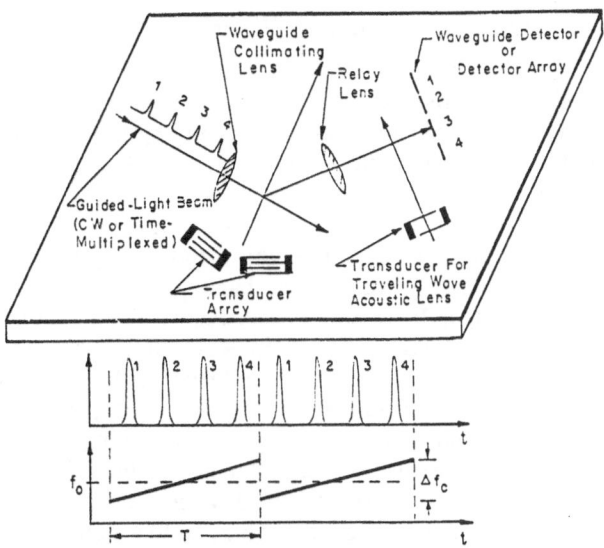

Fig:5.41. Integrated AO multiport switch and time-division demultiplexer

Some of the potential applications for the analog deflector described above include: (a) very high data rate multiport switching (Fig.5.41); (b) very high data rate optical writing and reading in applications such as printing and facsimile; (c) optical pulse compression of radar chirp signal [5.101]; (d) high-speed parallel to serial (spatial to temporal) readout for the RF spectrum of integrated optic spectrum analyzers [5.102]; and (e) time-demultiplexing of wide-band multichannel optical pulse trains [5.103]. The first two are self explanatory; the third and fourth are discussed in Sect.5.5.2. The fifth application is described here.

Referring to Fig.5.41, a guided-light beam consisting of time-multiplexed pulse trains with individual data channels can be time-demultiplexed and routed to a waveguide detector array using an analog deflector. This figure depicts a four-channel composite pulse train being routed to four detector elements. The traveling-wave acoustic lens [5.104] located in front of the detector array may be used to enhance the number of resolvable spots or ports. The resulting optical demultiplexer should be capable of processing a composite data stream of Gbit per second, and thus be useful in a single-mode multichannel integrated fiber-optic communication system which employs the optical pulse train from a mode-locked cw GaAlAs laser diode or a dye laser [5.105] as the carrier. In this connection it should be mentioned that standing-wave AO Bragg modulators using either bulk acoustic wave or surface acoustic wave have also been utilized to perform optical multiplexing and demultiplexing [5.106, 107]. In the former, multiplexing and demultiplexing of up to six channels per modulator at multigigabit per second has been accomplished [5.106].

c) Electronically Tunable Optical Wavelength Filtering

The working principle of bulk-wave tunable AO filters [5.108] is based on the characteristics of anisotropic AO Bragg diffraction. It is well known that in an optically anisotropic medium of certain crystal orientations the polarization of the Bragg-diffracted light is orthogonal to that of the incident light. Also, as a consequence of the wave-vector matching requirement for a given acoustic frequency, only a small range of optical wavelengths will be cumulatively diffracted into the orthogonal polarization. Thus, if the acoustic frequency is varied, the band of optical wavelengths of the diffracted (filtered) light is tuned accordingly. The possibilities of implementing such tunable optical filters in waveguide structures were explored first using guided-wave isotropic collinear [5.96,97] and anisotropic collinear [5.98] AO interactions. The above working principle for bulk-wave AO filters was found to be also applicable to their guided-wave counterparts. One additional feature of the latter is that the inherent waveguide dispersion facilitates the orthogonal polarization in the diffracted light through TE_0-TM_0 mode conversion. Thus, unlike bulk-wave AO filters, guided-wave AO filters can be constructed using isotropic waveguide substrates as well.

Guided-wave optical filtering that utilizes anisotropic noncollinear AO Bragg interaction in Y-cut $LiNbO_3$ waveguides was studied subsequently [5.99]. In the experimental study a guided-light beam from either a He-Ne laser at 6328 Å or an argon laser at around 5000 Å, propagating at an angle centering aroud 70° from the Z (c) axis of a titanium-indiffused $LiNbO_3$ waveguide, was Bragg diffracted by the SAW with 500 MHz center frequency and propagating at 16° from the X axis. An optical passband of 16 Å and an angular aperture of 9° were measured for the case in which the undiffracted and diffracted light propagated in the TE_0 mode and TM_0 mode, respectively. Furthermore, the filtered optical wavelength may be tuned from 8300 to 5000 Å by varying the acoustic frequency from 350 to 670 MHz. Thus this guided-wave noncollinear AO filter has been shown to be capable of simultaneously providing a combination of desirable performance characteristics: large angular aperture, high spectral resolution, distinct angular separation between the filtered and the unfiltered light, and moderate spectral tuning range. One other possible application of the guided-wave AO filters lies in the high-speed electronic tuning of the output of semiconductor lasers.

d) Wide-Band Optical Frequency Shifting

Light sources with electronically tunable frequency offset are needed as local oscillators in heterodyne integrated- and fiber-optic communication and fiber-optic sensor systems [5.109,110]. A number of schemes for producing such frequency-shifted light sources have been reported [5.111-121]. Among the schemes involving integrated-optic devices, the one utilizing spatially weighted coupling between two waveguide modes [5.111] was recently realized in a strip waveguide [5.113]. Earlier, the scheme that utilizes EO Bragg array in a $LiNbO_3$ planar waveguide [5.118] was also reported. This EO frequency shifting scheme has the unique features of baseband operation and fixed input and output light angles irrespective of the amount of frequency shift. It should be noted that the required beam width of the incident light and the resulting beam width of the frequency-shifted light are necessarily much larger than the core diameter of a single-mode optical fiber. As a result, separate waveguide lenses must be formed in the device substrate to facilitate efficient and robust coupling between the optical fibers and the planar waveguide at both input and output ports so that a compact integrated-optic module can be realized. Subsequently, a similar scheme was extended to channel waveguides [5.119].

An alternate integrated-optic scheme would be to utilize AO Bragg diffraction from a traveling SAW as the frequency of the diffracted light is shifted from that of the incident light by the acoustic frequency. Three distinct device configurations, namely those using the planar waveguide [5.1], the channel waveguide [5.117], and the spherical waveguide [5.121] have been utilized to facilitate this AO frequency-shifting scheme. These studies have shown that single-sideband supressed-carrier frequency shifting using AO Bragg diffraction from the traveling SAW is a viable approach and the resulting integrated modules can be readily and rigidly coupled with the single-mode optical fibers to provide simultaneously the unshifted light and the single- as well as multiple-frequency shifted light sources that are spatially separated. In particular, a compact integrated AO frequency shifter module that utilizes Bragg diffractions in cascade from two tilted and counterpropagating SAWs may be realized in a common LiNbO$_3$ substrate $0.2 \times 10.0 \times 1.5$ cm^3 in size [5.120] to provide electronically tunable wideband frequency shift. A detailed description of these integrated-optic frequency shifter modules will be given in Chap.8.

5.5.2 Optical Signal Processing

The block diagram of a typical optical signal processor [5.122] is shown in Fig.5.42 [5.123]. It consists of an electrical-to-optical transducer (input transducer) that converts a time window of a wide-band electrical signal into a spatial optical display, an optical system which acts on this display to generate the Fourier transform of the signal or the correlation/convolution of the signal with a reference signal, and an optical-to-electrical transducer (output transducer) which reads out this optical Fourier transform or correlation/convolution as an electrical analog signal or digital bit stream. Optical signal processors can process wide-band signals at very high speed, and offer a significant advantage in hardware performance (equivalent bits per second per dollar) over the digital electronic processors in certain areas of application [5.124].

One of the most important classes of optical signal processors is based on the use of coherent AO interactions. Note that in the resulting AO signal processors the input transducer, optical system, and output transducer are provided, respectively, by the acoustic transducer/delay line, AO Bragg cell, and photodetector. A number of bulk-wave AO signal

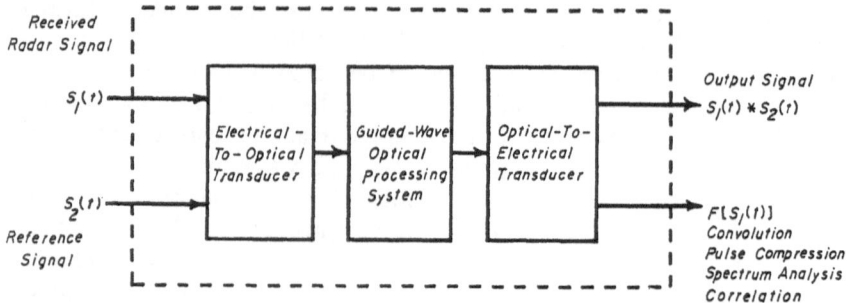

Fig.5.42. Block diagram of a basic guided-wave optical signal processor

processors have been studied and demonstrated [5.125]. Some of these bulk-wave AO signal processors are: spatial-integrating correlators, matched filters for chirp radar, convolvers, time-integrating correlators, snapshot PROM, and spectrum analyzers. Most of these bulk-wave AO processors for one-dimensional signal processing may be implemented in a guided-wave format [5.1] using the wide-band AO Bragg modulators and deflectors described in Sect.5.4.

As indicated earlier, these guided-wave versions possess a number of attractive features. However, it should be noted that an extension to two-dimensional signal processing is more difficult to implement in the guided-wave format than in the bulk-wave format. A number of these guided-wave AO signal processors are now described.

a) Spectral Analysis of Very Wide-Band RF Signals

As mentioned in the Introduction, development and implementation of an rf spectrum analyzer using hybrid integrated-optic technique as depicted in Fig.1.1 has already become a world-wide activity [5.126]. A detailed treatment of the LiNbO$_3$-based Integrated-Optic (RF) Spectrum Analyzer (IOSA) will be given in Chap.7. An update of the realization of the IOSAs on GaAs and other nonpiezoelectric substrates will be given in Chap.8. Only the principle of operation and some of the major parameters as well as a high-speed AO readout scheme for the IOSA are given here.

Some of the major parameters of IOSA. A schematic diagram of a monolithic integrated-optic RF spectrum analyzer is depicted in Fig.5.43. When a spectrum of RF signals is applied to the transducer, each spectral component generates a SAW which deflects the incident light beam in a corresponding direction. As shown in Sect.5.1, the deflection angle and the intensity of the Bragg-diffracted light are proportional to the frequency and power of the RF or the acoustic signal, respectively. Figure 5.44 shows the deflected light spots resulting from two independent RF signals as their frequency separation and relative RF power are varied. Thus, by measuring the linear positions and the intensities of the deflected-light spots at the focal plane of the transform lens, the frequency and the power spectral density distribution of the RF signal of interest may be determined. The number of frequency channels, the frequency

180

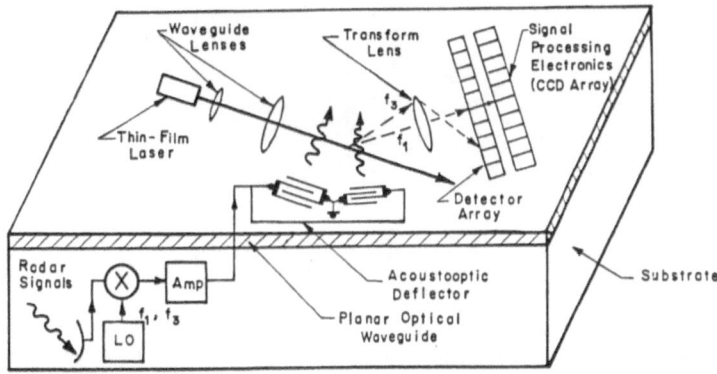

Fig.5.43. Integrated acousto-optic RF spectrum analyzer - an example of integrated optic circuitry

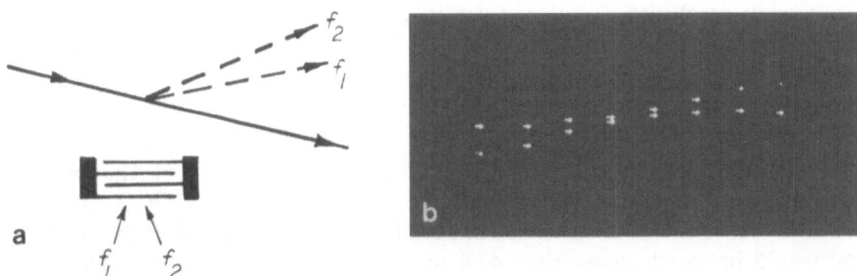

Fig.5.44. (a) Spectral analysis of RF signals using planar acousto-optic Bragg diffraction. (b) Deflected light spots from two spectral components of varying frequency separation and power level

resolution, and the dynamic range are among the major parameters in AO RF spectral analysis. The number of resolvable frequency channels N_R and the corresponding frequency resolution δf_R, based on Rayleigh criteria, are given by (5.12 a,b).

As described in Sect.5.4.2, 450 resolvable frequency channels and a frequency resolution of approximately 0.8 MHz were measured using a deflector of 358 MHz bandwidth and a uniform light beam aperture of 4.5 mm [5.14]. For the 680 MHz bandwidth deflector described in Sect. 5.4.2 the measured frequency resolution was 0.6 MHz for a truncated-Gaussian light beam of 6 mm aperture. This resolution was determined by measuring the half-power width of the deflected light spot [5.1,67]. Thus, based on the measured resolution this deflector, when used as a spectrum analyzer, would provide a maximum of 1130 resolvable frequency channels. Note that in practice the maximum number of resolvable frequency channels are further limited by both the focal spot size of the transform lens and the spacing of the photodetector array. The two deflected light spots (at the far field) resulting from two independent RF signals, with the frequency separation δf as a parameter, are shown in Fig.5.45. Clearly this deflector has provided deflected and undeflected light spots of very fine quality.

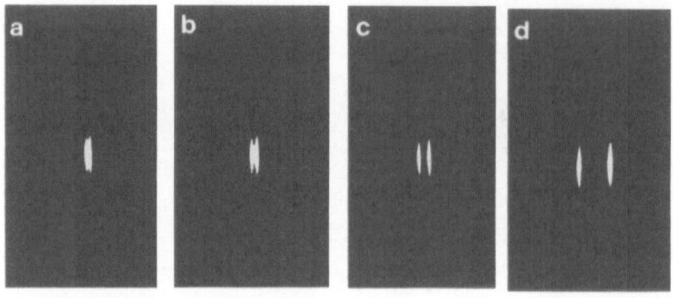

Fig.5.45. Deflected light spots as a function of frequency separation between two RF drive signals (aperture of incident light: 6 mm): (a) $\delta f = 0.7$ MHz, (b) $\delta f = 1.0$ MHz, (c) $\delta f = 2.0$ MHz, (d) $\delta f = 6.0$ MHz

In addition to the number of resolvable frequency channels and the frequency resolution, intermodulation and cross modulation between different frequency components resulting from multiple AO diffraction were measured using two independent RF signals of varying frequency separation [5.58]. The measured data show that even for a worst case with a combined diffraction as much as 43%, the intensity of the strongest intermodulation was 38 db down from those of the intensity of the primary diffracted spots which result from the two fundamental RF frequencies. For the practical cases in spectrum analysis, the total diffraction efficiency would be much lower than 43% and the corresponding inter- and crossmodulation would accordingly be even lower. Thus the dynamic range of guided-wave AO spectrum analyzers is likely to be limited by background noise (due to light scattering in the waveguide) rather than the inter- and crossmodulations due to multiple AO diffraction.

A high-speed acousto-optic readout scheme for IOSA. We now return to the fourth application of analog deflectors listed in Sect.5.5.1, namely high-speed AO readout for the RF spectra of the IOSA. In a conventional readout scheme, a waveguide lens is used to focus each of the deflected-light beams and a photodetector array used to detect the focused deflected light spots (Chap.7 and [5.126]). A CCD array is then used to perform the spatial-to-temporal conversion so that the RF spectrum is read out serially in time. The throughput rate of this scheme is relatively low because of the speed limitations of the existing CCD array. Also, the projected cost for a CCD photodetector array of very high speed and large number of elements, say 1000 elements or more, is understandably high. In the scheme (Fig.5.46) discussed here, neither a photodetector array nor a CCD array is employed. Instead a very-high-scan-rate analog deflector described in Sect. 5.5.1 is employed to convert the deflected light beams into fast-scanning focused-light spots. A single high-speed photodetector incorporating a narrow slit and located in the focal plane is then used to detect and convert the scanning focused light spots into an electrical spectrum which is already serial in time [5.102]. In contrast to the conventional scheme, this alternate scheme is inherently simple and is capable of

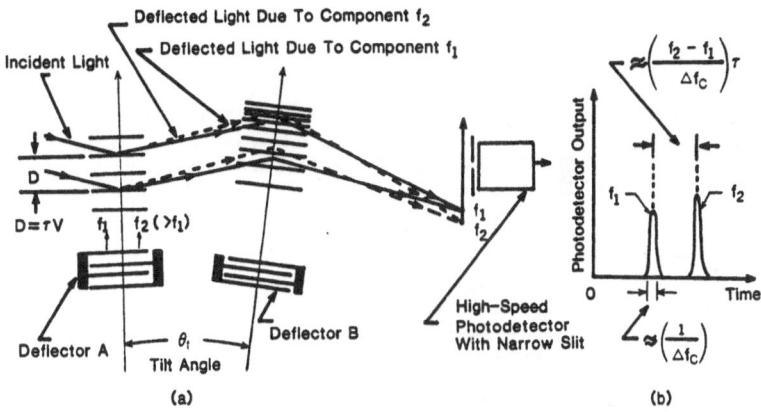

Fig.5.46. Very high-speed spatial-to-temporal conversion for integrated-optic RF spectral analysis

displaying the RF spectrum at a much higher and controllable throughput rate. A high throughput rate may be desirable for some applications which require greater temporal resolution of the received signals or to reduce reaction time. For example, a total cycle time approaching 1 μs is of interest in this situation, and this is compatible with the total scan time for this AO readout technique. Some preliminary experiments using SAW (centered at 300MHz) in a Y-cut LiNbO$_3$ waveguide and laser light at 6328 Å have been carried out [5.102]. A frequency resolution of 3 MHz and a bandwidth of 180 MHz have been demonstrated.

As stated previously, development and realization of the integrated-optic RF spectrum analyzers has become the focus of an international effort. Such integrated-optic RF spectrum analyzers, when fully developed, are expected to possess two major advantages: (i) increased performance and reduced cost over both currently employed technology and competing technologies, and (ii) reduced size and increased compactness. At present the development effort are focused on the hybrid module in LiNbO$_3$, as shown in Fig.1.1 of Chapter 1, in which the laser source and the photodetector array are edge-coupled to the input and output end faces of the LiNbO$_3$ substrate, respectively. An ultimate module will be realized in GaAs in which both the laser source and the photodetector array are incorporated in the same GaAs substrate [5.40].

b) Convolution of Wide-Band RF Signals

The capability of performing real-time convolution between two wideband RF signals is an important requirement in a radar system because this technique provides a great improvement in the signal-to-noise ratio (SNR). This improvement is equal to the time-bandwidth product of the convolver.

The basic AO interaction configuration for convolution is depicted in Fig.5.47a. One of the specific configurations employed for convolution experiment is shown in Fig.5.47b [5.128, 129]. Two end-to-end identical

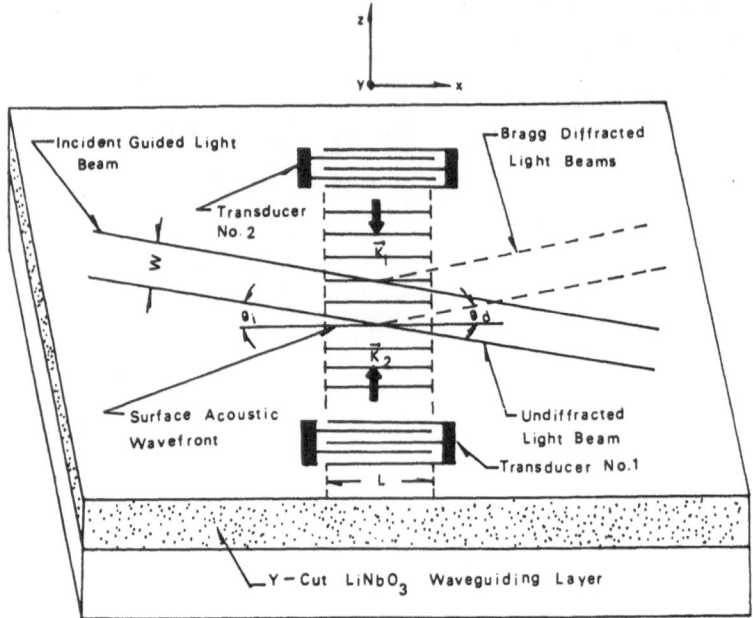

Fig.5.47a. Basic interaction configuration for convolvers using guided-wave acousto-optic Bragg diffraction

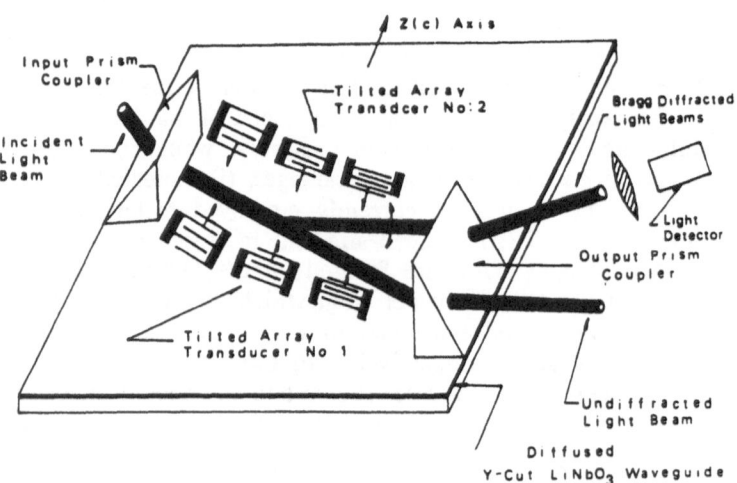

Fig.5.47b. Convolution of wideband RF signals using guided-wave acousto-optic Bragg diffraction

tilted-SAW transducers with center frequencies of 163, 194, and 230 MHz and separated at a distance of 1.3 cm were deposited on the top of a Ti-diffused Y-cut $LiNbO_3$ waveguide. One pulse-modulated RF signal (the reference signal) was applied to one transducer to generate a SAW and the other pulse-modulated RF signal (the radar signal to be processed) was applied to the other transducer to generate a second SAW propagating in the opposite direction. The two diffracted-light beams overlap and are collected by a lens and then mixed in a PIN photodiode detector. The component in the electrical output of the photodetector which corresponds to the convolution of the two RF signals is further processed by means of a heterodyne receiver and then displayed in a wide-band oscilloscope.

The measured performance figures [5.128, 129] include a time-bandwidth product of 305 with a time aperture of 2.85 μs and a bandwidth of 107 MHz, a dynamic range of approximately 50 db at a total RF power of 310 mW for maximum convolution output, and a frequency resolution of 1 MHz (defined at zero convolution output). A considerably larger time-bandwidth product can be achieved by increasing the center frequency of the SAW and/or the aperture of the light beam. For example, using the 1.5 cm light beam aperture mentioned and the 680 MHz bandwidth deflector described in Sect.5.4, a time-bandwidth product of 2950 may be obtained. The 50 db dynamic range of the convolver was obtained at an input light power (before being coupled into the waveguide) of 27 mW. The dynamic range can be greatly increased by using a higher light power and/or a more sensitive photodetector.

In summary, this guided-wave AO convolution experiment has demonstrated that multiple, tilted SAWs can be employed to obtain very good performance figures. Multiple, tilted SAWs have also been employed in a convolver without optical waveguide to achieve even larger time-bandwidth product [5.130].

c) Compression of RF Chirp Pulse

One of the commonly used RF pulses which can simultaneously provide a large range and a high range resolution in radar systems is the so-called linear FM or chirp pulse, in which the carrier frequency varies linearly within the pulse. By means of a signal-processing technique referred to as "pulse compression" at the receiver, the feeble radar echoes may be made sharp and strong. The important characteristics of the resulting "pulse compressor" are the width and the intensity of the compressed pulse. The pulse width Δt after compression is approximately equal to 1/B, B being the RF bandwidth of the chirp pulse. The pulse intensity is enhanced over that before pulse compression by a factor equal to TB, commonly called the compression ratio or the time-bandwidth product, where T is the original duration of the chirp pulse.

Acousto-optic pulse compressors of various configurations using bulk-wave AO modulators have been examined in recent years [5.125]. These AO pulse compressors have been shown to be capable of processing a wide variety of signal codes and waveforms. Such AO pulse compressors can also be implemented in guided-wave format. In one of the experimental guided-wave AO pulse compressor configurations, the chirp pulse received was used to excite a tilted-SAW transducer (Fig.5.48). The optical grating created by the SAW in the optical waveguide acted as a moving Fresnel zone lens. Thus the Bragg-diffracted light beam resulting from the incident collimated light beam was brought to a focus, and the

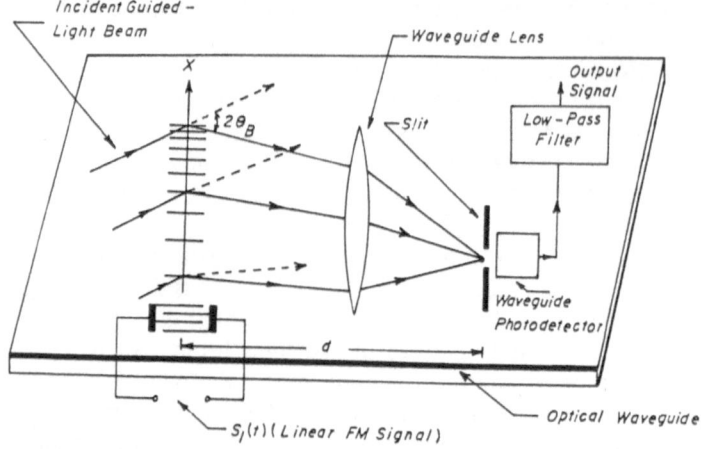

Fig.5.48. Guided-wave acousto-optic pulse compressor using one Bragg cell

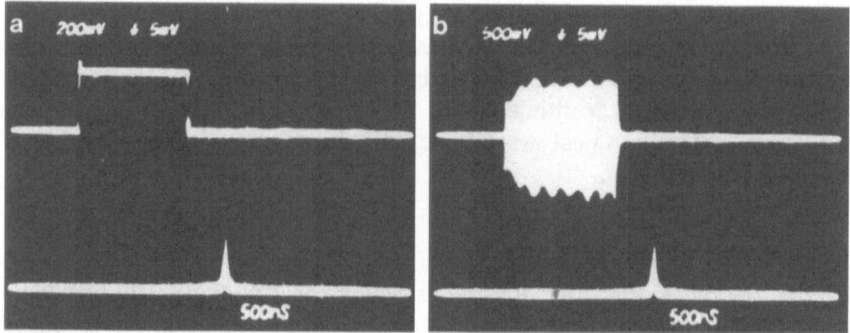

Fig.5.49. Optical pulse compression using guided-wave acousto-optic Bragg diffraction and RF chirp waveform [modulating pulse and RF chirp pulse (*upper traces*): 1.5 μs, compressed optical pulse (*lower traces*): 7.5 ns, compression ratio: 200]

focal spot swept in the focal plane at the acoustic wave velocity. Note that the above phenomenon has been discussed in Sect.5.5.1 in connection with analog light-beam deflection. A high-speed photodetector with a sufficiently narrow aperture would register an intense, compressed optical pulse when the focal spot sweeps across the aperture. Since the light energy from the entire time aperture T produces a current pulse of length Δt, the resulting compression ratio is $T/\Delta t$. A linear FM pulse of 430 MHz center frequency and 160 MHz bandwidth was applied to the tilted transducer referred to earlier. A collimated guide light beam from a He-Ne laser at 6328 Å was then Bragg-diffracted from the SAWs. Figure 5.49 shows the waveforms of the RF chirp pulse and the compressed optical pulse, indicating a compression ratio of 200 [5.101]. The best performance figures that have been measured are a compressed optical pulse of 5 ns width and a compression ratio of 300.

d) Correlation of Wide-Band RF Signals

The ability to perform real-time correlation of two analog signals, i.e. a received signal $S_1(t)+n(t)$, $n(t)$ being the additive noise, and a reference $S_2(t)$ supplied by the receiver, is one of the key requirements in radar and communication systems. As in the pulse compression system just discussed, the correlation operation enables achievement of a very narrow output pulse, even when the original signal is very long in time. The pulse width Δt is given by $1/B$, where B is the bandwidth of the signal being correlated. The correlation operation also provides an increase in the SNR over the original signal received, called correlation gain or processing gain. This gain is equal to the time-bandwidth product of the correlator. Thus two of the most important parameters of an AO correlator are the bandwidth and the time-bandwidth product discussed in Sect.5.2. The other two important parameters are the so-called range window and the time window. Range window is the allowable time error between the received and the reference signals for a correlation peak to be produced, and time window is the total duration of the correlated signal.

Like the bulk-wave AO correlators the guided-wave AO correlators can be classified into two major types: the spatial-integrating correlators and the time-integrating correlators [5.125]. A brief description of the first type and a more detailed description of the second type now follow.

Acousto-optic spatial-integrating correlator. Acousto-optic spatial-integrating correlators (AOSIC) [5.125] perform correlation by integrating the light diffracted by all parts of the signal(s) which are simultaneously present in the light beam aperture of the Bragg cell. A possible guided-wave version is shown in Fig.5.50. The received signal plus noise, $S_1(t)+n(t)$, to be correlated is fed into the first SAW transducer, producing a spatial display of a given time window of the received signal. This spatial display is multiplied by that produced by a time-reversed reference signal, $S_2(-t)$, which is applied to the second SAW transducer. The dif-

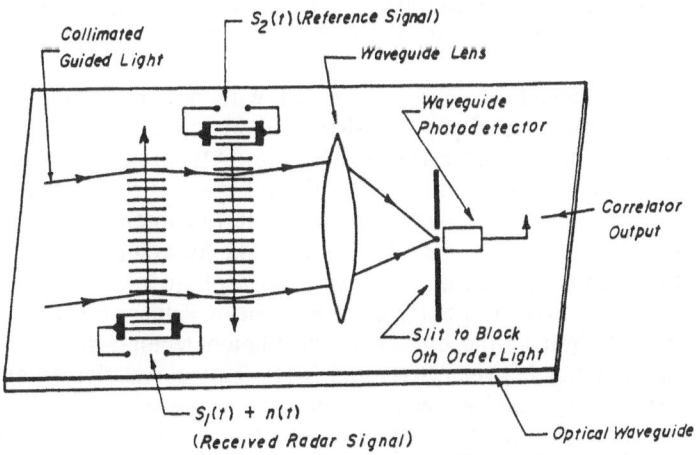

Fig.5.50. Guided-wave acousto-optic spatial-integrating correlator

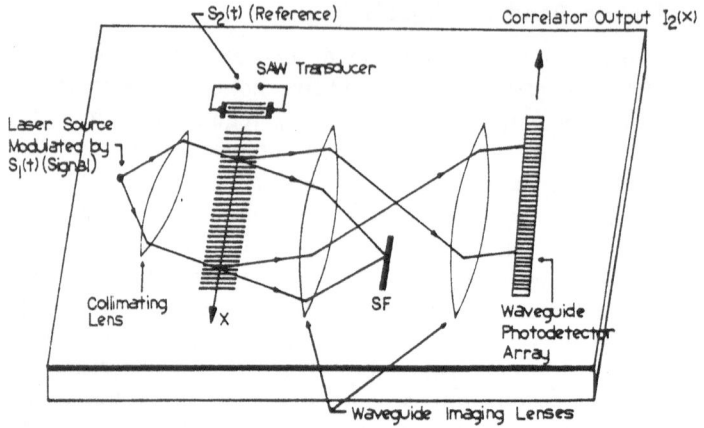

Fig.5.51. Guided-wave acousto-optic time-integrating correlator

fracted light is then spatially integrated onto a waveguide photodetector and results in the correlation signal. The advantage of using a second SAW as the reference signal is that a large variety of signal waveforms can be correlated by varying the reference signal appropriately. Note that this type of correlator has a large-range window but a limited time window and thus a limited time-bandwidth product. Using a uniform, guided-light beam aperture of 1.5 cm, the maximum time window achievable is 4.35 μs. Thus a time-bandwidth product of 4350 can be expected if a Bragg cell of 1 Ghz bandwidth is employed.

Acousto-optic time-integrating correlator. Acousto-Optic Time-Integrating Correlators (AOTIC) [5.131, 132] perform correlation by using a closely spaced photodetector array to integrate in time for each point within the Bragg cell. A fully integrated guided-wave version that utilizes isotropic Bragg diffraction is depicted in Fig.5.51. While such a monolithic integrated AO correlator is yet to be implemented, a hybrid correlator module that utilizes anisotropic Bragg diffraction has already been realized. A more detailed description of this hybrid module will be given in Chap.8.

The signal to be correlated, $S_1(t)$, is added with a bias voltage V_1 and used to modulate the intensity of a coherent light source. The modulated light is then collimated and diffracted by the SAW produced by an RF carrier which is amplitude-modulated by the reference signal $S_2(t)$. A proper choice of bias voltage V_2 would ensure that the intensity of the diffracted light is linearly proportional to $S_2(t)$. The diffracted light is then collected by a lens, filtered, and imaged onto a photodetector array. It can be shown that if $S_1(t)$ and $S_2(t)$ have zero mean values, the intensity of the diffracted light at the output of the photodetector array contains the correlation signal between $S_1(t)$ and $S_2(t)$. This correlation signal is displayed in space but can be read out in time using a CCD array. Since the correlation is performed in time rather than in space, this type of correlator is potentially capable of a very long processing time which is

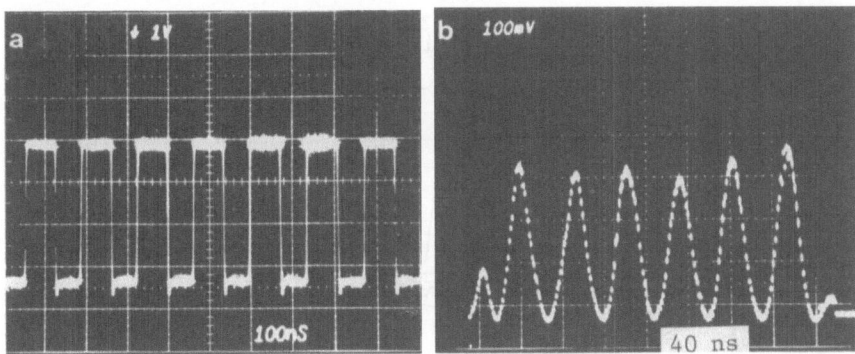

Fig.5.52a,b. Input and output waveforms of the guided-wave acousto-optic time-integrating correlator: (a) rectangular input pulse train to both modulator and deflector, (b) triangular autocorrelation output (Note different time scales)

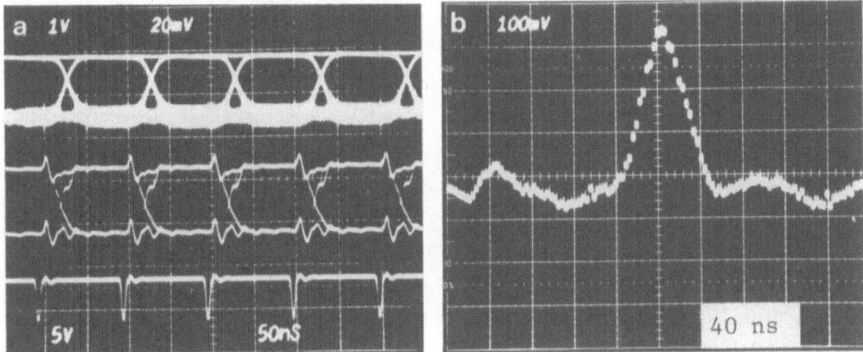

Fig.5.53a,b. Autocorrelation output of pseudorandom code with a chip rate of one per 100 ns: (a) top trace: output light from AO modulator, middle trace: input code to be correlated, bottom trace: clock pulses of pseudorandom code generator; (b) correlation output (Note different time scales)

determined by the time constant of the photodetector array. Furthermore, since both the coherent light source and the AO Bragg cell can be modulated at GHz bandwidth, this type of correlator is also potentially capable of very large bandwidths. Thus a very large time-bandwidth product is expected. The experiments that employed the SAW with the center frequency at 125 MHz in a Y-cut LiNbO$_3$ waveguide and a He-Ne laser (6328 Å), modulated by a bulk-wave AO Bragg cell and coupled into and out of the waveguide through a pair of rutile prisms, have demonstrated a processing time of 7 ms and a time-bandwidth product of $1.5 \cdot 10^5$ [5.133]. Figs.5.52,53 show the autocorrelation outputs of rectangular pulse train and pseudorandom code, respectively.

The GaAlAs waveguide structure is a potential substrate for a monolithic AOTIC module. In this case the light source takes a variety of convenient forms such as a distributed-feedback [5.134] or a Bragg-reflector laser [5.135]. Also, as shown in Chapter 8, both wideband AO Bragg cells [5.40] and waveguide lenses [5.136] have been realized on the

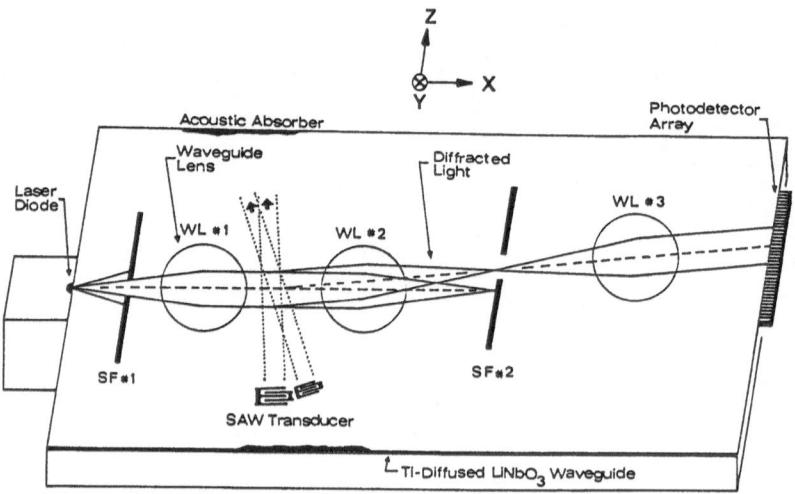

Fig.5.54. An acousto-optic time-integrating correlator using hybrid optical waveguide structure

GaAs waveguide most recently. Thus the prospects for ultimate realization of this monolithic module has been greatly enhanced. At present, thermally oxidized silicon, lithium niobate, and gallium arsenide can all be used for implementation of a hybrid AOTIC module [5.137]. In the first, the light source such as a GaAlAs DH laser diode is butt-coupled [5.138] to one edge of the Si substrate. In the second, both the GaAlAs DH laser diode and the photodetector array are butt-coupled to the edges of the $LiNbO_3$ substrate (Fig.5.54). Finally, in the third a photodetector array is butt-coupled to the output edge of the GaAs substrate. A model design has shown that a Y-cut $LiNbO_3$ plate having a substrate area of 2.5·8 cm² is sufficient to accommodate all passive and active components. In view of the fact that high-quality optical waveguides [5.3], planar waveguide lenses [5.75,82], and wide-band high-efficiency Bragg deflectors/modulators have been successfully fabricated in the $LiNbO_3$ substrate, and the continued advancement in the diode lasers, the photodetector arrays as well as their butt-coupling to the $LiNbO_3$ substrate, the hybrid structure as illustrated in Fig.5.54 has become the most common approach for the present. Based on the aforementioned preliminary results [5.133], this hybrid AOTIC should be capable of providing much better performance figures. As indicated previously, $LiNbO_3$ hybrid AOTIC that utilizes anisotropic Bragg diffraction [5.139] will be presented in Chap.8. The correlator that utilized integrated optical spatial light modulators of both AO and EO types has also been explored [5.140,141].

5.5.3 Optical Computing

Realization of optical computing functions in a waveguide substrate has long been considered one of the potential applications of integrated optics. Recently, the prospects for this realization are significantly advanced through the progress in guided-wave micro-optic components of both active and passive types in the LiNbO$_3$ substrate [5.85]. One of these components are the single-mode microlenses and microlens arrays fabricated by the Titanium-Indiffusion Proton-Exchange (TIPE) technique [5.75] (Chap.8). These microlenses and lens arrays have demonstrated a combination of desirable characteristics. A TIPE microlens array and a large-aperture integrating lens together with a channel waveguide array, a planar waveguide, and a SAW transducer were integrated on a LiNbO$_3$ substrate $0.2 \times 1.0 \times 2.0$ cm^3 in size [5.85, 141]. The resulting integrated AO modulator module has demonstrated efficient Bragg diffraction at 0.6328 μm from the SAW centering at 500 MHz.

The integrated AO modulator module should find applications in optical computing such as optical systolic array processing, as it can readily perform the two required basic operations "multiplication" and "addition". Multiplication is facilitated by AO Bragg diffraction, and addition by the integrating lens. Thus, by pulsing the analog or digital data sequences separately into the multiple input light beams (through the channel-waveguide array) and the SAW high-speed programmable filtering as well as matrix-vector and matrix-matrix multiplications [5.85, 141] can be performed. A detailed description of the integrated AO Bragg modulator module and the results obtained with optical computing experiments will be presented in Chap.8.

5.6 Summary

Numerous advancements have been made in planar guided-wave acousto-optics in recent years. These advancements include analytical treatment of complex interaction geometry, design and fabrication of wideband Bragg modulators and deflectors (Bragg cells), and demonstration of a number of novel applications. The continuing progress in the fabriction and performance of other components including optical waveguides, waveguide lenses, diode laser sources and photodetector arrays, and their integration has significantly advanced the prospects for realization of integrated acoustooptic device modules and circuits. The most notable examples of such modules realized earlier are the hybrid integrated optic RF spectrum analyzers, and correlators in LiNbO$_3$ substrate. The most recent advancements on realization of integrated acoustooptic device modules have been made in LiNbO$_3$ channel-planar and spherical waveguides and in GaAs planar waveguides. The most notable are the TIPE microlens-based multichannel integrated Bragg modulator modules in LiNbO$_3$ channel-planar composite waveguides. These integrated acoustooptic modules were shown to take up a small substrate dimension (up to 2.0cm) along the optical path and are also inherently of high modularity and vesatility. Finally, some of the applications of guided-wave acoustooptic devices for wideband multichannel integrated- and fiber-optic communication, signal processing, and computing that were demonstrated are listed in Table 5.6.

Table 5.6. Applications of guided-wave acoustooptic modulators and deflectors

---

I. *Communications*
   Light beam modulation and deflection
   Multiport switching
   Time- and wavelength-division multiplexing/demultiplexing
   Tunable optical wavelength filtering
   Optical frequency shifting and heterodyne detection

II. *Radio frequency signal processings*
   Spectral analysis or Fourier transform
   Pulse compression
   Convolution
   Time- and space-integrating correlations
   Adaptive filtering
   Ambiguity function

III. *Computing*
   Matrix-vector multiplication
   Matrix-matrix multiplication
   Programmable digital correlation/filtering
   Optical bistability

Acknowledgement. The author wishes to express his deep gratitude to Dr. Robert Adler of Zenith Corporation for his thorough review of the chapter and for many valuable suggestions.

References

5.1 C.S. Tsai: Guided-wave acousto-optic Bragg modulators for wideband integrated optic communications and signal processing. IEEE Trans. CAS-26, 1072 (1979)

5.2 I.P. Kaminow, J.R. Carruthers: Optical waveguiding layers in LiNbO$_3$ and LiTaO$_3$. Appl. Phys. Lett. 22, 326-329 (1973)

5.3 R.V. Schmidt, I.P. Kaminow: Metal-diffused optical wave guides in LiNbO$_3$. Appl. Phys. Lett. 25, 458-460 (1974)
   R.D. Stanley, V. Ramaswamy: Nb-diffused LiTaO$_3$ optical waveguides: Planar embedded strip guides. Appl. Phys. Lett. 25, 711 (1974)

5.4 L. Kuhn, M.D. Kakss, P.F. Heidrich, B.A. Scott: Deflection of and optical guided wave by a surface acoustic wave. Appl. Phys. Lett. 17, 265ff (1970)

5.5 Y. Omachi: Acousto-optical light diffraction in thin films. J. Appl. Phys. 44, 3928-3932 (1973)

5.6 D. Marcuse: TE modes of graded-index slab waveguides. IEEE J. QE-9, 1000-1006 (1973)

5.7 E.M. Conwell: Modes in optical waveguides formed by diffusion. Appl. Phys. Lett. 26, 328-329 (1973)

5.8 P.K. Tien, S.Riva-Sanseverino, R.J. Martin, A.A. Ballman, H. Brown: Optical waveguide modes in single crystalline LiNbO$_3$-LiTaO$_3$ solid-solution films. Appl. Phys. Lett. 24, 503-506 (1974)

5.9 H. Kogelnik: Theory of dielectric waveguides, in *Integrated Optics*, ed. T. Tamir, Topics Appl. Phys., Vol.7 (Springer, Berlin, Heidelberg 1975) Chap.2

5.10 G.B. Hocker, W.K. Burns: Modes in diffused optical waveguides of arbitrary index profile. IEEE J. QE-11, 270-276 (1975)

5.11 R.V. Schmidt, I.P. Kaminow: Acousto-optic Bragg deflection in LiNbO$_3$ Ti-diffused waveguides. IEEE J. QE-11, 57-59 (1975)

5.12 C.S. Tsai, Le T. Nguyen, S.K. Yao, M.A. Alhaider: High-performance guided light beam device using two tilted-surface acoustic waves. Appl. Phys. Lett. 26, 140-142 (1975)
C.S. Tsai: Wideband guided-wave acousto-optic Bragg devices and applications, in Proc. of the 1975 Ultrasonic Symposium, IEEE Cat. 75CHO-994-4SU, pp.120-125

5.13 R.V. Schmidt: Acousto-optic interactions between guided optical waves and acoustic surface waves. IEEE Trans. SU-23, 22-23 (1976)

5.14 C.S. Tsai, M.A. Alhaider, Le T. Nguyen, B. Kim: Wideband guided-wave acousto-optic Bragg diffraction and devices using multiple tilted-surface acoustic waves. Proc. IEEE 64, 318-328 (1976)

5.15 B. Kim, C.S. Tsai: High-performance guided-wave acousto-optic scanning devices using multiple-surface acoustic waves. Proc. IEEE 64, 788-796 (1976)

5.16 E.G. Lean, J.M. White, C.D.W. Wilkinson: Thin-film acousto-optic devices. Proc. IEEE 64, 779-788 (1976)

5.17 T.G. Giallorenzi: Acousto-optical deflectin in thin-film waveguides. J. Appl. Phys. 44, 242-253 (1973)

5.18 T.G. Giallorenzi, A.F. Milton: Light deflection in multimode waveguides using the acousto-optic interaction. J. Appl. Phys. 45, 1762-1774 (1974)

5.19 R. Normandin, V.C.-Y. So, N. Rowell, G.I. Stegeman: The scattering of guided optical beams by surface acoustic waves in thin films. J. Opt. Soc. Am. 69, 1153-1165 (1979)

5.20 C.S. Tsai, I.W. Yao, B. Kim, Le T. Nguyen: Wideband guided-wave anisotropic acousto-optic Bragg diffraction in LiNbO$_3$ waveguides. at Int'l. Conf. on Integrated Optics and Optical Fiber Commun. Tokyo, Japan (July 1977) in Digest Techn. Papers, pp.57-60

5.21 See, for example, V.V. Proklov: Acousto-optic interactions in planar waveguide, Pt.I. Proc. Int'l Symp. on Surface Waves in Solids and Layered Structures, Novosibirsk (1986), Vol.1, pp.I48-I63
D.V. Petrov: Acousto-optic interactions in a planar optical waveguide, Pt.II (Ibid.) pp.I64-I75

5.22 P.K. Tien, Bell Telephone Lab. Report, MM-61-124-8 (1961)

5.23 C.F. Quate, C.D.W. Wilkinson, D.K. Winslow: Interaction of light and microwave sound. Proc. IEEE 53, 1604-1623 (1965)

5.24 H.V. Hance, J.K. Parks: Wideband modulation of a laser beam using Bragg-angle diffraction by amplitude-modulated ultrasonic waves. J. Acoust. Soc. Am. 38, 14-23 (1965)

5.25 A.J. DeMaria, G.E. Danielson Jr.: Internal laser modulation by acoustic lens-like effects. IEEE J. QE-2, 157-164 (1966)

5.26 A. Korpel, R. Adler, P. Desmares, W. Watson: A television display using acoustic deflection and modulation of coherent light. Proc. IEEE 54, 1429-1437 (1966)

5.27 M.G. Cohen, I.E. Gordon: Acoustic beam probing using optical techniques. Bell Systems Techn. J. 44, 693-721 (1965)
E.I. Gordon: A review of acousto-optic deflection and modulation. Proc. IEEE 54, 1391-1401 (1966)

5.28 R. Adler: Interactions of light and sound. IEEE Spectrum 4, 42-54 (May 1967)

5.29 R.W. Dixon: Acoustic diffraction of light in anisotropic media. IEEE J. QE-3, 85 (1967)

5.30 R.W. Dixon: Photoelastic properties of selected materials and their relevance for applications to acoustic light modulators and scanner. J. Appl. Phys. **38**, 5149-5153 (1967)

D.A. Pinnow: Guidelines for the selection of acousto-optic materials. IEEE J. QE-6, 223-238 (1970)

5.31 N. Uchida, N. Niizeki: Acousto-optic deflection materials and techniques. Proc. IEEE **61**, 1073-1092 (1973)

I.C. Chang: Acousto-optic deviceds and applications. IEEE Trans. SU-23, 2-22 (1976)

5.32 See, for example, W.S.C. Chang: Acoustooptical deflections in thin films. IEEE J. QE-7, 167-170 (1971), V.V. Proklov, E.M. Korablev: Acousto-optic measurements of effective reflective indices of guided modes in planar waveguides. Electron. Lett. **19**, 238 (1983)

D.V. Petrov: Acoustooptic and electrooptic guided wave conversion to leaky waves in an anisotropic optical waveguide. Opt. Commun. **50**, 300 (1984)

D.V. Petrov, J. Ctyroky: Kvatovaya Elecktronika (USSR) **12**, 987 (1985)

E.A. Kolosovsky, D.V. Petrov, I.B. Yakovkin: Spatial spectrum of leaky waves generated by scattering of a guided wave from an acoustic wave. Opt. Commun. **60**, 280-286 (1986)

5.33 G.W. Farnell: Properties of elastic surface waves, in *Physical Acoustics* **6**, 109-166 (Academic, New York 1970)

5.34 R.N. Spaight, G.G. Koerber: Piezoelectric surface waves on LiNbO$_3$. IEEE Trans. SU-18, 237-238 (1971)

5.35 A.J. Slobodnik, E.D. Conway, R.T. Delmonico (eds.), *Microwave Acoustics Handbook*, Vol.1A: *Surface Wave Velocities*, AFCRL-TR-73-0593 (1 Oct. 1973)

5.36 B.A. Auld: *Acoustic Fields and Waves in Solids*, Vol.II, (Wiley, New York 1973)

5.37 R.J. Pressley (ed.): *Handbook of Lasers with Selected Data of Optical Technology* (Chemical Rubber, Cleveland, Ohio 1971)

5.38 K.W. Loh, W.S.C. Chang, W.R. Smith, T. Gudkowski: Bragg coupling efficiency for guided acoustic interaction in GaAs. Appl. Opt. **15**, 156-166 (1976)

5.39 O. Yamazaki, C.S. Tsai, M. Umeda, L.S. Yap, C.J. Lii, K. Wasa, J. Merz: Guided-wave acousto-optic interactions in GaAs-ZnO composite structure. in 1982 IEEE Ultrasonics Symposium Proc., IEEE Cat. No. 82CH1823-4, pp.418-421

A.A. Ilyich, S.M. Kikkarin, D.V. Tsarev, I.B. Yakovkin: A comparison of acoustooptic interaction in LiNbO$_3$ and GaAlAs waveguides. Opt. Commun. **56**, 161-166 (1985)

5.40 C.J. Lii, C.S. Tsai, C.C. Lee: Wide-band guided-wave acousto-optic Bragg cells in GaAs-GaAlAs waveguide. IEEE J. QE-22, 868-872 (1986)

A. Abdelrazek, C.S. Tsai: High-performance acoustooptic Bragg cells in ZnO-GaAs waveguide at GHz frequencies. Optoelectronics - Devices and Technologies **4**, 33-37 (1989)

5.41 J. Kushibiki, H. Sasaki, N. Chubachi, N. Mikoshiba, K. Shibayama, "Thickness dependence of acousto-optica diffraction efficiency in ZnO-film optical waveguides. Appl. Phys. Lett. **26**, 362-364 (1975)

5.42 N. Chubachi: ZnO films for surface acousto-optic devices on nonpiezoelectric substrates. Proc. IEEE **64**, 772-774 (1976)

5.43 J. Kushibiki, N. Chubachi, K. Shibayama: Improvement of diffraction efficiency in surface-acoustic-optic devices by means of multi-layered surface. Appl. Phys. Lett. **29**, 333-335 (1976)

5.44 N. Chubachi, H. Sasaki: Surface acousto-optic interaction in ZnO thin films. Wave Electronics **2**, 379 (1976)

5.45 D. Mergerian, E. Malarkey, B. Newman, J. Lane, R. Weinert, B.R. McAvoy, C.S. Tsai: Zinc oxide transducer arrays for integrated optics, in Proc. 1978 Ultrasonics Symp., IEEE Cat. No. 78-CH1344-1SU, pp.64-69

5.46 S.K. Yao, R.R. August, D.B. Anderson: Guided-wave acousto-optic interaction on nonpiezoelectric substrates. J. Appl. Phys. 49, 5728-5730 (1978)

5.47 K. Setsune, T. Mitsuyu, O. Yamazaki, K. Wasa: Discrete frequency Bragg deflector at 2 GHz range using ZnO/sapphire substrate. Proc. 1983 IEEE Ultrasonics Symp.

5.48 R.W. Smith, H.M. Gerard, J.H. Collins, T.M. Reeder, H.J. Shaw: Design of surface wave delay lines with interdigital transducers. IEEE Trans. MTT-17, 865-873 (1969)

5.49 T.M. Bristol: Analysis and design of surface acoustic wave transducers, in *Proc. Conf. on Component Performance and System Application of Surface Wave devices*, (IEEE, Stevenage, England 1973) pp.115-129

5.50 R.H. Tancrell, M.G. Holland: Acoustic surface wave filters. Proc. IEEE 59, 393-409 (1971)
C.S. Hartmann, D.T. Bell, R.C. Rosenfeld: IEEE Trans. SU-20, 80 (1973)

5.51 W.R. Smith, H.M. Gerard, W.R. Jones: Analysis and design of dispersive interdigital surface-wave transducers. IEEE Trans. MTT-20, 458-471 (1972)

5.52 H. Matthews (ed.): *Surface Wave Filters - Design, Construction, and Use* (Wiley, New York 1977)

5.53 D. Maydan: Acousto-optical pulse modulators. IEEE J. QE-6, 15-24 (1970)

5.54 G.L. Matthaei, L. Young, E.M.T. Jones: *Microwave Filters, Impedance Matching Networks and Coupling Structures* (McGraw-Hill, New York 1964)

5.55 G.L. Matthaei, D.Y. Young: Some techniques for interdigital acoustic surface wave filter synthesis, in Proc. 1973 Ultrasonics Symposium, IEEE Cat. No. 73CH0807-8SU, pp.427-435
O.L. Matthaei, D.J. Wong, P.B. O'Shaughnessy: IEEE Trans. SU-22, 105 (1975)

5.56 T.M. Reeder, W.R. Shreve, P.L. Adams: A new broadband coupling network for interdigital surface wave transducers. IEEE Trans. SU-19, 466 (1972)

5.57 D.L. Hecht: Broadband acousto-optic spectrum analysis, in Proc. 1973 IEEE Ultrasonics Symposium, IEEE Cat. No. 73CH0807-8SU, pp.98-101

5.58 C.S. Tsai, Le T. Nguyen, B. Kim, I.W. Yao: Guided-wave acousto-optic signal processors for wideband radar systems, in *Effective Utilization of Optics in Radar Systems*, SPIE 128, 68-74 (1978)

5.59 E.G. Lean, C.C. Tseng: Nonlinear effects in surface acoustic waves. J. Appl. Phys. 41, 3912-3917 (1970)

5.60 B. Kim, C.C. Lee, C.S. Tsai: Design and fabrication of GHz bandwidth guided-wave acousto-optic Bragg deflectors in LiNbO$_3$ waveguides (unpublished)

5.61 C.S. Tsai, L.T. Nguyen, B. Kim: Wideband guided-wave acousto-optic Bragg diffraction using phased-surface acoustic wave array in LiNbO$_3$ waveguides, in Proc. 1975 IEEE Ultrasonics Symposium, IEEE Cat. No. 75CH0-994-4SU, pp.42-43

5.62 C.S. Tsai: Wideband guided-wave acousto-optic Bragg devices and applications, in Proc. 1975 IEEE Ultrasonics Symposium, IEEE Cat. No. 75CH0-994-4SU, pp.120-125

5.63 C.D.W. Wilkinson, R.M. de la Rue, G. Trantor: Light beam deflector using SAW optical interaction. presented at the 1975 Int'l. Electron Devices Meeting, Washington, DC

5.64 Le T. Nguyen, C.S. Tsai: Efficient wideband guided-wave acousto-optic Bragg diffraction using phased-surface acoustic wave array in LiNbO$_3$ waveguides. Appl. Opt. 16, 1297-1304 (1977)

5.65 G.A. Coquin, J.P. Griffin, L.K. Anderson: Wideband acousto-optic deflectors using acoustic beam steering. IEEE Trans. SU-17, 34-40 (1970)

5.66 D.A. Pinnow: Acousto-optic light deflection: design consideration for first-order beam steering transducers. IEEE Trans. SU-18, 209-214 (1971)

5.67 C.C. Lee, K.Y. Liao, C.L. Chang, C.S. Tsai: Wideband guided-wave acousto-optic Bragg deflector using a tilted-finger chirp transducer. IEEE J. QE-15, 1166-1170 (1979)

5.68 H.I. Smith, F.J. Bachner, N. Efremow: A high-yield photolithographic technique for surface wave devices. J. Electrochem. Soc. 118, 822-825 (1971)

5.69 R.L. Davis, F.S. Hickernell, R.L. Ward, F.V. Richard, S.H. Arneson: Design, fabrication, and performance of wideband guided-wave Bragg cells at frequencies above 1 GHz. 1982 Topical Meeting on Integrated and Guided-Wave Optics, Pacific Grove, CA, in Technical Digest of Post Deadline Papers, pp.PDP11-1 to 4

5.70 T. Suhara, H. Nishihara, J. Koyama: One-gigahertz-bandwidth demonstration in integrated optic spectrum analyzers. paper 29C5-5 presented at Int'l Conf. on Integrated Optics and Optical Fiber Communications, Tokyo, Japan, 27-30 June 1983

5.71 C. Stewart, G. Serivener, W.J. Stewart: Guided-wave acoustooptic spectrum analysis at frequencies above 1 GHz. 3rd Int'l Conf. on Integrated Optics and Optical Fiber Communication (1981), Techn. Digest p.122

5.72 M.K. Barnoski, B. Chen, H.M. Gerald, E. Marom, O.G. Ramer, W.R. Smith, G.L. Tangonan, R.D. Weglein: Design, fabrication, and integration of components for an integrated optic spectrum analyzer. in: *1978 Ultrasonics Symposium Proc.*, IEEE Cat. No. 78CH1344-1SU, pp.74-78

5.73 K.Y. Liao, C.L. Chang, C.C. Lee, C.S. Tsai: Progress on wideband guided-wave acousto-optic Bragg deflector using a tilted-finger chirp transducer. In *1979 Ultrasonics Symposium Proc.*, IEEE Cat. No. 79CH1482-9SU, pp.24-27

5.74 T.R. Joseph: Broadband chirp transducers for integrated optics spectrum analyzers. *1979 Ultrasonics Symposium Proc.*, IEEE Cat. No. 79CH1482-9SU, pp.28-32

5.75 P.K. Tien, R. Ulrich, R.J. Martin: Modes of propagating light waves in thin deposited semiconductor films. Appl. Phys. Lett. 14, 291-294 (1969)
J.H. Harris, R. Shubert: Beam coupling to films. URSI Spring Meeting, Washington, DC (1969), Conf. Abstracts, p.71
J.E. Midwinter, F. Zernike: Experimental studies of evanescent wave coupling into a thin-film waveguide. Appl. Phys. Lett. 16, 198-200 (1970)

5.76 D.Y. Zang, C.S. Tsai: Single-mode waveguide microlenses and microlens arrays fabrication using titanium-indiffused proton exchange technique in LiNbO₃. Appl. Phys. Lett. 48, 703-705 (1985)

5.77 V.M. Ristic, S.A. Jones, G.R. Dubois: Evaluation of an integrated acousto-optic receiver. Cdn. Electr. Eng. J. 8, 59-64 (1983)

5.78 T. Suhara, T. Shiono, H. Nishihara, J. Koyama: An integrated optic Fourier processor using an acousto-optic deflector and Fresnel lenses in As₂S₃ waveguide. J. Lightwave Technol. LT-1, 624-630 (1983)

5.79 D. Gregoris and V.M. Ristic: Wideband transducer for acousto-optic Bragg deflector. Cdn. J. Phys. 63, 195-197 (1985)

5.80 See, for example, I. Hayashi: Recent progress in semiconductor lasers - cw GaAs lasers are now ready for new applications. Appl. Phys. 5, 25 (1974)
A. Yariv, P. Yeh: *Optical Waveguides in Crystals* (Wiley, New York 1984) Chap.11
Y. Suematsu: Advances in semiconductor lasers. Phys. Today 32, 32-39 (May 1985)

5.81 I. Hayashi, M.B. Panish, P.W. Foy, S. Sumski: Junction lasers which operate continuously at room temperature. Appl. Phys. Lett. 17, 109 (1970)
M. Nakamura, A. Yariv, H.W. Yen, S. Somekh, H.L. Garvin: Optically pumped GaAs surface laser with corrugation feedback. Appl. Phys. Lett. 22, 515 (1973)
H. Yonezu, I. Sakuma, K. Kobayashi, T. Kamejima, M. Ueno, Y. Nannichi, "A

GaAs-Al$_x$Ga$_{1-x}$As double heterostructure plane stripe laser. Japan J. Appl. Phys. 12, 1485-1492 (1973)

D.R. Scifres, R.D. Burnham, W. Streifer: Highly collimated laser beams from electrically pumped SH GaAs/GaAlAs distributed-feedback lasers. Appl. Phys. Lett. 26, 48 (1975)

Y. Suematsu, M. Yamada, K. Hayashi: Integrated twin-guide AlGaAs laser with multiheterostructure. IEEE J. QE-11, 457-460 (1975)

C.C. Tseng, D. Botez, S. Wang: Optically pumped epitaxial GaAs waveguide lasers with distributed Bragg reflectors. IEEE J. QE-12, 549 (1976)

H.W. Yen, W. Ng, I. Samid, A. Yariv: GaAs distributed Bragg reflector lasers. Opt. Cummun. 17, 213 (1976)

F.K. Reinhart, A.Y. Cho: Al$_y$Ga$_{1-y}$As/Al$_x$Ga$_{1-x}$As laser structures for integrated optics grown by molecular-beam epitaxy. Appl. Phys. Lett. 31, 457 (1977)

J.L. Merz, R.A. Logan: Integrated GaAs-Al$_x$Ga$_{1-x}$As injection lasers and detectors with etched reflectors. Appl. Phys. Lett. 30, 530 (1977)

C.C. Ghizoni, J.M. Ballantyne, C.L. Tang: Theory of optical-waveguide distributed feedback lasers: A Green's function approach. IEEE J. QE-13, 843 (1977)

W.T. Tsang, R.A. Logan: GaAs-Al$_x$Ga$_{1-x}$As strip buried heterostructure lasers. IEEE J. QE-15, 451 (1979)

L.A. Coldren, K. Iga, B.I. Miller, J.A. Rentschler: GaInAsP/InP strip-geometry laser with a reactive-ion-etched facet. Appl. Phys. Lett. 37, 681 (1980)

M. Nakamura: Single-mode operation of semiconductor injection lasers. IEEE Trans. on Circuits and Systems CAS-26, 1055-1065 (1979)

5.82   See the references for waveguide lenses, listed in Chap.8

5.83   D.B. Ostrowsky, R. Poirier, L.M. Reibor, C. Deverdun: Integrated optical photodetector. Appl. Phys. Lett. 22, 263-464 (1973)

H. Stoll, A. Yariv, R.G. Hunsperger, G.L. Tangonan: Proton-implanted optical waveguide detectors in GaAs. Appl. Phys. Lett. 23, 664-665 (1973)

G.E. Stillman, C.M. Wolfe, I. Melgailis: Monolithic integrated In$_x$Ga$_{1-x}$As Schottky-barrier waveguide photodetector. Appl. Phys. Lett. 25, 36-39 (1974)

C.C. Tseng, S. Wang: Integrated grating-type Schottky-barrier photodetector with optical channel waveguide. Appl. Phys. Lett. 26, 632-636 (1975)

J.T. Boyd, C.L. Chen: Integrated optical silicon photodiode array. Appl. Opt. 15, 1389 (1976)

J.L. Merz, R.A. Logan, A.M. Sergent: GaAs integrated optical circuits by wet chemical etching. IEEE J. QE-5, 72-82 (1979)

J.C. Campbell, T.P. Lee, A.G. Dentai, C.A. Burrus: Dual-wavelength demultiplexing InGaAsP photodiode. Appl. Phys. Lett. 34, 401-402 (1979)

G.M. Borsuk, A. Turley, G.E. Marx, E.C. Malarkey: Photosensor array for integrated optical spectrum analyzer systems. Proc. SPIE 176, 109 (1979)

G.W. Anderson, A.E. Spezio: Photodetector approaches for acousto-optics spectrum analysis. Proc. SPIE 477, 161-164 (1984)

I. Melngailis: Laser sources and detectors for guided-wave optic signal processing. Proc. SPIE 185 (1979)

F.F. So, S.R. Forrest, H.L. Garvin, D.L. Jackson: A fast organic-on-inorganic semiconductor photodetector. 1988 OSA Topical Meeting on Integrated and Guided-Wave Optics, Santa Fe, NM, Techn. Digest Ser., Vol.5, pp.210-214

5.84   See, for example, W.T. Tsang, A.Y. Cho: Molecular beam epitaxial writing of patterned GaAs epilayer structures. Appl. Phys. Lett. 32, 491 (1978)

E. Garmire: Semiconductor components for monolithic applications, in *Integrated Optics*, 2nd edn., ed. by T. Tamir (Springer, Berlin, Heidelberg 1979)

5.85   H.F. Taylor, W.E. Martin, W.M. Caton: Channel waveguide electro-optic devices for communications and information processing. 1976 Topical Meeting on

Integrated Optics, Salt Lake City, Utah, in Digest of Technical Papers, Cat. No. 75CH1039-7QEC, pp.WA5-1 to 5

R. Shubert, J.H. Harris: Optical survace waves on thin films and their application to integrated data processors. IEEE Trans. MIT-16, 1048-1054 (1968)

H.F. Taylor: Application of guided-wave optics in signal processing and sensing. Proc. IEEE 75, 1524-1535 (1987)

C.S. Tsai, D.Y. Zang, P. Le: Guided-wave acousto-optic Bragg diffraction in a LiNbO$_3$ channel-planar waveguide with application to optical computing. Appl. Phys. Lett. 47, 549-551 (1985)

C.M. Verber and R.P. Kenan: Integrated optical circuits for numerical computation. Proc. SPIE 408, 57 (1983)

C.M. Verber: Integrated-optical approaches to numerical optical processing. Proc. IEEE 72, 942-953 (1984)

C.S. Tsai: Titatium-indiffused proton-exchanged microlens-based integrated optic Bragg modulator modules for optical computing. Proc. SPIE 634, 409-421 (1986)

C.S. Tsai: Integraged-optical device modules in LiNbO$_3$ for computing and signal processing. J. Mod. Opt. 35, 965-977 (1988)

5.86   T. Van Duzer: Lenses and graded films for focusing and guiding acoustic surface waves. Proc. IEEE 58, 1230-1237 (1970)

5.87   E.A.J. Marcatili: Dielectric rectangular waveguide and directional coupler for integrated optics. Bell Syst. Techn. J. 48, 2071 (1969)

H. Kogelnik, R.V. Schmidt: Switched directional couplers with alternating $\Delta\beta$. IEEE J. QE-12, 396 (1976)

R.C. Alferness: Guided-wave devices for optical communication. IEEE J. QE-17, 946 (1981)

F. Auracher, H.H. Witte: New Directional coupler for integrated optics. J. Appl. Phys. 45, 4997 (1974)

K. Tada, K. Hirose: A new light modulator using perturbation of synchronism between two coupled guides. Appl. Phys. Lett. 25, 261 (1974)

W.J. Tomlinson, H.P. Weber, C.A. Pryde, E.A. Chandross: Optical directional couplers and grating couplers using a new high-resolution photolocking material. Appl. Phys. Lett. 26, 303 (1975)

J.C. Campbell, F.A. Blum, D.W. Shaw: GaAs electro-optic channel-waveguide modulator. Appl. Phys. Lett. 26, 640 (1975)

F.J. Leonberger, J.P. Donnelly, C.O. Bozler: GaAsP$^+$n$^-$n$^+$ directional-coupler switch. Appl. Phys. Lett. 29, 640 (1975)

J.C. Shelton, F.K. Reinhart, R.A. Logan: GaAs-Al$_x$Ga$_{1-x}$As rib waveguide switches with MOS electrooptical control. Appl. Opt. 17, 2548 (1978)

A. Carenco, L. Menigaus, P. Delpech: Multiwave length GaAs rib waveguide directional coupler with stepped $\Delta\beta$ Schottky electrodes. J. Appl. Phys. 50, 5139 (1979)

H.A. Haus, C.G. Fonstad: Three-waveguide couplers for improved sampling and filtering. IEEE J. QE-17, 2321 (1981)

5.88   M. Papuchon, Am. Ray, D.B. Ostrowsky: Electrically active bifurcations: BOS. Appl. Phys. Lett. 31, 266 (1977)

Y. Ohmachi, J. Noda: Electro-optic light modulator with branched ridge waveguide. Appl. Phys. Lett. 27, 544 (1975)

C.S. Tsai, B. Kim, F.R. El-Akkari: Optical channel waveguide switch and coupler using total internal reflection. IEEE J. QE-14, 513 (1978)

F.R. El-Akkari, C.L. Chang, C.S. Tsai: Electro-optical channel waveguide matrix switch using total internal reflection. 1980 Topical Meeting on Guided-Wave and Integrated Optics, Incline Village, Nevada, in Technical Digest, OSA/IEEE Cat. No. 80CH1489-4QEA, pp.TuEf-1 to 4

C.L. Chang, F.R. El-Akkari, C.S. Tsai: Fabrication and testing of optical channel waveguide TIR switching networks. Proc. SPIE 239, 147 (1981)

C.L. Chang, C.S. Tsai: GHz bandwidth optical channel waveguide TIR switches and 4x4 switching networks. 1982 Topical Meeting on Integrated and Guided-Wave Optics, Pacific Grove, CA, in Technical Digest, IEEE Cat. No. 82CH1719-4, pp.TuD2-1 to 4

C.S. Tsai, C.C. Lee, Phat Le: A 8.5 GHz bandwidth single-mode crossed channel waveguide TIR modulator and switch in LiNbO$_3$. 1984 Topical Meeting on Integrated and Guided-Wave Optics, in Technical Digest of Post Deadline Papers, IEEE Cat. No. 84CH1997-6, pp.PD5-1 to 4

K. Wasa, H. Adachi, T. Kawaguchi, K. Ohji, K. Setsune: Optical TIR switches using PLZT thin film waveguides on sapphire. Int'l. Conf. on Integrated Optics and Optical Fiber Communications, Tokyo, Japan (1983), in Technical Digest, pp.356-357

A. Neyer: Electro-optic x-switch using single-mode Ti:LiNbO$_3$ channel waveguides. Electron. Lett. 19, 553 (1983)

5.89 P.K. Tien, S. Riva-Sanseverino, A.A. Ballman: Light beam scanning and deflection in epitaxial LiNbO$_3$ electro-optic waveguides. Appl. Phys. Lett. 25, 563-565 (1974)

5.90 I.P. Kaminow, L.W. Stulz: A planar electro-optic prism switch. IEEE J. QE-11, 633-635 (1975)

5.91 C.S. Tsai, P. Saunier: Ultrafast guided-light beam deflection/switching and modulation using simulated electro-optic prism structures in LiNbO$_3$ waveguides. Appl. Phys. Lett. 27, 248-250 (1975)

C.S. Tsai, P. Saunier: New guided-wave acousto-optic and electcro-optic devices using LiNbO$_3$, Ferroelectrics 10, 257-261 (1976)

5.92 H. Kotani, S. Namba, M. Kawabe: Electro-optic Bragg deflection modulators in corrugated waveguides. IEEE J. QE-15, 270-272 (1979)

5.93 N. Sasaki, R.M. de la Rue: Electro-optic Bragg deflection modulators in corrugated waveguides. IEEE J. QE-15, 270-272 (1979)

P. Saunier, C.S. Tsai, I.W. Tao, Le T. Nguyen: Electro-optic phased-array light beam deflector with application to analog-to-digital conversion. at Topical Meeting on Integrated and Guided-Wave Optics, Salt Lake City, Utah, in Technial Digest, Paper TuC2, pp.TuC2-4 (1978)

5.94 D.C. Flanders, H. Kogelnik, R.V. Schmidt, C.V. Shank: Grating filters for think-film optical waveguides. Appl. Phys. Lett. 24, 194-196 (1974)

A.C. Livanors, A. Katzir, A. Yariv, C.S. Hong: Chirped-grating demultiplexers in dielectric waveguides. Appl. Phys. Lett. 30, 519-521 (1977)

P.S. Cross, H. Kogelnik: Suppression in corrugated-waveguide filters. Opt. Lett. 1, 43-45 (1977)

H.W. Yen, H.R. Friedrich, R.J. Morrison, G.L. Tangonan: Planar Rowland spectrometer for fiber-optic wavelength demultiplexing. Opt. Lett. 6, 639-641 (1981)

T. Suhara, Y. Handa, H. Nishihara, J. Koyama: Monolithic integrated micro-gratings and photodiodes for wavelength demultiplexing. Appl. Phys. Lett. 40, 120-122 (1982)

E. Voges: Multimode planar devices for wavelength division multiplexing and demultiplexing. presented at 1983 Int'l. Conf. on Integrated Optics and Optical Fiber Communications, Tokyo, Japan, 27-30 June, Paper No. 29A1-5

G. Winzer: Wavelength multiplexing components--a review of singe-mode devices and their applications. J. Lightwave Techn. LT-2, 369-378 (1984)

5.95 R.C. Alferness, R.V. Schmidt: Tunable optical waveguide directional coupler filter. 1978 Topical Meeting on Integrated and Guided-Wave Optics, in Technical Digest, Paper TuA3

R.C. Alferness, P.S. Cross: Filter characteristics of co-directionally coupled waveguides with weighted coupling. IEEE J. QE-14, 843 (1978)

R.C. Alferness: Efficient electro-optic TE-TM mode converter/wavelength filter. Appl. Phys. Lett. 36, 513-515 (1980)

5.96 L. Kuhn, P.F. Heidrich, E.G. Lean: Optical guided wave mode conversion by an acoustic sufrace wave. Appl. Phys. Lett. 19, 428-430 (1971)

5.97 M.S. Chang: Tolerances to the phase matching condition for the integrated acousto-optic filter. Appl. Opt. 13, 1867 (1974)

5.98 Y. Ohmachi, J. Noda: LiNbO$_0$ TE-TM mode converter collinear acousto-optic interaction. IEEE J. QE-13, 43-46 (1977)

K. Yamonouchi, K. Higuchi, K. Shibayama: TE-TM mode conversion by interaction between elastic surface waves and a laser beam on a metal-diffused optical waveguide. Appl. Phys. Lett. 29, 75-77 (1976)

V.P. Hinkov, R. Opitz, W. Sohler: Collinear acousto-optical TM-TE mode conversion in proton exchanged Ti-LiNbO$_3$ waveguide structures. J. Lightwave Tech. 6, 903-908 (1988)

5.99 B. Kim, C.S. Tsai: Thin-film tunable optical filtering using anisotropic interaction. IEEE J. QE-15, 642-647 (1979)

5.100 H. Jerominek, J.Y.D. Pomerleau, P. Tremblay, C. Delisle: An integrated acousto-optic bistable device. Opt. Commun. 51, 6-10 (1984)

5.101 C.S. Tsai, Le T. Nguyen, C.C. Lee: Optical pulse compression and very high-speed light beam scanning using guided-wave acousto-optic Bragg diffraction. 1978 Topical Meeting on Integrated and Guided Wave Optics, Salt Lake City, Utah, in Technical Digest, pp.TuC3-1 to 4, OSA 78Ch1280-7 QEA.

5.102 C.C. Lee, C.S. Tsai: An acousto-optic readout scheme for integrated optic RF spectrum analyzer. 1978 Ultrasonics Symposium Proc., IEEE Cat. No.78CH1344-1SU, pp.79-81

5.103 M. Ross, P. Freedman, J. Abernathy, G. Matassov, J. Wolf, J.D. Barry: Space optical communications with the Nd:YAG laser. Proc. IEEE 66, 319-344 (1978)

5.104 L.C. Foster, C.B. Crumley, R.L. Cohoon: A high-resolution linear optical scanner using a traveling-wave acoustic lens. Appl. Opt. 9, 2154 (1970)

5.105 P.T. Ho, L.A. Glaser, E.P. Ippen, H.A. Haus: Picosecond pulse generation with a cw GaAlAs laser diode. Appl. Phys. Lett. 33, 241 (1978)

Z.A. Yasa, A. Dienes, J.R. Whinnery: Subpicosecond pulses from a cw double mode locked dye laser. Appl. Phys. Lett. 30, 24-26 (1977)

5.106 C.S. Tsai: The increase of Bragg diffraction intensity due to acoustic resonance and its application for demultiplexing and multiplexing in laser communication. Appl. Opt. 10, 215-218 (1971)

S.K. Yao, C.S. Tsai: Acousto-optic Bragg diffraction with application to ultra-high data rate laser communications systems. Pt.I Theoretical considerations of the standing wave ultrasonic Bragg cell. Appl. Opt. 16, 3032-3043 (1977)

C.S. Tsai, S.K. Yao: Pt.II Experimental results of the SUBC and the acousto-optic multiplexer/demultiplexer terminals. Appl. Opt. 16, 3044-3060 (1977)

5.107 C.S. Tsai, S.K. Yao, M.A. Alhaider, P. Saunier: High-speed guided-wave acousto-optic and electro-optic switches. 1974 Int'l. Electronic Devices Meeting, in Technical Digest, pp.85-87

5.108 S.E. Harris, R.W. Wallace: Acousto-optic tunable filter. J. Opt. Soc. Am. 59, 744-747 (1969

I.C. Chang: Noncollinear acousto-optic filter with large angular aperture. Appl. Phys. Lett. 25, 370 (1974)

M. Khoshnevisam, E. Sovero, P.R. Newman, J. Tracy: Development of a cryogenic infrared acoustooptic tunable spectral filter. Proc. SPIE 245, 63 (1980)

J.D. Feichtner, M. Gottlieb, J.J. Conroy: A tunable collinear acousto-optic filter for the intermediate infrared using crystal Ti$_3$AsSe$_3$. CLEO'75, Washington, DC

T. Yano, A. Watanabe: Acousto-optic TeO$_2$ tunable filter using far-off-axis anisotropic Bragg diffraction. Appl. Opt. 15, 2250-2258 (1976)

5.109 J.L. Davis, S. Ezekiel: Closed-loop, low-noise fiber-optic rotation sensor. Opt. Lett. 6, 505 (1981)

5.110 R.F. Cahill, E. Udd: Phase-nulling fiber-optic laser gyro. Opt. Lett. 4, 93 (1979)

5.111 F. Heismann, R. Ulrich: Integrated-optical single-sideband modulator and phase shifter. IEEE J. QE-18, 767 (1982)

5.112 B. Culshaw, M.G.F. Wilson: Integrated optic frequency shifter modulator. Electron. Lett. 17, 135 (1981)

5.113 F. Heismann, R. Ulrich: Integrated-optical frequency translator with strip waveguide. Appl. Phys. Lett. 45, 490 (1984)

5.114 M. Izutsu, S. Shikama, T. Sueta: Integrated optical SSB modulator and frequency shifter. IEEE J. QE-17, 2225 (1981)

5.115 K. Nosu, S.C. Rashleigh, H.F. Taylor, J.F. Weller: Acousto-optic frequency shifter for birefringent fiber. Electron. Lett. 19, 816 (1983)

5.116 W.P. Risk, R.C. Youngquist, G.S. Kino, H.J. Shaw: Acousto-optic frequency shifting in birefringent fiber. Opt. Lett 9, 309 (1984)
J.N. Blake, B.Y. Kim, H.E. Engan, H.J. Shaw: All-fiber acoustooptic frequency shifter using two-mode fiber. Proc. SPIE 179, 92 (1986)

5.117 C.S. Tsai, C.L. Chang, C.C. Lee, K.Y. Liao: Acousto-optic Bragg deflection in channel optical waveguides. 1980 Topical Meeting on Integrated and Guided-Wave Optics, in Technical Digest of Post-Deadline Papers, IEEE Cat. No. 80CH1489-4QEA, pp.PD7-1 to 4
C.S. Tsai, C.T. Lee, C.C. Lee: Efficient acousto-optic diffraction in crossed channel waveguides and resultant integrated optic module. 1982 IEEE Ultrasonics Symposium Proc., IEEE Cat. No. 82CH1823-4, pp.422-425

5.118 R.H. Kingston, R.A. Becker, F.J. Leonberger: Broadband guided-wave optical frequency translator using an electro-optical Bragg array. Appl. Phys. Lett. 42, 759 (1983)

5.119 L.M. Johnson, R.A. Becker, R.H. Kingston: Integrated-optical channel-waveguide frequency shifter. 1984 Topical Meeting on Integrated and Guided-Wave Optics, IEEE Cat. No.84CH1947-6, Techn. Digest, Paper WD4-1

5.120 C.S. Tsai, Z.Y. Cheng: Novel guided-wave acousto-optic frequency shifting scheme using Bragg diffractions in cascade. Appl. Phys. Lett. 54, 1616-1618 (1989)

5.121 C.S. Tsai, Q. Li: Wideband optical frequency shifting using acousto-optic Bragg diffraction in a LiNbO$_3$ spherical waveguide (to be published in J. Appl. Phys.)

5.122 J.W. Goodman: Introduction to Fourier Optics (New York, McGraw-Hill 1968)
L.J. Cutrona, E.N. Leith, C.J. Palermo, L.J. Porcello: Optical data processing and filtering systems. IRE Trans. IT-6, 386-400 (1960)
A. van der Lugt: Coherent optical processing. Proc. IEEE 62, 1300-1319 (1974)

5.123 R.A. Sprague et al.: Optical signal processing for terminal defense radars. Itek Corp. Report 76-9528-1, for the project sponsored by Ballistic Missile Defense Advanced Technology Center (1976)

5.124 K. Preston Jr.: A comparison of analog and digital techniques for pattern recognition. Proc. IEEE 60, 1216-1231 (1972)

5.125 R.W. Damon, W.T. Maloney, D.H. McMahon: In Physical Acoustics, Vol.7, ed. by W.D. Mason, R.N. Thurston (Academic, New York 1970)
A. Korpel: In Optical Information Processing, ed. by Y.E. Nestenikhin et al. (Plenum, New York 1975)
R.A. Sprague: A review of acousto-optic signal correlators. Opt. Eng. 16 467-478 (1977)
A. van der Lugt (ed.), Special section on acousto-optic signal processing, Proc. IEEE 69, 48-92 (1981)

D. Casasent: Acousto-optic transducers in iterative optical vector-matrix processors. Appl. Opt. 21, 1958 (1982)

D. Psaltis: Two-dimensional optical processing using one-dimensional input devices. Proc. IEEE 72, 962 (1984)

5.126 M.C. Hamilton, D.A. Wille, W.J. Meceli: An integrated optical RF spectrum analyzer. Proc. 1976 Ultrasonics Symposium, IEEE Cat. No. 74CH1120-5SU, pp.218-222

D.B. Anderson, J.T. Boyd, M.C. Hamilton, R.R. August: An integrated-optical approach to the Fourier transform. IEEE J. QE-13, 268 (1977)

C.S. Tsai: Guided-wave acousto-optic Bragg modulators for wideband integrated optic communications and signal processing. IEEE Trans. CAS-26, 1072-1098 (1979)

M.K. Barnoski, B. Chen, T.R. Joseph, J.Y.M. Lee, O.G. Ramer, "Integrated-optic spectrum analyzer. IEEE Trans. CAS-26, 1113-1124 (1979)

E. Marx, L.D. Hutcheson, A.L. Keller: Operational integrated-optic RF spectrum analyzer. Appl. Opt. 19, 3033-3034 (1980)

R.L. Davis, F.S. Hickernell: An integrated optic spectrum analyzer with tin film lenses. 1981 Int'l. Conf. on Integrated Optics and Optical Fiber Communications, San Francisco, CA, Paper No. WE-6

V. Neuman, C.W. Pitt, L.M. Walpita: An integrated acousto-optic spectrum analyzer using grating components. Proc. 1st Europ. Conf. Integrated Optics, London, England (1981) pp.89-92

T. Suhara, H. Nishihara, J. Koyama: A folded-type integrated optic spectrum analyzer using butt-coupled chirped grating lenses. IEEE J. QE-18, 1057-1059 (1982)

V.M. Ristic, S.A. Jones: Theoretical considerations related to development of an integrated acousto-optic receiver. Can. Electr. Eng. J. 7, 7-18 (1982)

R.L. Davis, F.S. Hickernell: Application of wideband Bragg cells for integrated optic spectrum analyzer. Proc. SPIE 321, 141-148 (1982)

V.M. Ristic, S.A. Jones, G.R. Dubois: Evaluation of an integrated acousto-optic receiver. Can. Electr. Eng. J. 8, 59-64 (1983)

M. Kanazawa, T. Atsumi, M. Takami, T. Ito: High resolution integrated optic spectrum analyzer. 1983 Int'l. Conf. on Integrated Optics and Optical Fiber Communications, Tokyo, Japan, Paper No. 30B3-5

S. Valette, J. Lizet, P. Mottier, J.P. Jadot, S. Renard, A. Fournier, A.M. Grouillet, P. Gidons, H. Denis: Integrated optical spectrum analyzer using planar technology on oxidized silicon substrate. Electron. Lett. 19 883[885 (1983) and IEEE Proc. H 131, 325-331 (1984)

T. Suhara, T. Shiono, H. Nishihara, J. Koyama: An integrated-optic Fourier processor using an acousto-optic deflector and Fresnel lenses in As_2S_3 waveguide. J. Lightwave Technol. LT-1, 624-630 (1983)

C. Stewart, G. Serivener, W.J. Stewart: Guided-wave acousto-optic spectrum analysis at frequencies above 1 GHz. 1981 Int'l Conf. on Integrated Optics and Optical Fiber Communications, in Technical Digest, p.122

5.127 E.T. Aksenov, N.A. Esepkina, A.A. Lipovskii, A.V. Pavenko: Prototype integrated acousto-optic spectrum analyzer. Soviet Tech. Lett. 6, 519-520 (1980)

5.128 I.W. Yao, C.S. Tsai: Signal processing using guided-wave acousto-optic Bragg diffraction in $LiNbO_3$ waveguides. 1976 Int'l. Microwave Symposium, in Technical Digest, IEEE Cat. No. 876CH1087-7MITT, pp.21-23

5.129 W.S.C. Chang, C.S. Tsai, R.A. Becker, I.W. Yao: Convolution using guided acousto-optic interaction in thin-film waveguides. IEEE J. QE-13, 208-215 (1977)

5.130 C.J. Kramer, M.N. Araghi, P. Das: Real-time convolution using acousto-optic diffraction from surface waves. Appl. Phys. Lett. 25, 180 (1974)

N.J. Berg, B.J. Udelson: Large time-bandwidth acousto-optic convolver. 1976 Ultrasonics Symposium Proc., IEEE Cat. No. 76CH1120-5SU, pp.183-188

5.131 R.A. Sprague, K.L. Koliopaulos: Time-integrating acousto-optic correlator. Appl. Opt. **15**, 89 (1976)

5.132 T.M. Turpin: Time-integrating optical processors. SPIE **154**, 196-198 (1978)

5.133 I.W. Yao, C.S. Tsai: A time-integrating correlator using guided-wave acousto-optic interactions. 1978 Ultrasonic Symposium Proc., IEEE Cat. No. 78CH1344-ISU, pp.87-90

5.134 H. Kogelnik, C.V. Shank: Stimulated emission in a periodic structure. Appl. Phys. Lett. **18**, 152-154 (1971)

M. Nakamura et al.: Cw operation of distributed feedback GaAs-GaAlAs diode lasers at temperatures up to 300 K. Appl. Phys. Lett. 27, 403-504 (1975)

A. Yariv: *Introduction to Optical Electronics*, 2nd. ed. (Holt, Rinehart, Winston, New York 1976)

5.135 S. Wang: Principles of distributed feedback and distributed Bragg reflector lasers. IEEE J. QE-10, 413-427 (1974)

F.K. Reinhart et al.: GaAsAl$_x$Ga$_{1-x}$As injection lasers with distributed Bragg reflectors. Appl. Phys. Lett. 27, 45-48 (1975)

5.136 T.Q. Vu, J.A. Norris, C.S. Tsai: Planar waveguide lenses in GaAs by using ion milling. Appl. Phys. Lett. **54**, 1098-1100 (1989)

5.137 C.S. Tsai, J.K. Wang, K.Y. Liao: Acousto-optic time-integrating correlators using integrated optics technology. Proc. SPIE **180**, 160-163 (1979)

5.138 R.G. Hunsperger, A. Yariv, A. Lee: Parallel end-butt coupling for optical integrated circuits. Appl. Opt. **16**, 1026-1032 (1977)

J.D. Zino et al.: Coupling of an injection laser diode to planar waveguide. at 1979 SPIE East, 18 April, Washington, DC

5.139 K.Y. Liao, C.C. Lee, C.S. Tsai: Time-integrating correlator using guided-wave anisotropic acousto-optic Bragg diffraction and hybrid integration. 1982 Topical Meeting on Integrated and Guided-Wave Optics, Pacific Grove, CA, in Technical Digest, IEEE Cat. No. 82CH1719-4, p0.WA4-1 to 4

C.C. Lee, K.Y. Liao, C.S. Tsai: Acousto-optic time-integrating correlator using hybrid integrated optics. Proc. 1982 IEEE Ultrasonics Symposium, IEEE Cat. No. 82CH1823-4, pp.405-407

5.140 C.M. Verber, R.P. Kenan, J.R. Busch, "Correlator based on an integrated optical spatial light modulator. Appl. Opt. 20, 1626 (1981)

C.S. Tsai, D.Y. Zang, P. Les: High-packing density multichannel integrated-optic device modules in LiNbO$_3$ for programmable correlation of binary sequences. Opt. Lett. 14, 889-891 (1989)

5.141 C.S. Tsai: LiNbO$_3$-based integrated-optic device modules for communication, computing, and signal processing. CLEO'76, in Technical Digest, IEEE Cat. No. 86CH2274-9, pp.44-46

D.Y. Zang, C.S. Tsai: Titanium-indiffused proton-exchanged waveguide lenses in LiNbO$_3$ for optical information processing. Appl. Opt. 25, 2264-2271 (1986)

C.S. Tsai: Titanium-indiffused proton-exchanged microlens-based integrated optic Bragg modulator modules for optical computing. Proc. SPIE 634, 409-421 (1986)

C.S. Tsai: Integrated-optical device modules in LiNbO$_3$ for computing and signal processing. J. Mod. Opt. 35, 965-977 (1988)

6. Guided-Wave Acousto-Optic Interaction in a ZnO Thin Film on a Nonpiezoelectric Substrate

Nobuo Mikoshiba

With 26 Figures

The possibility of Surface Acousto-Optic (SAO) devices such as deflectors, modulators, and mode converters playing an important role in integrated optics led to the extensive study of the Bragg diffraction of Guided Optical Waves (GOW) by Surface Acoustic Waves (SAW). Since *Kuhn* et al. [6.1] first demonstrated SAW-GOW deflectors and modulators, numerous experiments have been carried out on various thin-film configurations. An extensive list of references on prior works has been provided in [6.2] and Chap.5. All the experiments performed thus far are classified by device structure into two categories, i.e., if the substrate on which the optical waveguide is constructed is a piezoelectric material, it is in the first category, while if it is a nonpiezoelectric material, it is in the second category. Since many kinds of devices in integrated optics must be constructed on nonpiezoelectric substrates, it is desirable to construct SAO devices with the second category. With this category a SAW can be launched by means of a piezoelectric thin film such as ZnO deposited on the substrate. ZnO, a memeber of the hexagonal wurtzite class, has a low dielectric constant and a high electromechanical coupling factor making it a useful material for SAW device technology. Moreover, ZnO is an interesting material for GOW because it is transparent from 0.4 to 2 μm and should be an optically low-loss material. Therefore, a ZnO thin film is one of the most promising materials for SAO devices.

In this chapter, we discuss the present state of the art of the various fabrication techniques for ZnO thin films. In addition, we describe the properties of ZnO thin-film waveguides constructed on various substrates and the excitation of a SAW by a ZnO thin film. A detailed description is given for various SAO interactions in ZnO thin film fabricated on fused-quartz substrates.

6.1 Fabrication of ZnO Thin Film

Various kind of fabrication techniques for ZnO thin films have been proposed and demonstrated, which include dc and RF sputtering, chemical vapor deposition (CVD), planar magnetron sputtering, etc. The following subsections give a brief description of these techniques.

6.1.1 DC Sputtering

The direct sputtering of a ZnO compound in both an inert argon atmosphere and a reactive oxygen atmosphere to obtain oriented ZnO films

Springer Series in Electronics and Photonics, Vol. 23
Guided-Wave Acoustooptics   Editor: C. S. Tsai
© Springer-Verlag Berlin, Heidelberg 1990

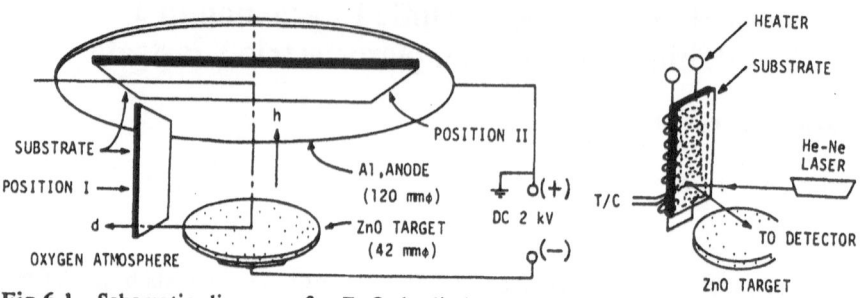

Fig.6.1. Schematic diagram of a ZnO dc diode sputtering apparatus [6.5]

with thicknesses from 0.1 to 10 μm was first demonstrated by *Rozgonyi* and *Polito* [6.3]. The ZnO source material was fabricated into a cylindrical cathode by hot-pressing small single crystals of high-purity ZnO into a mold. The cathode-to-anode spacing was 2 to 3 cm. The sputtering system achieved pressures of 10^{-9} torr after an overnight bake. A dc sputter condition was initiated after isolating the reaction chamber, with a volume of 2 liters, from the pump and backfilling it with 20 to 100 μm of spectral-grade argon. A plasma became excited between the cathode and anode. Positive ions from the plasma bombarded the target (cathode), thus sputtering off ZnO. This material was transported to the substrate (anode) where it was deposited. Deposition rates varied from 5 to 13 nm/min., depending on the argon and oxygen partial pressure and the power supplied to the glow discharge. A typical cathode potential was 2 kV, giving a cathode current density of 2 mA/cm^2. Film thicknesses were determined by a light section microscope and a Hobson-Tallysu RF stylus. The films were identified by Reflection Electron Diffraction (RED) as hexagonal ZnO oriented with the c axis perpendicular to the substrate. Substrate materials such as sapphire, pyrex glass, gold plated sapphire, and quartz rods were used. The performance of the ZnO film transducers thus obtained was examined by *Foster* and *Rozgonyi* [6.4].

A redesigned dc diode sputtering system used by *Chubachi* and others [6.5,6] is shown in Fig.6.1, where the substrate is suspended beside the target (Pos.I) so that the surface of the deposited film is kept out of the plasma column during the sputtering. In the conventional sputtering system, the substrate is suspended facing the target and exposed to the plasma (Pos.II). The optimum conditions under which highly c-axis oriented ZnO films were reproducibly obtained on various substrates were as follows [cathode potential: V = 1.7~2.2 kV, cathode current: I = 20~30 mA (cathode size: 42mm in diam.), pressure of oxygen atmosphere: p = 4.8~8.5·10^{-2} torr, substrate temperature: T_s = 180~260°C, and deposition rate of the film: 0.7~0.8 μm/h]. The resistivity of the ZnO film was greater than 10^6 Ω·cm. The spatial distribution of the c-axis orientation of the ZnO crystallites was investigated by X-ray rocking curve studies and pole-figure analysis. It was demonstrated that the distribution is Gaussian with a mean value m, the inclination angle of the c-axis orientation and a standard deviation σ [6.5,6]. It was also shown that the electromechanical

206

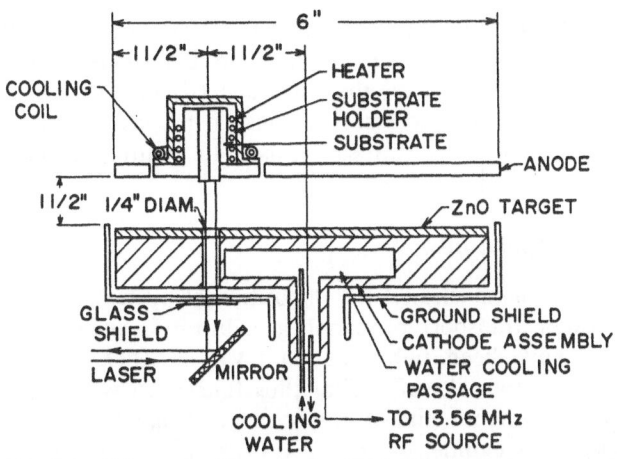

Fig.6.2. Schematic diagram of a ZnO RF sputtering apparatus [6.9]

coupling factors for longitudinal bulk waves, $k_t = 0.24 \sim 0.20$, were not affected remarkably by $\sigma < 6°$. It should be noted here that $k_t = 0.290$ for bulk ZnO crystals. Although the redesigned system has advantages compared to conventional sputtering systems, it has the difficult problem of obtaining a ZnO film with a uniform thickness over a large area.

The dc triode sputtering technique has been demonstrated by *Foster* [6.7], and *Hickernell* and *Brewer* [6.8].

6.1.2 RF Sputtering

Larson et al. [6.9] tried to fabricate a ZnO thin film on a gold film over sapphire substrates by RF diode sputtering in an 80% argon/ 20% oxygen atmosphere. The schematic diagram of the RF sputtering apparatus is shown in Fig.6.2. A plasma was excited between cathode and anode assemblies by a 13.56 MHz RF source. The ZnO target assumed a negative bias due to a blocking capacitor between cathode and RF source. Laser monitoring was employed to determine the film thickness during deposition. *Wagers* et al. [6.10] initially indicated the significance of the use of oriented gold films for the production of high-quality ZnO films. *Khuri-Yakub* et al. [6.11] carried out a detailed study of the optimum conditions for the growth of sputtered ZnO films on (111) gold, (111) silicon, (0001) sapphire, and fused quartz, by using an apparatus similar to Fig. 6.2. The best results were obtained with the following set of parameters (RF power: 160W, p = 4μm in an 80% argon/20% oxygen mixture, target-to-substrate distance: 3.81 cm, deposition rate: 100 ~ 13 nm/min., $T_s = 225 \sim 400°$C depending on the substrate). The value of $k_t = 0.25 \pm 0.01$ was obtained for the best samples. *Paradis* and *Shuskus* [6.12] fabricated sputtered ZnO films on sapphire for use in integrated optics. *Mitsuyu* et al. [6.13] produced epitaxial ZnO single-crystal films on sapphire substrates by RF sputtering. The sample configurations were $(11\bar{2}0)$ZnO/$(01\bar{1}2)$Al$_2$O$_3$ and (0001)ZnO/(0001)Al$_2$O$_3$ under the conditions of $T_s = 600°$C and a deposition rate of 3 nm/min.

The merits of dc and RF sputtering compared to CVD are as follows: (i) the substrate temperature is lower than 400°C so that there is less damage to devices already fabricated on the substrate; (ii) the surface of the film is so smooth that it is not necessary to polish it for use as optical waveguides; and (iii) laser monitoring is easily used to control the film thickness.

6.1.3 Chemical Vapor Deposition

Galli and *Coker* [6.14] first demonstrated the deposition of epitaxial layers of ZnO by chemical vapor transport on sapphire substrates. Since then, several papers [6.15-17] have been published on CVD of ZnO on sapphire. One of the serious drawbacks of a film thus fabricated is that the surface of the ZnO is not smooth and it is necessary to lap and polish them for use as an optical waveguide.

Ohnishi et al. [6.18], and *Shiosaki* et al. [6.19] showed a method for obtaining smooth surfaces on ZnO films grown on sapphire substrates using the CVD method by introducing a very thin ZnO sputtered layer as an intermediate layer. The schematic diagram of the reactor and its temperature profile are displyed in Fig.6.3. The temperature profile followed the broken line. The thermal equilibrium description of the reaction is as follows: $ZnO+H_2 \leftrightarrows Zn+H_2O$, $ZnO+H_2O \leftrightarrows Zn(OH)_2$ at the ZnO source exposed to the deoxidation gas mixture of $H_2+H_2O+N_2$, and $Zn(OH)_2 \leftrightarrows ZnO+H_2O$, $2Zn+O_2 \leftrightarrows 2ZnO$ at the substrate exposed to the oxidation gas mixture of O_2+N_2. The deposition rate of ZnO was as high as 0.1 to 1.2 μm/min. The most appropriate temperature was 975°C. Figures 6.4,5 show Scanning Electron Microscope (SEM) photographs of $(11\bar{2}0)ZnO$ $(CVD)/(1\bar{1}02) Al_2O_3$ and $(0001)ZnO(CVD)/(0001)Al_2O_3$ without the intermediate sputtered layers, respectively. Many isolated polycrystallites seen in Figs.6.4 and 5 disappeared when a saphire substrate with a ZnO sputtered layer was used.

The CVD method has the following merits: (i) a single-crystal film of high quality can be deposited on single crystals so that low-loss optical and acoustic devices can be constructed; and (ii) a ZnO film whose c-axis

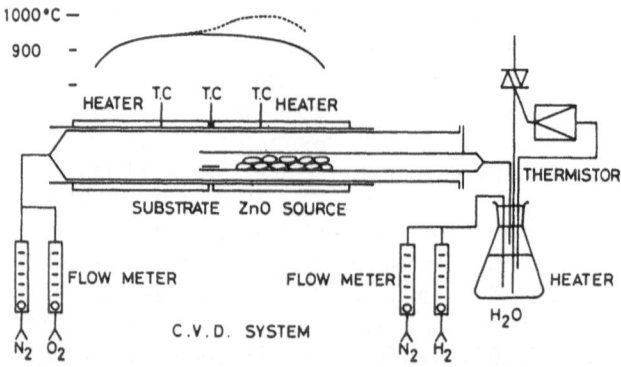

Fig.6.3. Schematic diagram of a CVD system for ZnO [6.19]

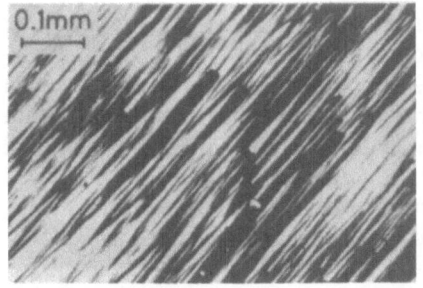
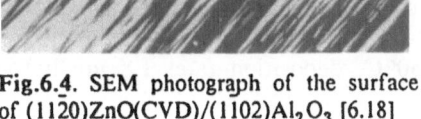

Fig.6.4. SEM photograph of the surface of (11$\bar{2}$0)ZnO(CVD)/(1$\bar{1}$02)Al$_2$O$_3$ [6.18]

Fig.6.5. SEM photograph of the surface of (0001)ZnO(CVD)/(0001)Al$_2$O$_3$ [6.18]

is in the film plane can be obtained so that a wider variety of SAO devices is possible compared to the sputtered film configurations. However, there are two problems - one is the high substrate temperature which may cause some damage on devices already fabricated on the substrate and the other is that some modification of the apparatus is necessary to use laser monitoring to control the film thickness.

6.1.4 Some Other Methods

One interesting approach to increase the sputtering-deposition rate of c-axis oriented ZnO film by one order of magnitude is to use planar magnetron equipment. One type of equipment is schematically shown in Fig. 6.6 [6.20]. In Fig.6.7 [6.20] there is a schematic configuration of the electric and the magnetic fields near the cathode. First, by introducing a planar magnet under the target holder, electrons are confined near the target. The confined electrons maintain the plasma discharge under lower background gas pressure so that the sputtered particles can reach the sub-

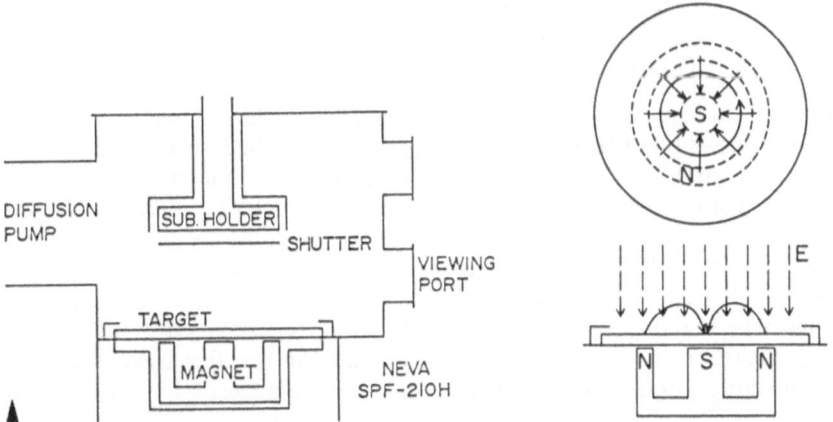

Fig.6.6. Schematic diagram of a planar magnetron sputtering equipment [6.20]

Fig.6.7. Schematic diagram of the electric and magnetic field configurations near the planar magnetron cathode [6.20]

strate more directly. Second, the secondary electrons are also trapped and fewer electrons reach the substrate so surface damage by electron bombardment and temperature rise of the substrate are suppressed. The film thus obtained has an electromechanical coupling factor comparable with that of excellent films obtained by conventional sputtering, and a standard deviation angle σ is approximately 1°. *Shiosaki* et al. [6.21] also reported the fabrication of epitaxical ZnO single-crystal films on sapphire using planar magnetron sputtering. Typical deposition parameters are as follows. Gas pressure: $5 \cdot 10^{-3}$ torr, $T_s = 280°C$, deposition rate: 1.7 μm/h for $(11\bar{2}0)ZnO/(01\bar{1}2)Al_2O_3$ and $T_s < 250°C$, deposition rate: 1.8 μm/h for $(0001)ZnO/(0001)Al_2O_3$.

Machida et al. [6.22] reported the fabrication of high-quality ZnO films on (0001) and ($1\bar{1}02$) sapphire, (111) silicon, fused quartz , and glass by using RF reactive ion plating [6.23]. The films obtained on sapphire were single crystals. *Takagi* et al. [6.24] also reported the fabrication of high-quality ZnO film on glasses by using a reactive ionized-cluster beam epitaxy [6.25].

6.1.5 Progress in ZnO Technology

Recent progress in ZnO technology has mainly been related to the progress of monolithic SAW convolvers and correlators, and has been reviewed by *Hickernell* [6.26], *Motamedi* et al. [6.27], *Minagawa* et al. [6.28], *Schwartz* et al. [6.29], and *Green* and *Kino* [6.30].

6.2 The ZnO Thin-Film Waveguide

The propagation losses of GOW in ZnO thin-film waveguides are summarized in the following subsections. It is shown that sufficiently low-optical-loss films of ZnO can be fabricated on various substrates.

6.2.1 Fused Quartz Substrate

Low-loss waveguides (less than 5 dB/cm for 0.6328 μm light) were obtained in dc diode sputtered ZnO films with polished surfaces deposited on fused quartz substrates [6.5,6]. It was found that the propagation loss increases with an increase in the electrical conductivity σ [6.5,6]. It was also found that the loss depends strongly on the substrate temperature, as shown in Fig.6.8 [6.31]. The as-grown ZnO film shows a minimum loss of about 7 dB/cm at $T_s = 220°C$. Fig.6.9 [6.31] exhibits the SEM photographs which indicate a smooth surface at $T_s = 220°C$. It is noted that fused quartz is a convenient substrate material for the basic study of acousto-optic interaction, but it may have a large acoustic loss in the GHz range.

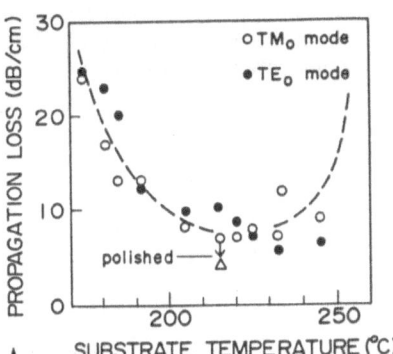

Fig.6.8. Substrate-temperature dependence of the optical propagation loss in dc diode sputtered ZnO films on fused-quartz substrates [6.31]

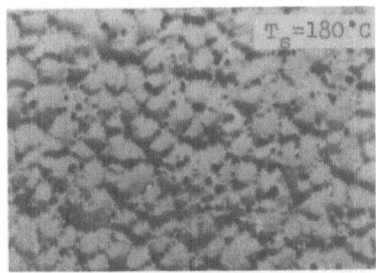

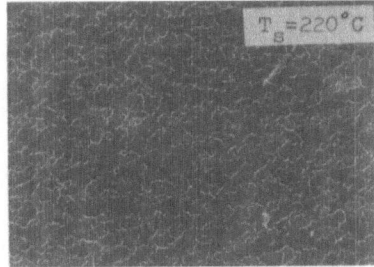

Fig.6.9. SEM photographs of the surface of dc diode sputtered ZnO films on fused-quartz substrates [6.31]

6.2.2 Sapphire Substrate

Optical waveguide losses less than 1 dB/cm were obtained in ZnO (CVD)/sapphire by *Hammer* et al. [6.15], and *Channin* et al. [6.16]. The c-axis of ZnO grown on (01$\bar{1}$2)-oriented Al_2O_3 lays in the plane of the film. The losses measured for the TE_1 mode as a function of optical wavelength λ_0 are shown in Fig.6.10, where straight lines represent the wavelength dependence of Rayleigh scattering by bulk inhomogeneities with characteristic dimensions small compared to λ_0. Sample A had a substantially rougher surface than that of sample B. It was shown that the losses can be understood as arising from two independent mechanisms. One independent mechanism is the bulk (Rayleigh) scattering from imperfections and the other is surface scattering at one or both film interfaces. The best sample of an RF sputtered ZnO film on sapphire obtained by *Paradis* and *Shuskus* [6.12] showed an optical loss as low as 2 dB/cm for a film thickness of 0.14 μm.

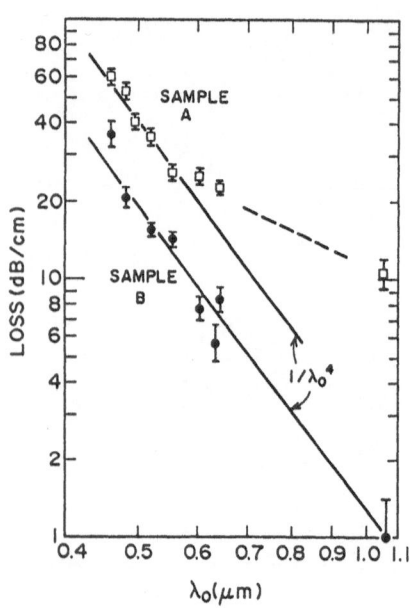

Fig.6.10. Optical propagation losses for TE_1 mode in $ZnO(CVD)/Al_2O_3$ as a function of optical wavelength λ_0 [6.15]

The chemical-vapor deposited ZnO film on an intermediately sputtered ZnO film on sapphire [6.18,19] had a smooth surface in which the 0.6328 μm light beam could be coupled to the as-grown film with a rutile prism without any post-deposition treatment. The measured optical losses are from 18.3 to 2.4 dB/cm for the seventh to second TE modes propagating perpendicular to the c-axis in a 2.57 μm thick ($1\bar{2}10$) ZnO film. As shown in Fig.6.11, the optical losses for the first- and zero-order modes deduced from the losses of higher-order modes were 1.5 and 0.7 dB/cm, respectively. It was also demonstrated that an optical strip waveguide is easily fabricated since a ZnO film grows selectively on a thin intermediate ZnO pattern. *Setsune* et al. [6.32] fabricated a ZnO/sapphire Bragg deflector in the GHz range. The ZnO film was grown by RF magetron sputtering and the propagation loss for TE_0 mode at 0.6328 μm was as low as 1.2 dB/cm.

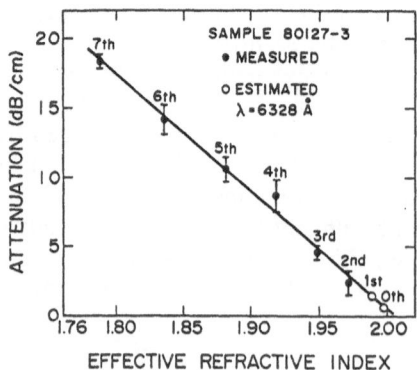

Fig.6.11. Optical propagation losses in $ZnO(CVD)/ZnO(sputtered)/Al_2O_3$ as a function of the effective refractive index of the TE mode [6.19]

It is noted that sapphire is one of the most interesting substrate materials, since there has been considerable interest and advancement in recent years in the development of epitaxical semiconductor films such as Si on sapphire for solid-state electronic device applications. Moreover, both optical and acoustic low-loss devices can be constructed on sapphire substrates because a single-crystal film can be easily deposited.

6.2.3 Si Substrate

Hickernell et al. [6.33] measured optical propagation losses in a dc triode sputtered ZnO film on thermally oxidized Si wafers together with the losses for Si_3N_4, Ta_2O_5, and 7059 glass. Typical losses in the ZnO films for the TE_0 mode of 0.6328 μm light were 1.0-2.0 dB/cm. The losses in Si_3N_4 (CVD) films were less than 1 dB/cm. It is noted here that *Stutius* and *Streifer* [6.34] fabricated similar Si_3N_4 (CVD) films on SiO_2/Si and measured losses less than 0.1 dB/cm. They showed that leakage into the silicon substrate is a major loss mechanism, especially at longer wavelengths and for higher mode numbers. Following their mathematical procedure *Sasaki* and *Mikoshiba* [6.35] calculated the leakage loss in a ZnO/ SiO_2/Si structure for the TE_0 mode of 0.6328 μm light, as shown in Fig.6.12.

The ZnO/SiO_2/Si system is particularly interesting from the view point of possible functional device development such as monolithic integration of acoustic, optical, and silicon devices on a single wafer of Si.

6.3 Excitation of SAW by ZnO Thin Films

The theory and experimental results for excitation of SAW on nonpiezoelectric materials by use of an interdigital electrode structure overlaid with a ZnO thin film are summarized in the following. It is shown that effective electromechanical coupling factors comparable to those presently obtained with the best $LiNbO_3$ delay lines can be realized. The propagation losses of SAW are also discussed.

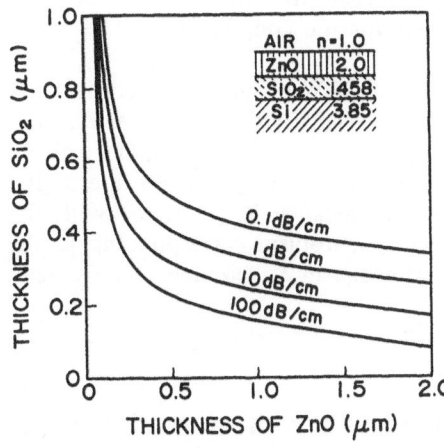

Fig.6.12. Leakage losses calculated in ZnO/SiO_2/Si system for the TE_0 mode of 0.6328 μm [6.35]

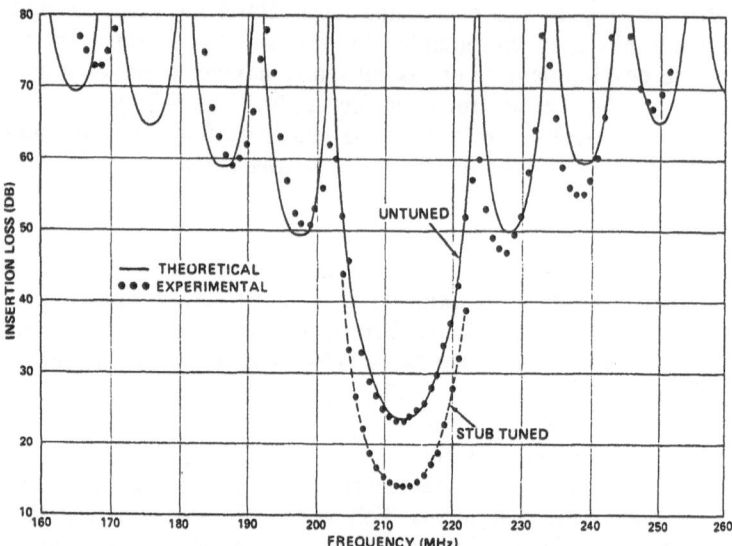

Fig.6.13. Insertion loss between adjacent transducers as a function of SAW frequency for a 4.3 μm ZnO film on oxidized silicon [6.8]

After the initial theoretical prediction that exceptionally high coupling factors are possible using piezoelectric film transducers [6.36,37] and low-frequency measurements (10-100 MHz) which showed that the theoretical relationship follows qualitatively [6.31-34], *Hickernell* and coworkers [6.8,42] carried out measurements of coupling factors, phase velocity, and propagation loss with dc triode sputtered ZnO films on SiO_2/Si and fused quartz at frequencies above 200 MHz. The interdigital electrode was on the substrate with the ZnO film overlaying the electrode. Two types of structures were used; a plated structure with a thin metal film evaporated on the ZnO film and an unplated structure with no metal film present. A typical example of the insertion loss as a function of frequency for a 4.3 μm ZnO film on a 7.8 μm SiO_2 layer on Si with an unplated structure is shown in Fig.6.13. From the best fit between theory and experiment, a value of 2.4% was obtained for the coupling factor k^2 above 200 MHz.

Detailed theories for the excitation of SAW on nonpiezoelectric materials by using an interdigital electrode overlaid with a piezoelectric film were developed by *Solie* [6.43], *Kino* and *Wagers* [6.44], *Sasaki* et al. [6.45], and *Inaba* et al. [6.46]. The series equivalent circuit for an interdigital transducer is described by an acoustic impedance Z_a in series with the capacity C_T between fingers [6.44]. At synchronism corresponding to the center frequency, the acoustic load impedance Z_a becomes purely real, i.e., R_a. It was shown that

$$R_a = (\pi/\omega C)(\Delta V_R/V_R') , \tag{6.1}$$

where ω is the angular frequency of the SAW, C is the capacitance per

214

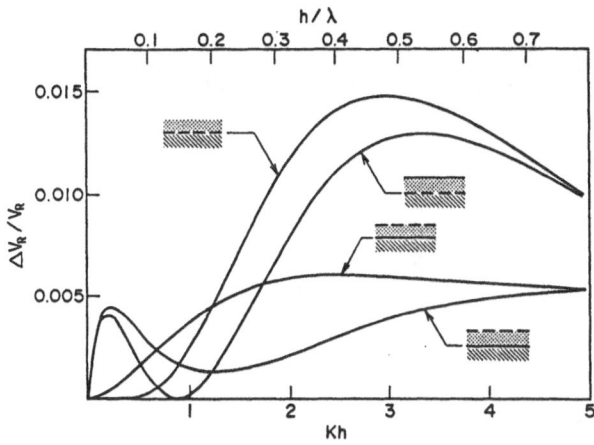

(ZX) ZnO / FUSED SILICA

Fig.6.14. Effective coupling factor $\Delta V_R/V_R$ for a (ZX)ZnO film on a fused-silica substrate, h being the thickness of the ZnO film and K the propagation constant of SAW [6.44]

finger pair, $\Delta V_R = V_R - V_R{}'$, V_R and $V_R{}'$ being the SAW velocity without and with a shorting plane, respectively. The effective coupling factor was defined as $k^2 = 2\Delta V_R/V_R$. Therefore, the values of $\Delta V_R/V_R$ and C completely specify the characteristics of the transducer. *Kino* and *Wagers* [6.44] calculated the values of $\Delta V_R/V_R$ and C in various configurations using the most reliable values for the crystal constants of ZnO given in Table 6.1 [6.47].

A typical example of the theoretical calculations of $\Delta V_R/V_R$ as a function of hK is shown in Fig.6.14, where h is the thickness of the film and K is the propagation constant of the SAW. In the figure, the results are shown in the four different configurations, i.e., (a) the interdigital electrode is located at the interface between the materials with a metal film on the top surface; (b) the same as (a) except without a metal film; (c) the interdigital electrode is located on the top surface with a metal film at the interface; and (d) the same as (c) except without a metal film. It should be remarked that there are two peaks for the cases a and c. The experiments on $\Delta V_R/V_R$ as a function of hK showed semiquantitative agreements with the theories. Theoretical calculations of capacitance for four combinations of film and substrate [6.44] are shown in Fig.6.15 when the interdigital electrode is at the interface. The status of research on ZnO thin-film SAW transducers was reviewed by *Hickernell* [6.48].

Hickernell and his co-worker [6.8,42] measured acoustic propagation losses in ZnO/SiO$_2$/Si system and obtained values of loss as low as 1.5 dB/μs at 215 MHz and 8.0 dB/μs at 630 MHz. The latter value was close to the calculated one of 6.0 dB/μs at 600 MHz in single crystal ZnO. *De Klerk* et al. [6.49] measured the loss of SAW propagated in the plane perpendicular to the c axis of single crystal ZnO and obtained the values of about 1 dB/cm at 100 MHz and about 25 dB/cm at 500 MHz. *Hickernell* et al. [6.33] also reported losses as low as 1.0 to 1.5 dB/cm at 200 MHz in

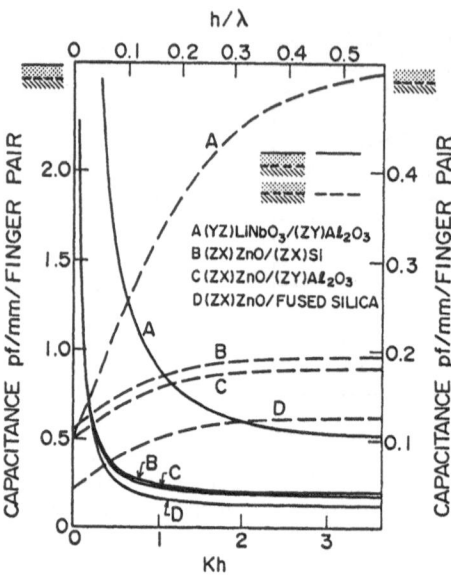

Fig.6.15. The capacitance of an interdigital transducer for four combinations of film and substrate [6.44]

A (YZ)LiNbO₃/(ZY)Al₂O₃
B (ZX)ZnO/(ZX)Si
C (ZX)ZnO/(ZY)Al₂O₃
D (ZX)ZnO/FUSED SILICA

a ZnO/SiO₂/Si system. All these results indicated that sufficiently low acoustic loss can be expected in ZnO/SiO₂/Si configurations. However, a report [6.50] showed that the loss in a system of ZnO/SiO₂/Si was extremely high, ranging from 40 dB/cm at 290 MHz to 118 dB/cm at 485 MHz and appeared to be independent of ZnO-transducer quality. The losses in the ZnO/SiO₂/Si may depend on (i) the quality of the ZnO; (ii) the relative thickness of the ZnO and the SiO₂ to the SAW wavelength; and (iii) the electronic states of the Si surface. Therefore a more detailed study will be required to resolve this apparent discrepancy on the losses of ZnO/SiO₂/Si configurations. Recently, *Minagawa* et al. [6.28] obtained a loss value of 7.5 dB/cm at 215 MHz in a ZnO/SiO₂/Si structure.

6.4 Photoelastic Effect in ZnO

The values of the photoelastic tensor elements in ZnO must be known in order to develop ZnO-thin-films SAO devices. The experimental methods used and the values of independent photoelestic tensor elements are briefly described in this section.

Sasaki et al. [6.51] measured the photoelastic constants of ZnO by comparative acoustooptic diffraction method [6.52] (reference material, fused quartz) at an acoustic frequency of 250 MHz and an optical wavelength of 0.6328 μm. Two crystals with different electrical conductivities – one highly conductive and the other highly resistive – were used as samples. Measured values of the independent photoelastic tensor elements related to nonpiezoactive strains, p_{1111}, p_{1112}, p_{3311}, p_{1313} taken with two samples are in agreement, while the values related to piezoactive strains,

p_{1133}, p_{3333}, and p_{1331}, are considerably different between the two samples (Table 6.2). This difference was interpreted as follows: The photoelastic effect in a piezoelectric (therefore electrooptic) material is described by

$$\Delta(1/\epsilon)_{ij} = p^E_{ij,kl}S_{kl} + r^S_{ij,k}E_k \tag{6.2}$$

where ϵ is the dielectric constant, p^E is the photoelastic (strain-optic) tensor measured at constant electric field, S is the acoustic strain, r^S is the electrooptic tensor measured at constant strain, and E is the electric field induced by the piezoelectricity of the crystal. The first and second terms of (6.2) represent direct and indirect photoelastic effects, respectively. The indirect effect is a two-step contribution to the photoelasticity through piezoelectric and electrooptic effects. The electric field which accompanies the acoustic wave within a piezoelectric semiconductor is given by [6.53]

$$E_r = -\frac{a_r a_s e_{s,kl} S_{kl}}{a_p \epsilon^S_{pq} a_q} \frac{1 + j(\omega/\omega_D)}{1 + j(\omega_c/\omega + \omega/\omega_D)} \tag{6.3}$$

where a is the component of the unit wave vector **a** of the acoustic wave, $\omega_c = \sigma/\epsilon^S$ (σ is the electrical conductivity) is the dielectric relaxation frequency, $\omega_D = qV^2/\mu k_B T$ (q is the electronic charge, V is the velocity of the acoustic wave, μ is the carrier mobility, k_B is the Boltzmann constant, and T is the absolute temperature) is the diffusion frequency. From (6.1, 2) we obtain

$$\Delta(1/\epsilon)_{ij} = (p^E_{ij,kl} + p^{ind}_{ij,kl})S_{kl} , \tag{6.4}$$

$$p^{ind}_{ij,kl} = -\frac{r^S_{ij,r} a_r a_s e_{s,kl}}{a_p \epsilon^S_{pq} a_q} \frac{1 + j(\omega/\omega_D)}{1 + j(\omega_c/\omega + \omega/\omega_D)} . \tag{6.5}$$

The variation of the indirect contribution with crystal conductivity σ is illustrated in Fig.6.16, where p_0^{ind} is the value of p^{ind} when $\sigma = 0$. As shown in Fig.6.16, the indirect contribution vanishes at the high conductivity limit because the electric field induced by the piezoelectricity is completely screened by free carriers. Therefore, the photoelastic tensor measured at the high resistivity limit is referred to as $p^D_{ij,kl}$ because the electric displacement D within the crystal is constant, and we obtain

$$p^D_{ij,kl} = p^E_{ij,kl} - (r_{ij,r} a_r a_s e_{s,kl}/a_p \epsilon^S_{pq} a_q) . \tag{6.6}$$

The indirect contribution at the high resistivity limit was calculated from the electrooptic data [6.54] ($r^S_{333} = +2.6 \cdot 10^{-12}$ m/V, $r^S_{113} = -1.4 \cdot 10^{-12}$ m/V) and from the piezoelectric and dielectric data (Table 6.1) ($e_{333} = +1.321$ C/m^2, $\epsilon^S_{33} = 9.03 \cdot 10^{-11}$ F/m) as follows: $p^{ind}_{1133} = -r^S_{113}e_{333}/\epsilon^S_{33} =$

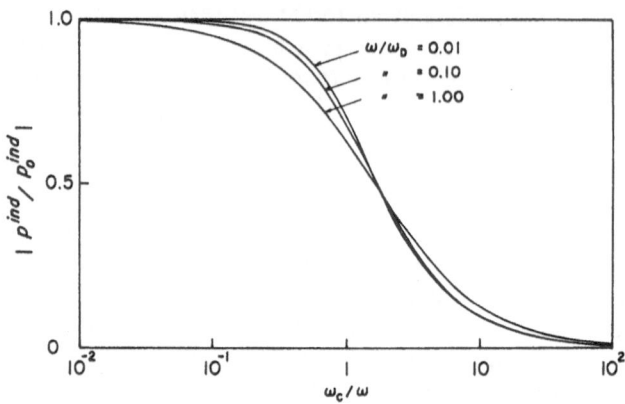

Fig.6.16. Variation of indirect photoelastic effect with crystal conductivity. ω_c is the dielectric relaxation frequency and ω_D is the diffusion frequency [6.51]

+0.0205 and $p^{ind}_{3333} = -r^S_{333}e_{333}/\epsilon^S_{33} = -0.038$. The signs of p_{1133} and p_{3333} in Table 6.2 were determined from the signs of the indirect contribution thus calculated. The measured indirect contributions were $p^{ind}_{1133} = +0.021$ and $p^{ind}_{3333} = -0.028$ in agreement with the calculated values. The measured and calculated values of p^{ind}_{1331} could not be compared because r^S_{311} was not known. It was concluded that the difference between the measured photoelastic tensor elements of the two samples is due to the difference between $p^E_{ij,kl}$ and $p^D_{ij,kl}$.

Next let us discuss the difference between p_{1331} and p_{1313} measured for the highly conductive sample. These tensor elements had been de-

Table 6.1. Elastic, piezoelectric, and dielectric constants of Li-doped ZnO single crystal (high resistivity, 800°C annealed crystal) [6.47]

c_{11}	2.096	s_{11}	7.824	$\epsilon_{11}{}^S$	7.57
c_{12}	1.205	s_{12}	-3.403	$\epsilon_{33}{}^S$	9.03
c_{13}	1.046	s_{13}	-2.196		
c_{33}	2.106	s_{33}	6.929		
				$\epsilon_{11}{}^T$	8.11
c_{44}	0.423	s_{44}	23.64		
c_{66}	0.445	s_{66}	22.46	$\epsilon_{33}{}^T$	11.19
		$\rho = 5.665$			
e_{15}	-0.480	d_{15}	-1.134	k_{15}	0.259
e_{31}	-0.573	d_{31}	-0.543	k_{31}	0.184
e_{33}	1.321	d_{33}	1.167	k_{33}	0.419
				k_t	0.290

Elastic constants $c_{ij}{}^E$ in 10^{11} N/m^2, elastic compliances $s_{ij}{}^E$ in 10^{-12} m^2/N, piezoelectric stress constants e_{ij} in C/m^2, piezoelectric strain constants d_{ij} in 10^{-11} C/N, density ρ in 10^3 kg/m^3, constant strain permittivities $\epsilon_{ij}{}^S$ in 10^{-11} F/m, constant stress permittivities $\epsilon_{ij}{}^T$ in 10^{-11} F/m, electromechanical coupling factor k_{ij}

Table 6.2. Measured photoelastic tensor elements of ZnO (optical wavelength: 0.633 μm). For the purpose of clarity, we have used the following shorthands: Prop. for propagation, Displ. for displacement, Piez. for piezoactive, Pol. for polarization, Photoel. tensor el. for photoelastic tensor elements, Hi.cond. for highly conductive sample (σ = 2.67 S/cm), and Hi.res. for highly resistive sample (σ < 10^{-8}S/cm)

Mode	Acoustic wave			Optical wave		$\dot{p}_{ij,kl}$	Photoel. tensor el.	
	Prop.	Displ.	Piez.	Prop.	Pol.		Hi.cond.	Hi.res.
L	X	X	no	$\simeq Y^a$	$\simeq X$	$\|\dot{p}_{1111}\|$	0.222	0.221
L	X	X	no	$\simeq Z$	Y	$\|\dot{p}_{1122}\|$	0.099	0.099
L	X	X	no	$\simeq Y$	Z	$\|\dot{p}_{3311}\|$	0.088	0.089
L	Z	Z	yes	$\simeq Y$	X	$\dot{p}_{1133}$	-0.111	-0.090
L	Z	Z	yes	$\simeq Y$	$\simeq Z$	$\dot{p}_{3333}$	-0.235	-0.263
S	X	Z	yes	θ_B^b to Y	Z or θ_B to X	$\dot{p}_{1331}$	-0.056	-0.061
S	Z	X	no	θ_B^c to Y	X or θ_B to Z	$\dot{p}_{1313}$	-0.061	-0.061

[a] $\simeq$ are used because the Bragg angle for longitudinal wave is less than 1°.
[b] Bragg angle (θ_B) for ordinary wave θ_0 and extraordinary wave θ_e are 15.44° and 17.05° for the highly conductive sample and 17.15° and 18.66° for the highly resistive sample, respectively.
[c] θ_0 = 13.95° and θ_e = 15.55° for the highly conductive sample, and θ_0 = 15.03° and θ_e = 16.61° for the highly resistive sample.

scribed by the same abbreviated notation (p_{55}) until the contribution of the rotation of the volume element to the photoelasticity was pointed out by *Nelson* and *Lax* [6.55]. From their theory, this contribution is written as

$$p_{(ij)[kl]} = \frac{1}{2} [(\epsilon^{-1})_{il}\delta_{kj} + (\epsilon^{-1})_{lj}\delta_{ik} - (\epsilon^{-1})_{ik}\delta_{lj} - (\epsilon^{-1})_{kj}\delta_{il}] .\qquad(6.7)$$

Using the refractive index measured for the highly conductive sample (n_0 = 1.9795 and n_e = 1.9957), $p_{(13)[13]}$ and $p_{(13)[31]}$ were calculated as $p_{(13)[13]}$ = -0.0021 and $p_{(13)[31]}$ = +0.0021. Therefore the difference between p_{1313} and p_{1331} becomes -0.0042. The signs of the measured photoelastic tensor elements p_{1313} and p_{1331} were determined from the sign of this contribution. The measured difference between p_{1313} and p_{1331} was -0.005 which was in agreement with the calculated value. For the highly resistive sample, the measured values of p_{1313} and p_{1331} were the same. This may be explained by the fact that the contribution of the volume rotation and the contribution of the indirect effect cancel out each other.

In order to determine the absolute signs of all the photoelastic tensor elements in ZnO, *Kushibiki* et al. [6.56] carried out an experiment based on the *Biegelsen* method [6.57]. All of the signs were determined to be negative so such a modification should be made in Table 6.2.

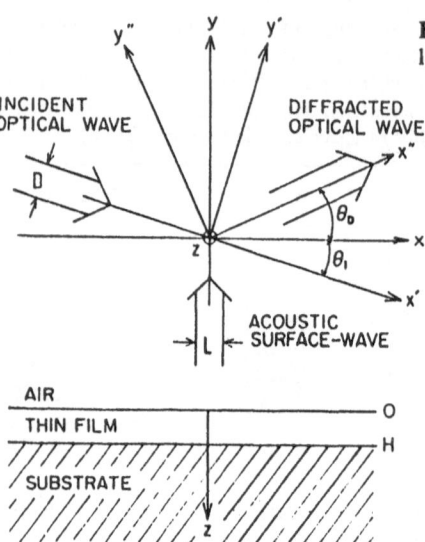

Fig.6.17. Coordinate system for the analysis of acoustooptic interaction [6.64]

INCIDENT
OPTICAL WAVE

DIFFRACTED
OPTICAL WAVE

ACOUSTIC
SURFACE-WAVE

AIR

THIN FILM

SUBSTRATE

6.5 Acousto-Optic Interaction in ZnO Thin Film

Theories of GOW Bragg diffraction in thin films by SAW have been developed [6.1,2,58-66] and treated in detail in Chap.5. In this section, we review the theory developed by *Sasaki* et al. [6.63,64] that is particularly useful to ZnO thin-film waveguides. In addition, we discuss the experiments carried out on ZnO-thin film optical waveguides deposited on fused-quartz substrates.

6.5.1 Diffraction Efficiency

Let us consider the coordinate system shown in Fig.6.17. Using a simplified matrix notation [6.67], the electric fields of the incident and diffracted GOW, and the strain of the SAW are written as

$$E_I = A_I(x) \, u_I(z) \, \exp[j(k_I x' - \omega_I t)] \; ,$$

$$E_D = A_D(x) \, u_D(z) \, \exp[j(k_D x'' - \omega_D t)] \; ,$$

$$S = A_s(x) \, u_s(z) \, \exp[j(Ky - \Omega t)] \; , \tag{6.8}$$

where boldface letters mean the matrices of vector and tensor, A is the amplitude, and u represents the spatial distribution. The explicit expressions for u can be derived by the solution of wave equations with suitable boundary conditions.

The periodic variation of the specific dielectric tensor induced by the SAW can be written as

$$\Delta\epsilon = -\epsilon\cdot\Delta B\cdot\epsilon = -\epsilon\cdot(p{:}S)\cdot\epsilon \tag{6.9}$$

where ΔB is the variation of impermeability, p is the photoelastic tensor (p^E or p^D depending on the conductivity). The polarization induced by the SAW and GOW is written as

$$P = \epsilon_0\Delta\epsilon\cdot E_I \tag{6.10}$$

where ϵ_0 is the vacuum dielectric constant, and E_I is the electric field of incident light. This polarization induces a polarization current given by

$$J = \partial P/\partial t . \tag{6.11}$$

When the phase-matching conditions (Bragg conditions),

$$k_D = k_I \pm K, \omega_D = \omega_I \pm \Omega \tag{6.12}$$

are satisfied, the average rate of work per unit volume done by the polarization current on the diffracted GOW, $\langle E_D{}^*\cdot J\rangle$, can be maximized. Let us consider the energy conservation.

$$\int_V \langle E_D{}^*\cdot J\rangle dv = \int_{\hat{s}} \langle E_D\cdot H_D{}^*\rangle d\hat{s} \tag{6.13}$$

where $\hat{s}$ is the area, v is the volume, and E_D and H_D are, respectively, the electric and magnetic fields of the diffracted light. Equation (6.13) means that the average rate of work done on the diffracted GOW is equal to the diffracted GOW power flowing out of the interaction volume. Applying (6.13) to the small region, the differential equation for A_D can be easily derived and is written as

$$\frac{\partial A_D}{\partial x} = \frac{A_I}{\cos\theta_D}\left(j\epsilon_0\omega_D A_s\int_0^\infty \Psi d\zeta\right)/4\int_{-\infty}^{+\infty} Y_D\,dz , \tag{6.14}$$

where Y_D is the spatial distribution function of the diffracted GOW power flow defined by

$$\langle E_D \times H_D{}^*\rangle\cdot n = \frac{Y_D A_D A_D{}^*}{2} \tag{6.15}$$

where n is the unit vector in the direction of the diffracted GOW power flow. In (6.14), we defined Ψ as

$$\Psi \equiv u_D{}^*\cdot\epsilon\cdot(p{:}u_s)\cdot\epsilon\cdot u_I . \tag{6.16}$$

Similar procedures lead to the differential equation for A_I:

221

$$\frac{\partial A_I}{\partial x} = \frac{A_D}{\cos\theta_I}\left(-j\epsilon_0\omega_I A_s^*\int_0^\infty \Psi^* dz\right)\Big/\left(4\int_{-\infty}^{+\infty} Y_I dz\right). \tag{6.17}$$

From (6.14, 17) with the boundary conditions $A_D = 0$ and $A_I = 1$ at $x = 0$, the diffraction efficiency, i.e., the ratio of the power of the diffracted GOW and that of the incident GOW is obtained as

$$\eta = \sin^2(\sqrt{\eta_0}) \tag{6.18}$$

where

$$\eta_0 = \frac{\pi^2}{2\lambda_0^2}\frac{LP_s}{H\cos\theta_I\,\cos\theta_D}\,Mk\Gamma\,, \tag{6.19}$$

$$M = n_I^3 n_D^3/\rho V_R^3\,, \tag{6.20}$$

$$\kappa = \int_0^H Y_s dz\Big/\int_0^\infty Y_I dz\,, \tag{6.21}$$

$$\Gamma = \frac{\epsilon_0\rho V_{Rv}^3 H\left|\int_0^\infty \Psi dz\right|^2}{n_I^3 n_D^3\int_0^H Y_s dz\int_{-\infty}^{+\infty} Y_I dz\int_{-\infty}^{+\infty} Y_D dz}\,. \tag{6.22}$$

Here, λ_0 is the optical wavelength in a vacuum, P_s and V_R are the power flow and the velocity of the SAW, respectively, ρ is the density of the film, n_I and n_D are the effective refractive indices for the incident and diffracted GOW, respectively. Finally, L and H designate the aperture and the penetration depth of the SAW, respectively. In deriving (6.18), the relation,

$$P_s = -L\int_{-\infty}^{+\infty}\langle(v\cdot T)\cdot n'\rangle\,dz \equiv \frac{L}{2}A_s A_s^*\int_0^\infty Y_s dz \tag{6.23}$$

was used, where v and T are the particle velocity and the stress of the SAW, respectively, and n' is the unit vector in the direction of SAW power flow. It should be remarked here that (i) in the small-signal limit, where the SAW power is sufficiently low, the diffraction efficiency is approximated by η_0; (ii) $M\kappa\Gamma$ corresponds to the figure of merit M_2 for the bulk acousto-optic interaction; (iii) $\kappa\Gamma$ represents the effect of spatial distribution of the GOW and SAW; and (iv) κ represents the ratio of the SAW power confined within the film to the total power.

When the Bragg condition is not completely satisfied so that there is a deviation Δk in the wave vectors, it is necessary to take account of a phase factor, $\exp(j\Delta kx)$, in (6.14, 17). In such a situation, we obtain

$$\eta = \eta_0 \left[\frac{\sin\sqrt{\eta_0 + (L\Delta k/2)^2}}{\sqrt{\eta_0 + (L\Delta k/2)^2}} \right]^2 \tag{6.24}$$

which is the same as the result obtained in the bulk acousto-optic interaction [6.68], and takes the same form as that given in Chap.5.

It is also noted that (6.18-22) hold under the condition of $L \ll D$, in which D designates the aperture of the incident light. In general, the effect of the interaction geometry must be taken into account. In the small-signal limit, we obtain

$$\eta = g\eta_0 \tag{6.25}$$

where

$$g = \begin{cases} 1-\alpha/3 & (\alpha < 1) \\ (3\alpha-1)/3\alpha^2 & (\alpha > 1) \end{cases}, \tag{6.26}$$

$$\alpha \equiv L \frac{\sin(\theta_I + \theta_D)}{D\cos\theta_D}. \tag{6.27}$$

This geometry effect is the same as that for the bulk-wave acousto-optic interaction [6.69].

For the case of the collinear interaction where the propagation directions of the incident and diffracted GOW and of the SAW are on the same straight line, the diffraction efficiency is given by (6.18) with

$$\eta_0 = \frac{\pi^2}{2\lambda_0^2} \frac{P_s}{LH} M\kappa\Gamma\ell^2 \tag{6.28}$$

where ℓ is the interaction length.

The spatial distributions of the electric fields and powers of the GOW calculated for a ZnO film whose c-axis is normal to the fused-quartz substrate are shown in Fig.6.18 [6.64]. Figure 6.19 [6.64] shows the spatial distributions of the strains and powers of the SAW. It should be noted that the field distributions of the SAW exhibit a marked change with the change in film thickness, while the distributions of the GOW show little change.

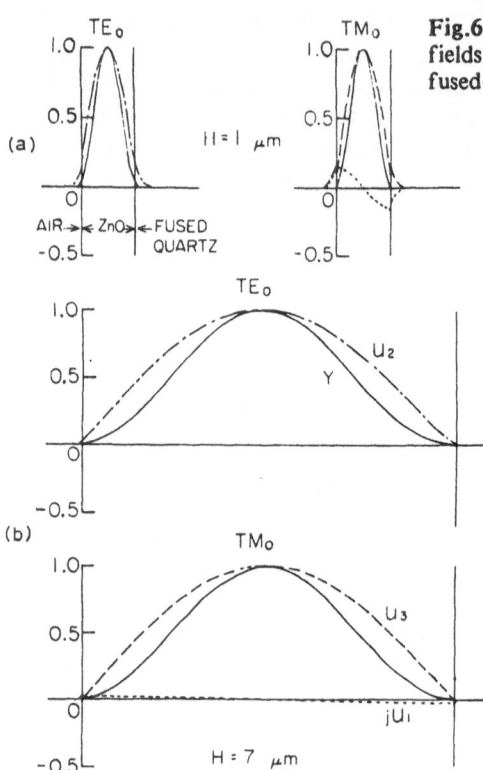

Fig.6.18. Spatial distribution of the electric fields and power of GOW in a ZnO film on fused-quartz substrate [6.64]

Let us now discuss the experiments [6.64,70,71] on Bragg diffraction between the same GOW modes ($TE_0 \rightarrow TE_0$, $TM_0 \rightarrow TM_0$) by a SAW of 130 MHz in a dc sputtered ZnO-film optical waveguide fabricated on a fused-quartz substrate. The experimental arrangement is schematically shown in Fig.6.20. From (6.12), the Bragg angle θ_B between the same GOW modes ($n_I = n_D = n$) is given by

$$|\sin\theta_B| = \lambda_0/2\Lambda n \quad (\theta_I = -\theta_D) , \tag{6.29}$$

where Λ is the wavelength of the SAW. Equation (6.29) corresponds to the isotropic diffraction in the case of bulk-wave acousto-optic Bragg diffraction. The film thickness of the optical waveguides used was 0.6-0.8 μm, while the film thickness of the SAW transducers was chosen to be H = 9.6 μm (KH = 3, K is the wavenumber of the SAW) for all samples to make the effective coupling factor maximum [6.45]. Interdigital electrodes were formed on the fused-quartz substrates by ordinary photolithographic techniques. The transducers had 20 finger pairs, a 4 mm aperture length, and a 20 μm period. A typical insertion loss of about 20 dB was measured when the distance between sending and receiving transducers was 15 mm. A 0.6328 μm HeNe laser beam was used for the light source and a pair of rutile prisms were employed for input and output couplers.

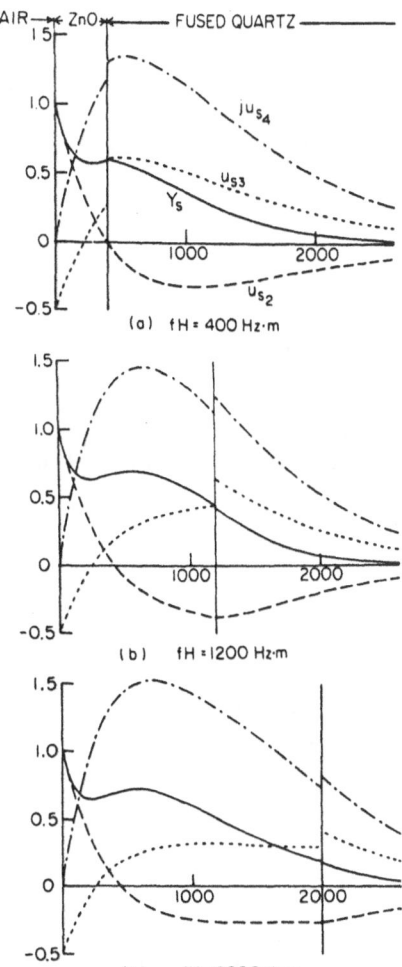

Fig.6.19. Spatial distribution of the strains and power of SAW in a ZnO film on fused-quartz substrate [6.64]

(a) $fH = 400\ Hz \cdot m$

(b) $fH = 1200\ Hz \cdot m$

(c) $fH = 2000\ Hz \cdot m$

Experimental results for $TE_0 \rightarrow TE_0$ diffraction in the case of a film thickness of 0.6 μm are shown in Fig.6.21 as functions of acoustic power (curve a) and the square root of acoustic power (curve b), respectively. For diffraction efficiencies less than 30%, the efficiency is proportional to the acoustic power in accordance with the theory. Also, the overall behaviour is close to the theoretical sine-squared dependence on the square root of the acoustic power, except for the maximum value of 90%. The discrepancy may be due to the fact that (i) the criterion for the Bragg regime was not completely satisfied; (ii) there was scattering of the GOW in the waveguide; and (iii) there was misalignment of the interaction.

Figure 6.22 shows the film-thickness dependence of the diffraction efficiency within the small signal limit where the efficiency is proportional to the acoustic power. It is observed that the efficiencies depend strongly on the film thickness and there exists a film thickness for each mode where the efficiency goes to zero. The thickness dependence of dif-

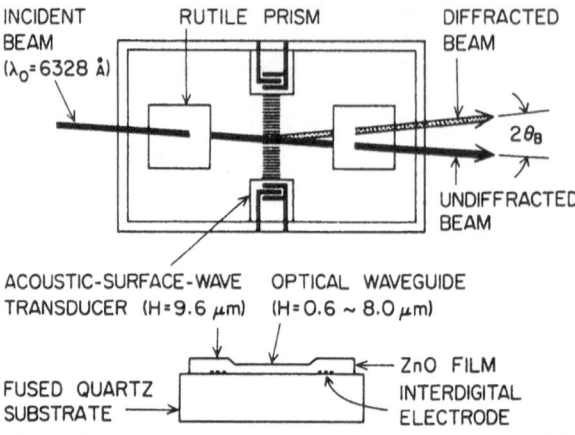

Fig.6.20. Experimental arrangement for the Bragg diffraction in a ZnO-film optical waveguide on a fused-quartz substrate [6.71]

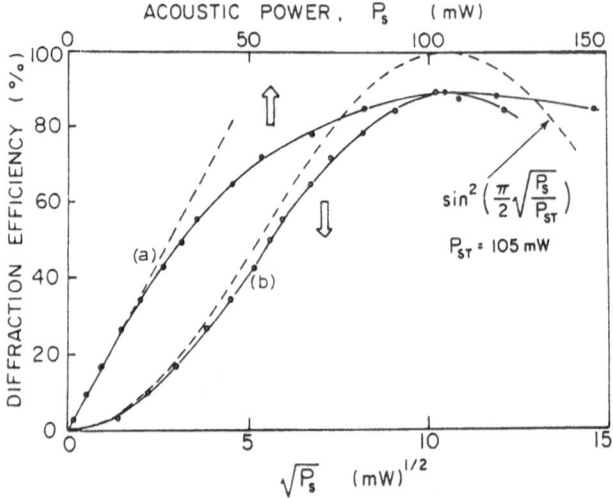

Fig.6.21. Experimental results of $TE_0 \rightarrow TE_0$ diffraction as a function of the acoustic power (curve *a*) and as a function of the square root of the acoustic power (curve *b*) [6.64]

fraction efficiency observed can be well explained by the theoretical calculations, although the theoretical curves for the TE_0 modes are shifted as a whole to a thicker region compared to the experimental curves. One of the reasons for such a discrepancy is due to the fact that the values of the material constants of ZnO single crystal used for the calculation differ from those of ZnO films employed in the experiments. It was shown that the essential origin of the observed thickness dependence is due to the variation of the spatial distributions of the acoustic strains with the ZnO-film thickness.

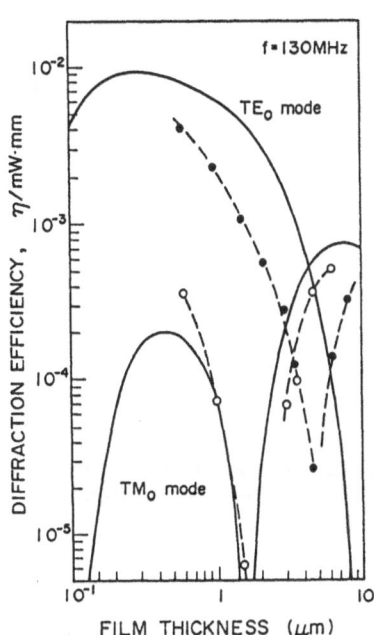

Fig.6.22. Experimental and theoretical results on the film-thickness dependence of the diffraction efficiency between the same optical modes in ZnO-film waveguides on fused-quartz substrates ([6.71]; theoretical curves were recalculated using the values of photoelastic constants given in Table 6.2)

An improvement of the diffraction efficiency in a ZnO-film waveguide by introducing a strain-controlling film of fused-quartz on a ZnO/(fused-quartz)system was proposed and demonstrated [6.72].

6.5.2 Bragg Condition

From (6.12), the generalized Bragg condition between different GOW modes is given by

$$\sin\theta_I = \pm \frac{\lambda_0}{2\Lambda n_I} [1 + (\Lambda/\lambda_0)^2 (n_I{}^2 - n_D{}^2)] \, ,$$

$$\sin\theta_D = \mp \frac{\lambda_0}{2\Lambda n_D} [1 + (\Lambda/\lambda_0)^2 (n_D{}^2 - n_I{}^2)] \, . \qquad (6.30)$$

Diffraction between different GOW modes, which correspond to anisotropic diffraction in the case of bulk-wave acousto-optic interaction can be divided into two categories: one is the diffraction without polarization rotation ($TE_m \leftrightarrows TE_n$, $TM_m \leftrightarrows TM_n$, $m \neq n$) and the other is the diffraction with polarization rotation ($TE_m \leftrightarrows TM_n$). It should be remarked that there exists a lower cut-off acoustic frequency f_c and a normal diffraction frequency f_n at which one of the Bragg angles becomes zero. The expressions for f_c and f_n are given by

$$f_c = (V_R/\lambda_0)|n_I - n_D| \qquad \text{and} \qquad (6.31)$$

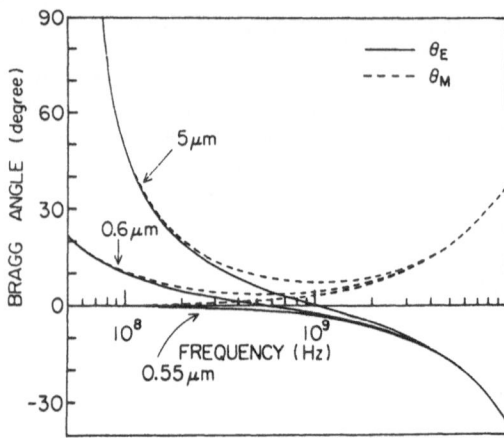

Fig.6.23. Calculated frequency dependence of Bragg angles for diffraction with the mode conversion between TE_0 and TM_0 modes. The parameter is ZnO-film thickness [6.64]

$$f_n = (V_R/\lambda_0)\sqrt{|n_I^2 - n_D^2|} ,\qquad(6.32)$$

respectively.

Some examples of the calculated frequency dependence of the Bragg angles for $TE_0 \leftrightarrows TM_0$ mode conversion are shown in Fig.6.23 [6.64] with the film thickness as a parameter. For the case of 0.55 μm film the condition becomes isotropic since the TE_0 and TM_0 modes are degenerate at this film thickness.

6.5.3 Mode Conversion; $TE \leftrightarrows TM$, $TM_i \leftrightarrows TM_j$

Experiments on the $TE \leftrightarrows TM$ mode conversion was carried out by *Sasaki* et al. [6.73]. The experimental arrangement was essentially the same as that in Fig.6.20. The TE and TM waves enter the rutile prism as extraordinary and ordinary waves, respectively. Thus, these two waves emerge from the output prism as two separated beams. The thickness of the ZnO film at the transducer was 5.3 μm to maximize the effective coupling factor for f(SAW) = 226 MHz [6.45]. The experimental results for the efficiency of the mode conversion between the TE_0 and TM_0 modes are shown in Fig.6.24 [6.73] as a function of the acoustic power. The parameter is the thickness of the ZnO film. Efficient mode conversion as high as 88% was obtained for a film thickness of 5 μm at an acoustic power of 350 mW. It is shown that the thickness dependence of the mode conversion efficiency is well explained by the theoretical calculation. The mode conversion between TE_n and TM_n modes was also observed in the 5 μm thick waveguide from n = 1 to n = 7.

Finally, let us discuss the collinear interaction which can be utilized for acoustooptic tunable filters [6.74]. The Bragg condition and the diffraction efficiency when there is a wave number mismatch Δk are given by

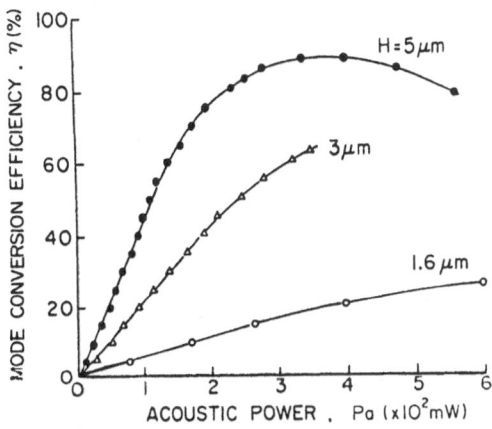

Fig.6.24. Experimental results of mode conversion between TE_0 and TM_0 modes, the parameter being film thickness of ZnO on fused-quartz substrate [6.73]

$$k_0(n_I - n_D) = K \tag{6.33}$$

and

$$\eta = \eta_0 \left[\frac{\sin\sqrt{\eta_0 + (L\Delta k/2)^2}}{\sqrt{\eta_0 + (L\Delta k/2)^2}} \right]^2 , \tag{6.34}$$

respectively, where k_0 is the optical wavenumber in vacuum and η_0 is given by (6.28). In the case where the c axis of the ZnO is normal to the fused-quartz substrates, the collinear interaction between TE and TM modes is impossible because off-diagonal elements of the photoelastic tensor needed for the TE-TM mode conversion vanish in this configuration. However, the collinear interaction is possible for the case of the mode conversion without polarization rotation, i.e., $TE_i \leftrightarrows TE_j$ and $TM_i \leftrightarrows TM_j$ mode conversions. In these cases, the incident and diffracted modes may be separated by the use of a prism coupler or a mode selective branching waveguide [6.75].

Using the data for dispersion of refractive indices of ZnO [6.76], the collinear-interaction frequency (acoustic tuning frequency) was evaluated from (6.33). The results are shown in Fig.6.25 as a function of optical wavelength. The parameter H is the thickness of the ZnO film. This figure shows the tuning characteristics of a ZnO-film acousto-optic filter. For example, tuning from 0.45 to 0.63 μm can be achieved by changing the SAW frequency from 240 to 335 MHz for H = 1 μm. When the SAW frequency is fixed and the optical wavelength is changed, the diffraction efficiency varies as shown in the inset in Fig.6.26. The resolution of the acoustooptic filter is defined by the wavelength difference between $\ell\Delta k = 0$ and $\pi(3)^{1/2}$ and is given by

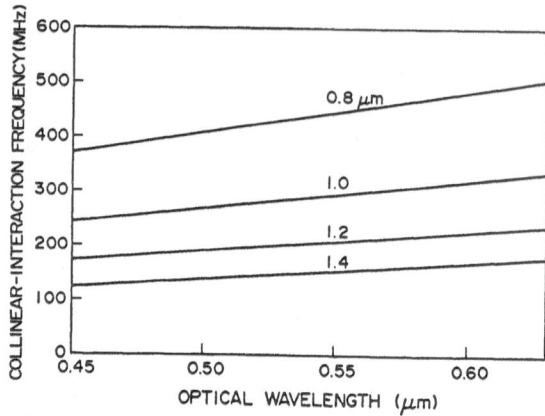

Fig.6.25. Collinear-interaction (i.e., acoustic tuning) frequency for TM_0-TM_1 mode conversion in ZnO/(fused-quartz) system. H is the thickness of the ZnO film [6.74]

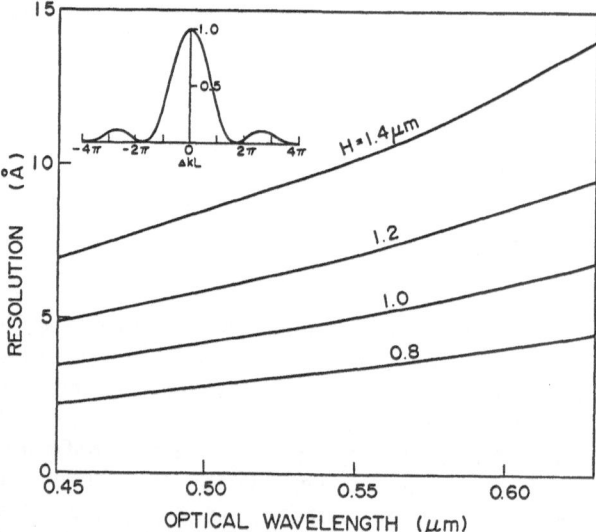

Fig.6.26. Resolution of tunable acoustooptic filters which utilize TM_0-TM_1 mode conversion in ZnO/(fused-quartz) system. Interaction length $\ell = 1$ cm [6.74]

$$\Delta\lambda_0 \simeq \sqrt{3} \, \frac{\lambda_0^2}{2\ell} \, \frac{1}{\Delta n - \lambda_0 \, (\partial\Delta n)/(\partial\lambda_0)} \, , \qquad (6.35)$$

where $\Delta n = n_I - n_D$. The resolution evaluated from (6.35) is shown in Fig.6.26 for $\ell = 1$ cm. The resolutions are several 0.1 nm for H = 1.0 μm. Small-signal diffraction efficiency was calculated as $\eta_0 \simeq 10$ (W·cm)$^{-1}$ at $\lambda_0 = 0.6328$ μm for H = 0.8 μm. This efficiency is comparable with that obtained in the TE_0-TM_0 collinear interaction in a Ti-diffused LiNbO$_3$ optical waveguide [6.77].

230

6.6 Combined Structure of ZnO and Other Thin Films

Structures that combine an independent optical-waveguiding film of high optical quality (less than 1 dB/cm loss) and a good piezoelectric ZnO film for the construction of practical SAO devices on nonpiezoelectric substrates have been investigated. One of the structures investigated was to fabricate the waveguide on the surface of F2 glass by an ion-exchange technique [6.78], and the other was to fabricate thin film waveguides of Si_3N_4, Ta_2O_5, and 7059 glass on a SiO_2/Si substrate [6.33, 79]. The latter is interesting in view of the compatibility with silicon technology as mentioned previously.

Although there is no clear-cut guideline for the choice of materials in the combined-structure SAO devices, the following items must be considered carefully: (1) Films with a high acousto-optic figure of merit are preferable. This means that the film must have a large refractive index n, large photoelastic constant p, and low acoustic velocity. Generally, a low acoustic velocity is not consistent with low acoustic propagation loss; (2) since the SAW usually penetrates into the substrate, both single crystal film and substrate are preferable for low-loss acoustic devices; (3) to obtain low optical losses, the following points must be considered: (3a) films with a high transparency for the wavelength of light are preferable; (3b) single crystalline films are preferable for weak Rayleigh scattering; (3c) a film with a smooth interface between the film and substrate are preferable for weak surface scattering; and (3d) when a buffer layer such as SiO_2 in a ZnO/ SiO_2/Si system is used, the leakage loss must be considered.

Before closing this chapter, I would like to make a comment that a ZnO film is not necessarily the best piezoelectric film. Other potentially useful films such as AlN, $LiNbO_3$, etc. should be investigated more intensively.

Acknowledgement. I am much indepted to Dr. H. Sasaki for his invaluable help in preparing this manuscript. I am also indepted to Professor C.S. Tsai for the time and energy devoted to editing the manuscript.

References

6.1   L. Kuhn, M.L. Dakss, P.F. Heidrich, B.A. Scott: Appl. Phys. Lett. 17, 265 (1970)
       L. Kuhn, P.F. Heidrich, E.G. Lean: Appl. Phys. Lett. 19, 428 (1971)
6.2   C.S. Tsai: IEEE Trans. CAS-64, 1072 (1979)
6.3   G.A. Rozgonyi, W.J. Polito: Appl. Phys. Lett. 8, 220 (1966)
6.4   N.F. Foster, G.A. Rozgonyi: Appl. Phys. Lett. 8, 221 (1966)
6.5   N. Chubachi: Proc IEEE 64, 772 (1976)
6.6   N. Chubachi, M. Minakata, Y. Kikuchi: Jpn. J. Appl. Phys. 13, (Suppl.2, Pt.1) 737 (1974)
6.7   N.F. Foster: J. Appl. Phys. 40, 4202 (1969)
6.8   F.S. Hickernell, J.W.Brewer: Appl. Phys. Lett. 21, 389 (1972)
6.9   J.D. Larson, D.K. Winslow, L.T. Zitelli: IEEE Trans. SU-19, 18 (1972)

6.10 R.S. Wagers, G.S. Kino, P. Galle, D.K. Winslow: Proc. 1972 Ultrasonics Symp. p.194

6.11 B.T. Khuri-Yakub, G.S. Kino, P. Galle: J. Appl. Phys. **46**, 3266 (1975)

6.12 E.L. Paradis, A.J. Shuskus: Thin Solid Films **38**, 131 (1976)

6.13 T. Mitsuyu, S. Ono, K. Wasa: J. Appl. Phys. **51**, 2464 (1980)

6.14 G. Galli, J.E. Coker: Appl. Phys. Lett. **16**, 439 (1970)

6.15 J.M. Hammer, D.J. Channin, M.T. Duffy, J.P. Wittke: Appl. Phys. Lett. **21**, 358 (1972)

6.16 D.J. Channin, J.M. Hammer, M.T. Duffy: Appl. Opt. **14**, 923 (1975)

6.17 M. Kasuga, S. Ishihara: Jpn. J. Appl. Phys. **15**, 1835 (1976)

6.18 S. Ohnishi, Y. Hirokawa, T. Shiosaki, A. Kawabata: Jpn. J. Appl. Phys. **17**, 773 (1978)

6.19 T. Shiosaki, S. Ohnishi, Y. Hirokawa, A. Kawabata: Appl. Phys. Lett. **33**, 406 (1978)

6.20 T. Shiosaki: Proc. 1978 IEEE Ultrasonics Symp. p.100

6.21 T. Shiosaki, S.Ohnishi, Y.Murakami, A.Kawabata: J. Crystal Growth **45**, 346 (1978)

6.22 M. Machida, M. Shibutani, T. Murai, Y. Murayama: Jpn. J. Appl. Phys. Suppl. **20-3**, 141 (1981)

6.23 Y. Murayama: Jpn. J. Appl. Phys. **13**, Suppl.2, Pt.1, 459 (1974)

6.24 T. Takagi, K. Matsubara, H. Takaoka, I. Yamada: Thin Solid Films **63**, 41 (1979)

6.25 T. Takagi, I. Yamada, A. Sasaki: J. Vac. Sci. Tech. **12**, 1128 (1975)

6.26 F.S. Hickernell: 1979 IEEE Trans. SU-32, 621 (1985)

6.27 M.E. Motamedi, M.K. Kilcoyne, R.K. Asatourian: IEEE Trans. SU-32, 663 (1985)

6.28 S. Minagawa, T. Okamoto, T. Niitsuma, K. Tsubouchi, N. Mikoshiba: IEEE Trans. SU-32, 670 (1985)

6.29 S.S. Schwartz, R.L. Gunshor, R.F. Pierret: IEEE Trans. SU-32, 707 (1985)

6.30 J.B. Green, G.S. Kino: IEEE Trans. SU-32, 734 (1985)

6.31 J. Kushibiki, H. Sasaki, N. Chubachi, Y. Kikuchi: Preprints Spring Meeting Jpn. Soc. Appl. Phys. (April 1974) p.A1 (in Japanese)

6.32 K. Setsune, T. Mitsuyu, O. Yamazaki, K. Wasa: Proc 1983 IEEE Ultrasonics Symp. p.467

6.33 F.S. Hickernell, R.L. Davis, F.V. Richard: Proc. 1978 IEEE Ultrasonics Symp. p.60

6.34 W. Stutius, W. Streifer: Appl. Opt. **16**, 3218 (1977)

6.35 H. Sasaki, N. Mikoshiba: unpublished (1978)

6.36 L.P. Solie: Appl. Phys. Lett. **18**, 111 (1971)

6.37 W.R. Smith: J. Appl. Phys. **42**,3016 (1971)

6.38 C.P. Sandbank, M.B.N. Butler: Electron. Lett. **7**, 499 (1971)

6.39 D.R. Evans, M.F. Lewis, E. Patterson: Electron. Lett. **7**,557 (1971)

6.40 F.S. Hickernell: In *Acoustic Surface Wave and Acousto-Optic Devices*, ed. by T.Kallard (Optosonic, New York 1971) p.31

6.41 F. Pizzarello: J. Appl. Phys. **43**, 3627 (1972)

6.42 F.S. Hickernell: J. Appl. Phys. **44**, 1061 (1973)

6.43 L.P. Solie: J. Appl. Phys. **44**, 619 (1973)

6.44 G.S. Kino, R.S. Wagers: J. Appl. Phys. **44**, 1480 (1973)

6.45 H. Sasaki, N.Chubachi, Y.Kikuchi: Electron. Lett. **9**, 92 (1973)

6.46 R. Inaba, K. Kajimura, N. Mikoshiba: J. Appl. Phys. **44**, 2495 (1973)

6.47 R.T. Smith, V.E. Stubblefield: J. Acoust. Soc. Am. **45**, 105 (1969)

6.48 F.S. Hickernell: Proc. IEEE **64**, 631 (1976)

6.49 J.de Klerk, R.W. Weinert, B.R. McAvoy: Proc. 1978 IEEE Ultrasonics Symp. p.667

6.50 D. Mergerian, E.C. Malarkey, B.A. Newman, J.R. Lane, R.W. Weinert, B.R. McAvoy, C.S. Tsai: Proc. 1978 IEEE Ultrasonics Symp. p.64

6.51 H. Sasaki, K. Tsubouchi, N. Chubachi, N. Mikoshiba: J. Appl. Phys. **47**, 2046 (1976)

6.52 R.W. Dixon, M.G. Cohen: Appl. Phys. Lett. **8**, 205 (1966)

6.53 D.L. White: J. Appl. Phys. **33**, 2547 (1962)

6.54 D.A. Pinnow: In *Handbook of Lasers with Selected Data of Optical Technology*, ed. by R.J. Pressley (Chemical Rubber, Cleveland, Ohio 1971) p.480

6.55 D.F. Nelson, M. Lax: Phys. Rev. B3, 2778 (1971)

6.56 J. Kushibiki, N. Chubachi, N. Mikoshiba: J. Appl. Phys. **49**, 4961 (1978)

6.57 D.K. Biegelsen: Appl. Phys. Lett. 22, 221 (1973)

6.58 Y. Ohmachi: J. Appl. Phys. **44**, 3928 (1973)

6.59 C.S. Tsai, M.A. Alhaider, L.T. Nguyen, B. Kim: Proc. IEEE **64**, 318 (1976)

6.60 B. Kim, C.S.Tsai: Proc. IEEE **64**, 788 (1976)

6.61 R.V. Schmidt: IEEE Trans. SU-23, 22 (1976)

6.62 E.G.H. Lean, J.M. White, C.D.W. Wilkinson: Proc IEEE **64**, 779 (1976)

6.63 H. Sasaki, J. Kushibiki, N. Chubachi, Y. Kikuchi: Trans. Inst. Electron. & Commun. Eng. (Japan) 57-C.419 (1974)

6.64 N. Chubachi, H. Sasaki: Wave Electron. 2,379 (1976)

6.65 L.T. Nguyen, C.S. Tsai: Appl. Opt. 16, 1297 (1977)

6.66 C.S. Tsai: Proc. Soc. of Photo-Optical Instr. Eng. **139**, 132 (1978)

6.67 R. Holland: IEEE Trans. SU-14, 18 (1967)

6.68 P. Phariseau: Proc. Indian Acad. Sci. 44A, 165 (1956)

6.69 D.H. McMahon: IEEE Trans. SU-16, 41 (1969)

6.70 J. Kushibiki, H. Sasaki, N. Chubachi, N. Mikoshiba, K. Shibayama: 1974 Ultrasonics Symp. Proc. (1974) p.85

6.71 J. Kushibiki, H. Sasaki, N. Chubachi, N. Mikoshiba, K. Shibayama: Appl. Phys. Lett. **26**, 362 (1975)

6.72 J. Kushibiki, N. Chubachi, K. Shibayama: Appl. Phys. Lett. **29**, 333 (1976)

6.73 H. Sasaki, J. Kushibiki, N. Chubachi: Appl. Phys. Lett. **25**, 476 (1974)

6.74 H. Sasaki, N. Mikoshiba: Proc. 1979 IEEE Ultrasonics Symp. p.18

6.75 H. Yajima: Proc. Symp. Optical & Acoustical Micro-Electronics (New York, 1974) p.339

6.76 N. Sakagami: Dissertation, Tohoku University (1977)

6.77 K. Yamanouchi, K. Higuchi, K. Shibayama: Trans. Inst. Electron. & Commun. Eng. (Japan) J 61-C, 47 (1978)

6.78 M. Kondo, Y. Ohta, F. Saito, H. Sasaki, N. Chubachi, K. Koizumi: Electron. Lett. **10**, 518 (1974)

6.79 S.K. Yao, R.R. August, D.B. Anderson: Proc. 1977 IEEE Ultrasonics Symp. p.439

7. Spectrum Analysis with Integrated Optics

Michael C. Hamilton and Anthony E. Spezio

With 18 Figures

The potential efficiency, bandwidth, small size and low cost of Integrated-Optical (IO) circuits makes them attractive for a variety of applications. *Schubert* and *Harris* [7.1] first proposed the use of IO technology for one-dimensional analog optical processing. Coherent analog optical processors have not found wide application because of excessive size, cost and environmental susceptibility. At least for one-dimensional applications, IO technology may remove those obstacles.

Spectrum analysis of Radio Frequency (RF) signals using acousto-optic Bragg diffraction is an example of an analog optical processor which could benefit from IO technology. The dependence of the diffraction angle on acoustic wavelength makes it possible for an acousto-optic Bragg cell to channelize received signals according to frequency. This permits unambiguous determination of the amplitude and temporal characteristics of simultaneous signals by post-detection processing. Many applications which would benefit from an acousto-optic spectrum analyzer place a premium on small size and low cost. The utilization of IO technology in place of "bulk" optics technology should help in the realization of those objectives. While Chap.5 treats guided-wave acousto-optic Bragg cells, this chapter deals with the operating principles of a resulting Integrated-Optical Spectrum Analyzer (IOSA). The technology of the components will be reviewed in detail. The IOSA is perhaps the simplest form of an analog optical processor. Variations of the IOSA and similar configurations for performing other signal processing functions such as correlation and convolution will be briefly reviewed. A more detailed treatment on correlation and convolution as well as an update on related IO modules are given, respectively, in Chaps.5 and 8.

7.1 Acousto-Optic Spectrum Analysis

Although the idea of applying acousto-optic Bragg diffraction to RF spectrum analysis is old, it has not recieved serious attention until recent years during which bandwidths and efficiencies have reached attractive levels. An excellent review of the features of acousto-optic spectrum analysis has been written by *Hecht* [7.2]. The heart of the Acousto-Optic Spectrum Analyzer (AOSA) is the Bragg cell in which an acoustic beam produces a traveling optical phase grating via the photoelastic effect and the electro-optic effect, if the cell material is piezoelectric. The interac-

Springer Series in Electronics and Photonics, Vol. 23
Guided-Wave Acoustooptics   Editor: C. S. Tsai
© Springer-Verlag Berlin, Heidelberg 1990

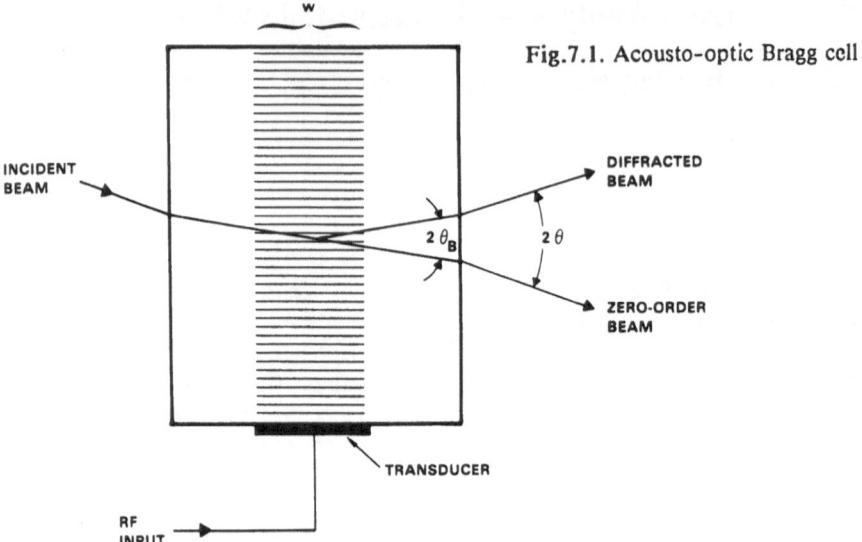

Fig.7.1. Acousto-optic Bragg cell

tion is depicted in Fig.7.1. If the acoustic beamwidth L is large enough such that

$$L \geq \frac{2n_B \Lambda^2}{\lambda_0} , \tag{7.1}$$

then essentially all the diffracted light appears at an angle θ (external to the cell) relative to the direction of the input beam. Here, n_B is the refractive index of the Bragg cell, Λ is the acoustic wavelength and λ_0 is the free space optical wavelength. The angle θ is given by

$$\theta = 2\sin^{-1}\left[\frac{n_B}{n}\sin\theta_B\right] = 2\sin^{-1}\left[\frac{\lambda_0 f}{2nV}\right] \simeq \frac{\lambda_0 f}{nV} , \tag{7.2}$$

where f is the acoustic frequency, v is the acoustic velocity, n is the refractive index external to the Bragg cell, and θ_B is the angle of incidence (Bragg angle) between the optical beam and the acoustic wave fronts for which diffraction is maximized. Efficient diffraction occurs over the bandwidth Δf provided that the acoustic-beam spread is at least as great as the variation in θ_B across the band; that is,

$$L \leq \frac{2n_B V^2}{\lambda_0 f_m \Delta f} , \tag{7.3}$$

where f_m is the high-frequency extreme of Δf. Since the diffraction efficiency is proportional to L, a compromise must be struck between bandwidth and efficiency.

236

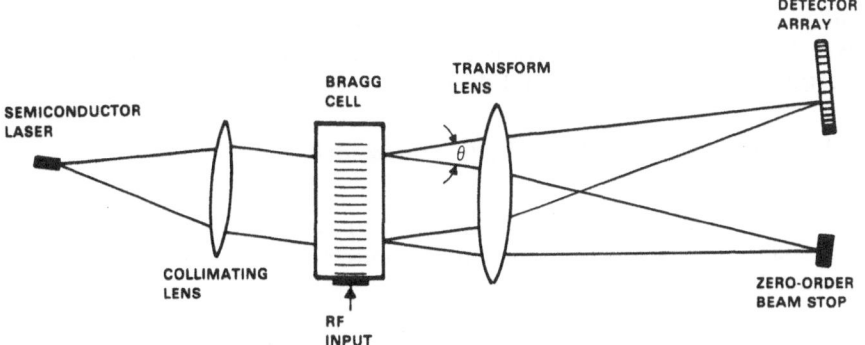

Fig. 7.2. Acousto-optic RF spectrum analyzer

The overall optical configuration of an AOSA is shown in Fig. 7.2. Light from a laser is collimated and passed through an acousto-optic Bragg cell. An RF signal of frequency f applied to the transducer results in a diffracted beam at the angle θ with respect to the incident collimated beam which is focused onto the detector array at position $x = \theta F$, where F is the transform-lens focal length. It is assumed that the signal has been amplified and frequency shifted, if necessary, to fall within the processing band. A very important feature of the diffraction process is that for low efficiencies, the interaction is linear to the first order so that multiple, simultaneous signals are processed in parallel over a wide instantaneous band.

At a given instant of time, the intensity distribution $I(K)$ at the detector array is proportional to the square of the Fourier transform of the function

$$q(x') = r(x')s(x')t(x') , \qquad (7.4)$$

where x' is the coordinate along the direction of acoustic propagation. The function $r(x')$ is the instantaneous amplitude of the acoustic representation of the signal in the Bragg cell, $s(x')$ represents the exponential attenuation of the acoustic beam and $t(x')$ accounts for the apodization of the optical beam, typically Gaussian-shaped. Thus,

$$I(K) = \left| \int_{-\infty}^{+\infty} r(x')s(x')t(x')e^{-jkx'}dx' \right|^2 , \qquad (7.5)$$

where K $(=2\pi f/V)$ is the spatial frequency component that produces diffraction at an angle θ. The relationship between the position x in the detector array plane and K is

$$x = \frac{\lambda_0 K F}{2\pi n} . \qquad (7.6)$$

237

Since the amplitude of the diffracted light is proportional to the acoustic wave's amplitude, the intensity distribution at the detector array is proportional to the power spectral density of the RF input.

7.1.1 Frequency Resolution and Bandwidth

In general, acoustic attenuation and optical beam truncation cause the focused-spot (representing a single frequency) cw input to have finite width and side lobes. As an example, for a uniform optical beam of width d, and neglecting acoustic attenuation, the -4 db spot width is

$$\delta x = \frac{\lambda_0 F}{nd} .$$ (7.7)

This corresponds to a frequency range δf of

$$\delta f = \frac{nV}{\lambda_0 F}\delta x = \frac{V}{d} = \frac{1}{\tau} ,$$ (7.8)

where τ is the acoustic transit time across the optical beam. The quantity δf is the frequency resolution if one signal is present and the detector periodicity is equal to the spot width. For a bandwidth Δf the number of resolvable frequencies such that adjacent spots have a 1 db dip between them is the time-bandwidth product

$$\frac{\Delta f}{\delta f} = \tau \Delta f .$$ (7.9)

If the optical beam is not uniform, but weighted to reduce sidelobes, the spot width is increased and resolution reduced. The discussion to this point has assumed the presence of only one signal. If it is required to resolve two adjacent spots, then the detector periodicity must be no greater than half the spot width.

For a simple transducer, the peak diffracted light intensity I_d is given by

$$I_d = I_0 \sin^2 \sqrt{\frac{\pi^2}{2\lambda_0^2} \frac{L}{h} M_2 P_a} ,$$ (7.10)

where I_0 is the incident light intensity, L/h is the aspect ratio of the acoustic beam, M_2 is an acousto-optic figure-of-merit and P_a is the acoustic power. The RF frequency response of I_d depends on the band shape of the acousto-optic interaction, the band shape of the transducer's electro-acoustic conversion and the band shape of the electrical matching circuitry. Typical of results obtained with bulk $LiNbO_3$ Bragg cells are an overall bandwidth of 1.0 GHz with a peak diffraction efficiency of 5% per W of RF input power at an optical wavelength of 0.83 μm.

7.1.2 Dynamic Range

Dynamic range can be defined in different ways depending on whether single or simultaneous signals are anticipated. For a single signal, the dynamic range is the range between the signal level at which an acceptable false alarm rate results and the signal level corresponding to either detector saturation or maximum achievable diffracted signal, whichever is least. Pulses of duration less than the transit time across the optical beam cause a reduction in dynamic range because (i) they intercept less than the entire optical beam, and (ii) their broader inherent spectrum causes spreading of the light across a greater number of detector positions.

Since the ability to handle simultaneous signals is a prime attribute of an AOSA, a more meaningful definition of dynamic range is the ratio of the strongest signal to the weakest signal that can be simultaneously received and detected. Weak signals tend to be obscured by scattering, sidelobes, intermodulation products and detector noise. Scattering is more significant in the integrated-optics configuration and is discussed in Sect. 7.2.3. Detector noise sources are discussed in Sect. 7.3.5. Sidelobes and intermodulation products are adressed in the remainder of this section.

If the light intensity across the aperture is uniform, then the first sidelobe is 13 db down from the signal level. This can be substantially improved if the light amplitude is weighted by an appropriate function. Since single-mode laser beams are typically Gaussian in shape, or near to it, this is a good case to consider. In [7.2] it has been shown that if a Gaussian-weighted beam is truncated at 1.5 times the e^{-2} width in intensity, the highest side lobe is 34 db down. The price paid in doing this is about a 40% increase in the spot width. Referring to (7.4), note that since the attenuation factor $s(x')$ has an exponential dependence, the product $s(x')t(x')$ is Gaussian in shape if $t(x')$ is Gaussian. Attenuation has the effect of reducing the efficiency but not changing the spot width or sidelobes.

It can be seen from (7.10) that for small diffraction efficiencies, the relationship between diffracted light intensity and acoustic power is nearly linear. Nevertheless, second- and third-order interaction terms must be considered. Sum and difference frequencies (2nd-order terms) can be excluded from consideration by restricting the bandwidth to less than one octave. The third-order intermodulation products $2f_1 - f_2$ and $2f_2 - f_1$ are always present and have an intensity dependent on the intensity of the individual signals. In [7.2] it is seen that for two signals of equal, and small, diffraction efficiency η_0, the relative level ϵ of the third-order intermodulation products is $\eta_0^2/36$. As an example, for $\eta_0 = 0.05$, ϵ is down by 41 db. It is therefore desirable to operate an AOSA at a small diffraction efficiency to suppress these spurious signals. A second reason is to prevent strong signals from causing excessive cross-modulation of simultaneously occurring signals which would hamper precise determination of signal amplitude. In practice, for the very high bandwidths that are desirable, it is difficult to achieve more than a few percent diffraction efficiency at reasonable RF drive levels.

239

7.2 The Integrated Optical Spectrum Analyzer

The potential advantages of an integrated-optical spectrum analyzer (IOSA) were first recognized by *Hamilton* and *Wille* [7.3] in 1974. Since then the concept has been discussed by *Hamilton* et al. [7.4], *Anderson* et al. [7.5], *Barnoski* et al. [7.6], and *Spezio* and *Orsino* [7.7]. In the preceding section, the bulk AOSA was discussed in detail because the principle of operation and the constraints on performance are very similar for an IOSA. An IOSA can be viewed as a very thin AOSA in which light is confined by optical waveguides rather than lenses. This leads to both advantages and disadvantages, as will become apparent in the following subsections.

7.2.1 IOSA Features

A schematic diagram of an IOSA is shown in Fig.7.3. It is functionally the same as the AOSA of Fig.7.2 except that here the entire optical path is through optical waveguides. The waveguides only confine light in the direction transverse to the substrate so that within the plane of the surface, light is free to propagate in any direction. Because of this, the equations governing the design of an IOSA are essentially the same as those for an AOSA if the so-called effective refractive index n_g of the guided mode is used in place of the bulk refractive index. In the IOSA, light diffraction is produced by surface acoustic waves (SAWs) which, traveling at the surface, interact very efficiently with guided optical waves. It is important that the waveguide supports only a single mode or frequency resolution will suffer because of different modes being diffracted at different angles as a consequence of their unequal effective refractive indexes. The waveguides must also have low total throughput loss, and low in-plane scatter that would appear as optical noise at the detector array.

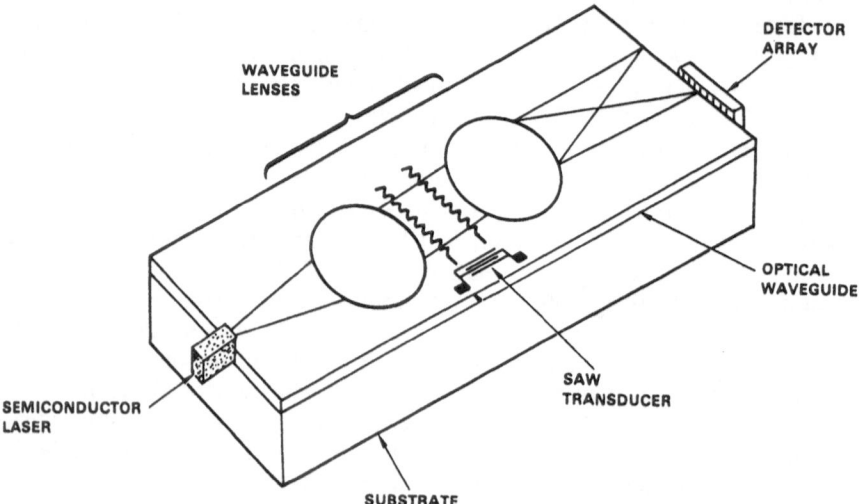

Fig.7.3. Integrated optical spectrum analyzer

240

Since small size is one of the principal advantages of an IOSA, a semiconductor laser is clearly the preferred choice for a light source. It must operate continuously in a stable, single mode that can be efficiently coupled to the waveguide. The linewidth must be·sufficiently narrow not to degrade the frequency resolution. This is ensured by operation in a single longitudinal mode. Spontaneous emission must be suppressed or it will produce an optical background spread across the detector array, similar to scattering.

Collimation and focusing are accomplished with waveguide lenses which must be diffraction limited, have low scatter, and have high throughput. Furthermore, the fabrication process for the lenses must be highly reproducible so that they may be placed with precision, in advance, relative to the laser and detector array.

7.2.2 Size Constraints

It is of interest to relate the physical dimensions of the IOSA to the relevant performance parameter, frequency resolution. In the direction of SAW propagation, the device size is determined by the optical beam width needed for a specified frequency resolution δf. For a Gaussian beam truncated at 1.5 times the e^{-2} width in intensity,

$$\delta f \simeq \frac{1.4 V_R}{d} ,$$
(7.11)

where V_R designates the velocity of the SAW. For a titanium diffused lithium niobate (Ti:LiNbO$_3$) waveguide with $V_R \simeq 3500$ m/s, 1 MHz frequency resolution requires that d = 5 mm. For better resolution, the beam size approaches the point for which the physical size of waveguide lenses and the tolerance for diffraction limited performance present severe fabrication problems.

The principal determinant of the length of the optical train is the focal length F of the Fourier transform lens. The overall length is obviously about twice F. For F/numbers, F′, less than about 10, the frequency resolution will be limited by the minimum periodicity achievable in a detector array rather than the optical spot size. In that case the single-frequency resolution will be

$$\delta f = \frac{V_R s n_g}{F \lambda_0} ,$$
(7.12)

where s is the periodicity of the detector array. Again, using as an example Ti:LiNbO$_3$ ($n_g \simeq 2.2$), it is found that 2 MHz resolution is achievable with a focal length of 4.5 cm and a detector periodicity of 10 μm. Five centimeters is probably the maximum allowable focal length for substrates of realistic size, unless beam folding with waveguide reflectors is employed.

The value of F' for which both (7.11, 12) are satisfied is

$$F' = \frac{F}{d} = \frac{sn_g}{1.4\lambda_0} \ . \tag{7.13}$$

For Ti:LiNbO$_3$ and s = 10 μm, F' is about 18. If the optics are designed for a lower F/number, frequency resolution is not improved; however, sidelobe rejection is enhanced since more of the stronger, nearby sidelobes are detected in the signal cell rather than adjacent cells.

7.2.3 Dynamic Range

The single-signal dynamic range may be estimated in the following way. The power P_d reaching the detector array for the strongest permissible signal is

$$P_d = P_L - (\alpha L + \eta_A + \eta_B + \eta_C + \eta_D) , \tag{7.14}$$

(P_L: laser output power [dbm], α: waveguide loss [db/cm], L: optical path length [cm], η_A: lens loss [db], η_B: maximum diffraction efficiency [db], η_C: laser-to-waveguide coupling loss [db], and η_D: waveguide-to-detector coupling loss [db]). If we take as realistic values for these parameters P_L = 15 dbm, α = 0.5 db/cm, L = 6 cm, η_A = 2 db, η_B = 13 db, η_C = 5 db and η_D = 2 db, then we find P_d = -10 dbm. State-of-the-art silicon detector arrays with 1 MHz bandwidth are capable of about -60 dbm sensitivity. This gives a projected dynamic range of 50 db with a signal-to-noise ratio of one. Anticipated improvements in laser output, detector sensitivity and throughput losses may improve dynamic range another 10 db or so.

When the dynamic range is defined in terms of the ability to detect weak signals in the presence of strong ones, consideration must be given to numerous noise sources such as in-plane scattering, sidelobes, laser spontaneous emission and detector noise. In-plane scattering refers to scattered light that remains trapped in the waveguide. Sidelobes were discussed above; spontaneous emission and detector noise will be treated below. Except for scattering, the magnitude of these problems is about the same for an IOSA as for an AOSA. Scattering is more of a problem for an IOSA because it occurs throughout the optical path.

The effect of scattering is illustrated in Fig. 7.4. A signal f_n causes a diffracted spot to fall at the detector position n. At the same time, light falls at the detector position m which is scattered from both the zero-order beam and the diffracted beam, and which tends to mask a signal f_m falling at that position. Note that the intensity distribution of the scattered light in a signal beam is the same as that of the zero-order beam multiplied by the diffraction efficiency and shifted in the detector plane. Thus, understanding the effect of scattering from a strong signal on a nearby weak signal permits assessment of the overall problem.

Boyd and *Anderson* [7.8] have analyzed the effect on dynamic range of scattering from the diffracted beam in the region between the lens and

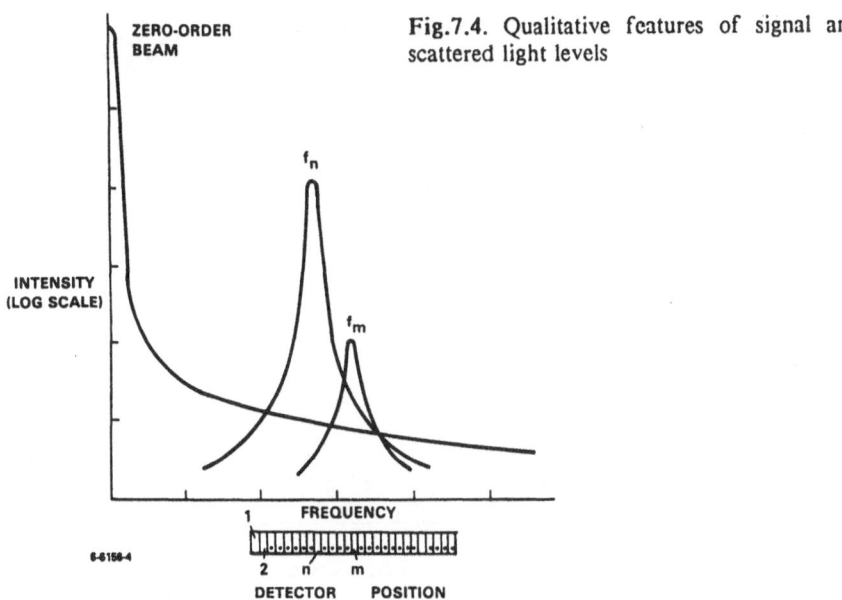

Fig.7.4. Qualitative features of signal and scattered light levels

the detector array. They defined dynamic range as the ratio of the signal light intensity at detector position n to the scattered light intensity at position n+1. It was assumed that the primary source of scattering was surface roughness described by the autocorrelation function

$$R(\mu) = \sigma^2 \exp(-|\mu|/z) , \qquad (7.15)$$

where μ is the separation of two points on the waveguide, and σ^2 is the mean-squared surface roughness. The surface-roughness correlation length z is given by

$$z = \frac{\lambda}{2\pi\Delta\eta} , \qquad (7.16)$$

where λ is the optical wavelength, and $\Delta\eta$ is a parameter describing the spatial frequency spectrum of the surface roughness. The angular distribution of the in-plane scattering is determined by z such that the larger its value, the more sharply peaked is the scattering in the forward direction.

Although in some waveguides scattering from inhomogeneities may make a significant, or even dominant, contribution to the in-plane scatter, it is expected that, qualitatively, the findings of *Boyd* and *Anderson* will apply. Their results may be summarized as follows:

(a) Scattering that occurs close to the detector array is the most important one since the capture angle subtended by a detector approaches zero for scattering occuring at a large distance;

(b) A smaller detector gives a larger dynamic range because the angle subtended is smaller;

243

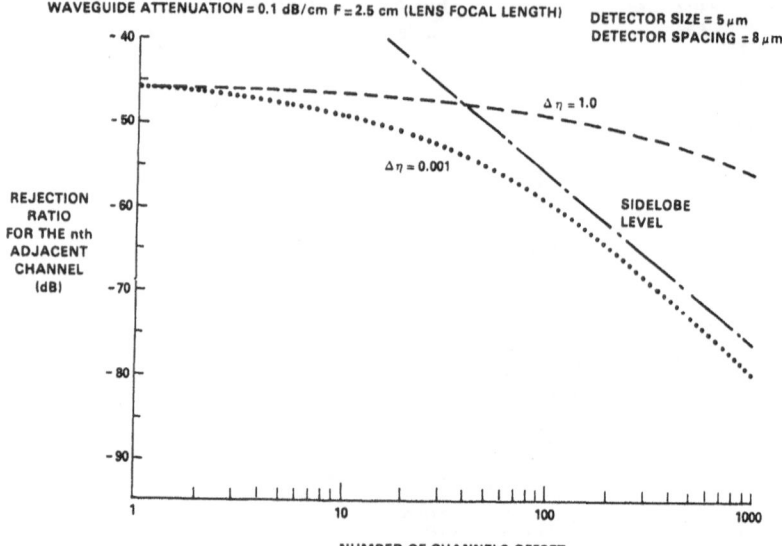

Fig.7.5. Rejection ration in db for the nth adjacent channel. Also shown is the locus of sidelobe maxima for a truncated Gaussian optical beam. Reprinted with permission from [7.8]

(c) The larger the overall attenuation, the smaller the dynamic range;

(d) The rate at which the scattering falls for elements far removed from a signal channel depends primarily on the correlation length. Small correlation lengths cause wide-angle scattering which affects the more distant detector elements to a greater extent.

Calculations in [7.8] predict a scattering level at the first adjacent detector that is at or below the expected sidelobe level for total scattering attenuation less than 1 db/cm. The fall-off in scattering level for detectors further removed is illustrated in Fig.7.5, which gives the rejection ratio (scattered light relative to a signal) for a 0.1 db/cm waveguide as a function of detector position for different correlation lengths. Note that if the correlation length is very small, the scattering level may fall at a slower rate than the sidelobes. It is this effect, along with diffraction efficiency, that determines the contribution of the zero-order beam. Excluding the effect of the zero-order beam, for the conditions given, the rejection ratio is over 45 db for all detector positions. There is some justification for ignoring the effect of the zero-order beam because its scattering contribution is constant in time and can be subtracted out by post-detection processing.

No comparable analysis has been done to determine the effect of scattering occurring in front of the transform lens. Fortunately, this contribution to the overall scattering level can be simulated fairly well experimentally. This is done by coupling a collimated beam into, then out of, the waveguide, and focusing the beam with an external lens rather than a waveguide lens. The intensity distribution in the focal plane of the external lens is equivalent to that portion of the total scattered light in an

244

IOSA that originated along a length of waveguide of a particular length used in the experiment. For good waveguides it was found by *Vahey* [7.9] and *Brandt* [7.10] that scattered light levels fall by about 50 db at 1° off-axis, corresponding to an RF frequency difference of about 200 MHz. The rate of fall-off of the scattering suggests that the zero-order beam contribution will not be significant.

Vahey et al. [7.11] have studied in-plane scattering in waveguides formed by titanium diffusion into $LiNbO_3$. They concluded that the typically 3.0 nm or less rms surface roughness could not explain the observed scattering. The dominant scattering contribution was attributed to the presence of subsurface Li-Ti-O compounds and could be reduced by post-diffusion polishing. Using an experimentally observed scattering distribution, an analysis of an IOSA similar to those dicussed in Sect.7.4 led to a prediction of 27 db dynamic range. It was argued though that the same distribution of scattering centers in a waveguide with a surface refractive index change of 0.005 and a diffusion depth of 3.3 μm should give 40 db dynamic range.

7.2.4 Materials

The component functions required for the IOSA are lasing, SAW transduction, acousto-optic diffraction, light focusing and detection. The obvious choice of a material system for a fully integrated circuit is one of the ternary or quaternary alloys of the III-V semiconductors such as $Al_x Ga_{1-x} As$ or $In_x Ga_{1-x} As_{1-y} P_y$. Since these materials are not piezoelectric, SAW transduction requires the deposition of a piezoelectric overlay such as ZnO. A review on thin-film ZnO SAW transducers and an update on Guided-wave acousto-optic Bragg cells in $Al_x Ga_{1-x} As$ waveguides are provided in Chaps.6 and 8, respectively. If the concept of an integrated laser is abandoned, then silicon is an attractive substrate because of its high level of development for electronic components and detectors. However, it also requires the deposition of a piezoelectric overlay. If a substrate is chosen with emphasis on the acoustic properties required for wide-band operation, then $LiNbO_3$ is the best choice. The properties of primary interest are a high piezoelectric coupling constant and low SAW propagation loss. Lithium niobate excels in both respects and is already available in good quality. Because the application of RF spectrum analysis does indeed place a premium on wide-band operation with good frequency resolution, present development activities are emphasizing this material.

It would seem to be a desirable objective to implement an IOSA in a fully integrated configuration in order to take maximum advantage of planar processing technology. Yet the dimensions are such that only one or, at best, a few circuits per substrate are possible. Consequently, the economic benefits of full integration will be somewhere between those of hand, or machine assembly of discrete parts and large-scale batch processing. Furthermore, the complexity of double heterostructure semiconductor lasers and electronically scanned detector arrays might result in such a low yield, when combined on a single substrate, that full integration would be economically unattractive. Because of these considerations, a hybrid configuration is attractive (at least in the near term) in which the laser, the detector array, or both, are external to the primary waveguide. This adds more impetus to development activities employing LiNbO$_3$.

7.3 Components

In this section we will review the progress made in the semiconductor lasers, optical waveguide devices, and detector arrays that are needed for implementation of the IOSA concept. The emphasis will be on hybrid configurations, with special attention given to the Ti:LiNbO$_3$ system.

7.3.1 Laser and Laser/Waveguide Coupling

Since small size is one of the strongest motivations for developing an IOSA, semiconductor diode lasers are attractive as laser sources because of their small overall size and the very small size of the light emitting region (about 0.5 μm x 5 μm). This makes it possible to butt-couple the laser with no intervening optical elements, as would be required for any other kind of laser. In this subsection we discuss the characteristics of these lasers and the conditions to be met for efficient butt-coupling to an optical waveguide. For more detailed discussions of diode lasers, the reader is referred to texts by *Casey* and *Panish* [7.12], and *Kressel* and *Buttler* [7.13].

a) Semiconductor Diode Laser Characteristics

A representative diode laser structure is illustrated in Fig.7.6. The region of light emission is indicated by cross-hatching. Room temperature, cw operation requires that current flow be restricted to the lasing volume and that the stimulated emission be confined to the waveguiding region (GaAs). Current confinement in Fig.7.6 is accomplished by the oxide mask and light confinement is provided by enclosing sections of GaAlAS, whose refractive indices are less than that of GaAs. There are numerous variations of this structure, some which have been reviewed by *Nakamura* [7.14].

Light generation occurs when electrons and holes recombine at the p-n junction under forward bias. Optical feedback for lasing is provided by the dielectric discontinuity at the cleaved edges of the diode. Lasers made with the alloy system GaAs/GaAlAs typically emit light at about 830 nm, which is well matched to silicon detectors. Typical threshhold currents are about 20 mA. They are capable of providing over 10 mW of power in a single transverse and longitudinal mode. Single-transverse-mode operation is important because it makes possible efficient coupling to the waveguide and collimation without need for a spatial filter. A single longitudinal

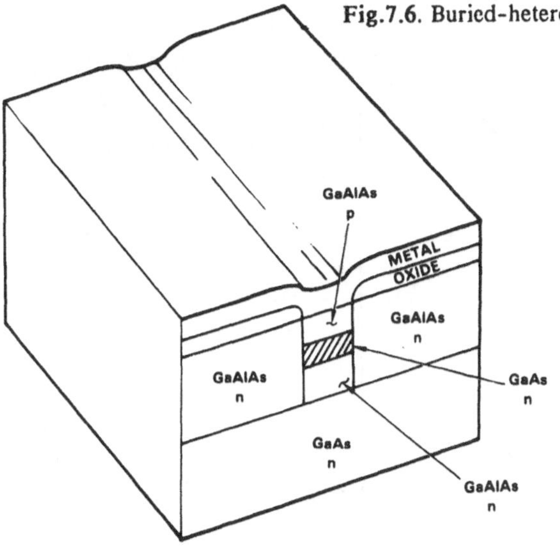

Fig.7.6. Buried-heterostructure semiconductor laser

GaAlAs
p

METAL
OXIDE

GaAlAs
n

GaAlAs
n

GaAs
n

GaAs
n

GaAlAs
n

mode assures a narrow linewidth; otherwise, RF resolution would be degraded. The lifetime of diodes surviving an initial burn-in period has
been estimated to be over 100,000 hours.

Normally, diode lasers are polarized in the plane of their junction
which, in butt-coupling, causes excitation of a TE mode in the optical
waveguide. This is the preferred mode for efficient acousto-optic diffraction on a y-cut $LiNbO_3$ substrate. The beam divergence of these lasers is
rather large, typically about $10°$ x $30°$. The spread in the direction transverse to the junction requires that the diode be positioned very close to
the waveguide for efficient coupling. The beam spread in the plane of the
junction is advantageous since it permits collimation of a beam of a prescribed width by proper positioning of a single lens.

There are other features of the laser diode spectrum that potentially
can degrade the IOSA's performance. These are multimode operation,
wavelength shift with temperature and broadband spontaneous emission.
To see this, consider that a wavelength spread or shift $\Delta\lambda$ in the laser
wavelength λ will cause a spread, or shift Δf, in the measured RF frequency f such that

$$\frac{\Delta f}{f} = \frac{\Delta\lambda}{\lambda} . \tag{7.17}$$

For λ = 830 nm and f = 1 GHz, a $\Delta\lambda$ of 1 nm results in a Δf of about
1.2 MHz. A typical multimode linewidth of 2-3 nm would clearly degrade
the RF resolution. On the other hand, the linewidth of a single-mode
laser (<0.1 nm) is of no concern. More troublesome is the 0.2-0.3 nm/°C
wavelength shift with temperature. This shift introduces an error in the
frequency calibration of the IOSA which must be compensated for by
temperature stabilizers or post-detection processing.

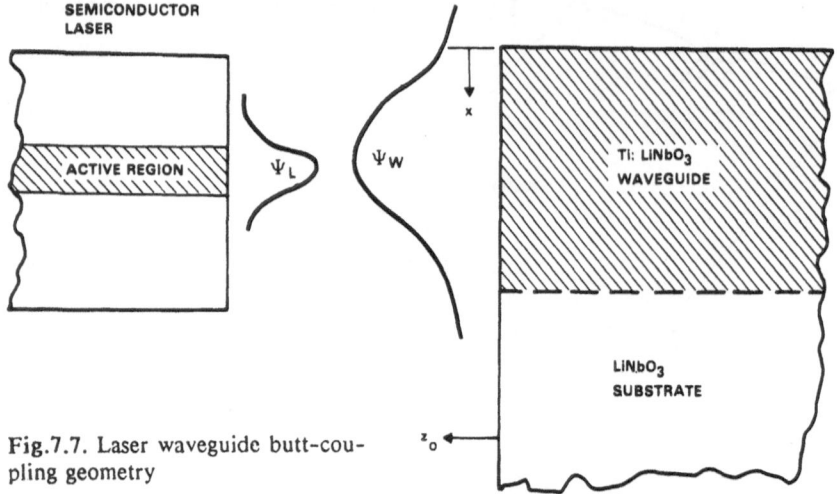

Fig. 7.7. Laser waveguide butt-coupling geometry

In addition to stimulated emission, the laser beam contains a component of broad-band, incoherent, spontaneous emission. Though this contribution to the output has not been extensively studied, its level can be 1-10% of the total output in a spectral band of up to a few tens of nanometers. The actual values depend on the details of the laser structure and the operating level. Since this emission is largely codirectional with the laser beam, it is coupled into the IOSA, diffracted, and appears as a low-level smear surrounding a signal spot. The effect is to degrade the dynamic range for stimulated signals. *Burns* and *Moeller* [7.15] have measured the spontaneous emission content of buried heterostructure lasers and inferred that this component of a signal diffracted beam would be down by about 40 db at the 4th nearest-neighbor detector. It was assumed that the RF resolution was 2 MHz and the processing band was 400 MHz, centered at 600 MHZ, using a LiNbO$_3$ substrate.

b) Butt-Coupling

The geometry for butt-coupling is illustrated in Fig. 7.7. Because of the large inherent beam spread of the diode laser, it must be positioned with a separation z_0 of the order of a few micrometers. The coupling efficiency, neglecting reflection losses, has been shown by *Mueller* et al. [7.16] to be

$$A = \frac{\left| \int_{-\infty}^{+\infty} \psi_{inc} \psi_w^* \, dx \right|^2}{\int_{-\infty}^{+\infty} \psi_{inc} \psi_{inc}^* \, dx \int_{-\infty}^{+\infty} \psi_w \psi_w^* \, dx}, \tag{7.18}$$

where ψ_{inc} is the laser beam electric field distribution incident on the waveguide, and ψ_w is the field distribution of the waveguide mode. Of course, the field distribution ψ_L of the laser at its output facet equals ψ_{inc} for $z_0 = 0$.

Since A is a normalized overlap integral of the fields ψ_{inc} and ψ_w, the coupling efficiency will be greatest when the laser and waveguide mode widths are comparable and z_0 is very small. Extensive calculations appear in [7.16] regarding the effect on coupling efficiency of varying the separation z_0, transverse misalignments and angular misalignments. It was found that, typically, the coupling efficiency is within 3 db of its maximum value for z_0 up to 15 μm, for transverse misalignments of about ±0.8 μm, and angular misalignments of about ±15°. If a laser of optimum mode width ($\simeq 2\mu$m) were used with a Ti: LiNbO$_3$ waveguide mode-depth of 1.35 μm, the coupling efficiency would be about 95%. Mode-widths more typical of commercially available lasers are under 1 μm, leading to efficiencies of 50% or less. Using a laser diode of mode-width 0.8 μm and a Ti:LiNbO$_3$ waveguide of mode-width $\simeq$ 3 μm, *Hall* et al. [7.17] measured a butt-coupling efficiency of 50%.

7.3.2 Waveguides

The properties of optical waveguides and techniques for forming them have been reviewed by *Tamir* [7.18], as well as in Chap.3. Two material systems (silicon and LiNbO$_3$) will be discussed here with emphasis on LiNbO$_3$ because of its high piezoelectric coupling constant and low SAW propagation loss.

a) Silicon

Silicon offers the opportunity to include integrated detectors in the IOSA. Its use requires the establishment of a low refractive index layer to optically isolate the waveguide from the highly absorbing substrate. This is readily accomplished by thermal oxidation of the surface to a depth of at least 1 μm. The oxide growth process has a surface smoothing effect which yields an excellent quality substrate. The low refractive index of the oxide makes consideration of many different materials for a waveguide possible. Sputtered thin films of a number of glasses [7.19,20] and oxides such as Ta$_2$O$_5$ [7.21] and Nb$_2$O$_5$ [7.22] have measured losses below 1 db/cm. *Stutius* and *Streifer* [7.23] deposited Si$_3$N$_4$ by chemical vapor deposition on oxidized silicon with a loss as low as 0.1 db/cm. Chalcogenide glasses are attractive for their high acousto-optic figure-of-merit, but they tend to be optically lossy. *Dutta* et al. [7.24] have used laser annealing to reduce the losses of sputtered glass waveguides to as low as 0.01 db/cm.

b) Lithium Niobate

Schmidt and *Kaminow* [7.25] showed that a number of transition metals can be deposited on LiNbO$_3$ and diffused in to form waveguides. The resulting refractive index profile is given by

$$n(y) = n_s + \Delta n(0) \cdot f(y/D) , \qquad\qquad (7.19)$$

where y is the depth below the surface, n_s is the substrate refractive index, $\Delta n(0)$ is the refractive index change at the surface, and D is the diffusion depth. The function $f(y/D)$ represents the refractive index profile. The principle waveguide parameters are determined by the prediffusion metal film thickness, the diffusion time and the temperature. *Hocker* and *Burns* [7.26] have presented a method for numerical calculation of the effective refractive indexes for the modes of diffused waveguides of arbitrary $f(y/D)$. They provide universal charts relating mode index, diffusion depth, surface index change and number of modes.

The most commonly used, and best understood, diffusant is titanium. *Naitoh* et al. [7.27] were able to fabricate single-mode waveguides for an optical wavelength of 0.873 μm by depositing 10-50 nm of Ti and diffusing it for seven hours at 970°C into y-cut plates of LiNbO$_3$. The resulting concentration profile and, hence, the index profile were Gaussian. *Burns* and *Hocker* [7.28] showed that a Gaussian field distribution is a good approximation where a Gaussian index profile exists. The effective mode depth of Ti:LiNbO$_3$ is about 2 μm which permits efficient interaction with SAWs for frequencies up to 1 GHz before efficiency begins falling off because of a decrease in the overlap integral factor.

There is a complication that arises if x- or y-cut plates are used in the IOSA. *Sheem* at al. [7.29] have found that anisotropic coupling occurs between TE and TM modes resulting in power leakage to the substrate at a rate dependent on the propagation angle θ with respect to the z-axis. This loss is substantial for $\theta > 10°$, but reduces to about 3 db/cm at $\theta \simeq 75°$, corresponding to an F/1.9 optical system oriented along the x (or y) axis. *Vahey* et al. [7.30] have shown that this results in the introduction of an additional weighting factor on the optical beam such that the more highly converging rays leaving the transform lens suffer the greatest loss.

The net attenuation of Ti:LiNbO$_3$ waveguides is under 0.5 db/cm at a wavelength of 0.8 μm. Measurements of absorption and the photorefractive effect have been reported by *Glass* et al. [7.31]. The effect of titanium in-diffusion on the composition of the waveguiding layer has been studied by *Burns* et al. [7.32] and *Sugii* et al. [7.33].

7.3.3 Waveguide Lenses

A great variety of methods for implementing waveguide lenses exists. Refracting lenses are possible because of the sensitivity of a waveguide mode's effective refractive index n_g to changes in the waveguide thickness and bulk refractive index. This type of lens was proposed by *Schubert* and *Harris* [7.1], and was demonstrated by *Schubert* and *Harris* [7.34], and *Ulrich* and *Martin* [7.35]. Examples are shown in Fig.7.8a,b. A recently reported variation on this type of lens is the titanium-indiffused proton-exchanged (TIPE) lens in which a lens-shaped proton-exchanged region is formed in a titanium in-diffused LiNbO$_3$ planar waveguide [7. 36]. The planar nature of waveguides provides greater flexibility in lens

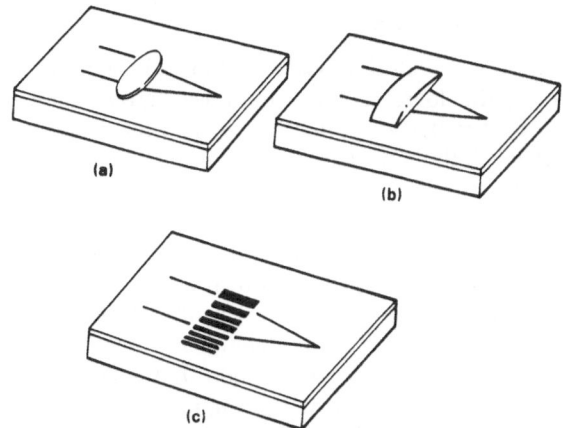

Fig.7.8a–c. Examples of waveguide lenses: (a) constant thickness with shaped edges, (b) tapered thickness with parallel edges, and (c) diffraction grating

fabrication than is possible with "bulk" optics. For example, a lens with a continuously varying n_g can be made by continuously varying the waveguide thickness in the lens region. A waveguide Luneburg lens made in this fashion was reported by *Zernike* [7.37]. Diffraction lenses can be implemented by introducing a diffraction grating in the waveguide of an appropriate geometry, as illustrated in Fig.7.8c. The grating structure can be formed by changing the waveguide thickness or applying overlays for absorption or phase-shifting.

A third type of waveguide lens, which does not have a counterpart in "bulk" optics, is the so-called geodesic lens which takes the form of a quasi-spherical depression or protrusion of the waveguide. The focussing effects of such a structure are well known to microwave engineers. The name "geodesic lens" arises because when Fermat's principle is applied in the analysis of these structures, it is found that the light ray paths follow geodesic curves of the surface (i.e., the geometric path difference along the curve between any two points is the least of all possible curves on the surface connecting the two points). The first demonstration of an optical waveguide geodesic lens was reported by *Righini* et al. [7.38].

The application to acousto-optic spectrum analysis requires that the lenses used be diffraction limited with respect to both spot size and side-lobes. The Fourier transform lens must be diffraction limited over a field-of-view corresponding to the acousto-optic bandwidth. For a bandwidth of 1.0 GHz, this is about 6°. For maximum sensitivity and dynamic range, the lenses must have high throughput and low in-plane scattering. Since small size is an important consideration, lenses with low F/numbers are desirable. This inevitably means the fabrication tolerances are tougher to meet. On the other hand, low production cost is also important. The development of a lens technology that satisfies all of these conditions has been the subject of a great deal of research. In the remainder of this section we review progress on some of the more promising approaches for high quality waveguide lenses.

a) Generalized Waveguide Luneburg Lens

The classic Luneburg lens is one which focuses a collimated beam to a point on the circumference of the lens. A generalized waveguide Luneburg lens is a variable-index, circularly symmetric, refracting structure which reimages two concentric circles on each other, one of which may be at infinity. This class of lens is free of all aberrations except field curvature, which is tolerable for sufficiently small fields-of-view. Luneburg derived an integral equation which gives the refractive index distribution for which a perfect image of an infinite object is formed. *Southwell* [7.39] has reported an analytical approximation to Luneburg's integral equation from which the index profile for a generalized Luneburg lens may be extracted with sufficient accuracy to provide diffraction limited performance. *Sochacki* [7.40] has derived a closed-form approximation for Luneburg's integral for waveguide lenses with focal distances exceeding twice the lens radius. Using the dispersion equation for a given waveguide structure, it is possible to calculate the required thickness profile of the waveguide layer.

A waveguide Luneburg lens can be deposited by sputtering the lens material through an appropriately shaped mask as shown in Fig.7.9. The lens material may be the same as the waveguide or different, as in the example shown where Ta_2O_5 is sputtered onto a 7059 glass waveguide on oxidized silicon. *Yao* and *Anderson* [7.41] have developed an iterative procedure for optimizing a sputtering mask consisting of conical sections to produce lenses of a specified aperture and focal length. Using the layered structure described above, they achieved a diffraction-limited spot size of 1.6 μm with an F/4 lens of 8 mm focal length.

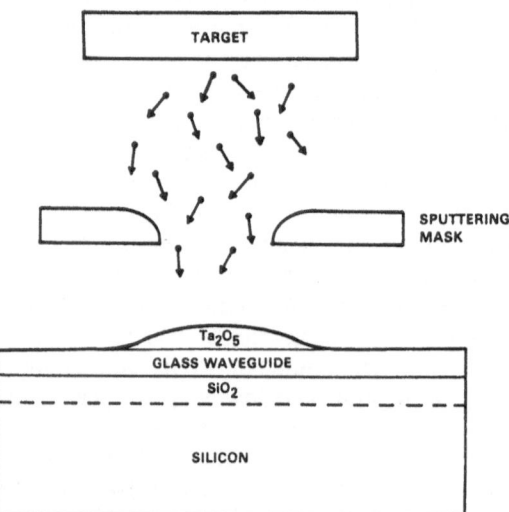

Fig.7.9. Geometry for sputter deposition of a waveguide Luneburg lens

b) Diffraction Lenses

The focussing effect of an acousto-optic Bragg cell with a frequency chirped RF input is well known. This same effect can be utilized to create a waveguide lens, as illustrated in Fig.7.8. The lens consists of a waveguide diffraction grating with a linear chirp rate K of its spatial frequency across the aperture d. For large F/numbers, it can be shown that the focal length f is given by $n_g/\lambda K$, where n_g is the waveguide effective refractive index and λ is the free-space wavelength. *Yao* and *Thompson* [7.42] have made such a lens by plasma etching the grating pattern into a glass waveguide on a thermally oxidized silicon substrate. The grating period varied from 3.4 to 6.8 μm over an aperture of 2 mm. The resulting focal length was 32 mm. A diffraction limited spot of 6.1 μm was measured with 90% throughput. *Forouhar* et al. [7.43] have discussed the performance and limitations of chirped grating lenses on Ti:LiNbO$_3$ waveguides. They achieved over 80% diffraction efficiency, but with a field-of-view less than 4° (corresponding to about 700 MHz acousto-optic bandwidth). The field-of-view can be improved by reducing the lens thickness at the expense of diffraction efficiency.

Other versions of the diffraction type lens have been reported by numerous researchers [7-44-51]. Most recently, negative-index change diffraction-type microlenses and lens arrays have been implemented in both LiNbO$_3$ [7.50] and GaAs [7.51] waveguides by using ion-milling. Ion-milling has proved to be a simple and versatile method, and represents the first technique for fabrication of waveguide lenses in GaAs. The same scheme is applicable to any other substrate. All of these lenses and reflectors have the desirable feature of being easily mass-produced once a master mask is available. Furthermore, their focal properties are dependent only on the mask geometry and thus, are extremely reproducible. Unfortunately, at present there are some problems associated with these lenses that tend to counterbalance the advantages. Low F/numbers require large diffraction angles and, consequently, very small periodicities which are difficult to achieve. Achieving high throughput with a wide field-of-view is a problem because the thicker a Bragg grating is, the more sensitive it is to angular misalignment. Finally, waveguide grating lenses tend to generate more scattered light than other types of lenses.

c) Geodesic Lenses

The profile of a spherical depression goedesic lens is illustrated in Fig. 7.10. Note that in actuality the waveguide thickness is negligible so that the optical properties of the lens can be predicted by considering only the surface shape. *Van Duzer* [7.52] has shown that the paraxial focal length f of a spherical depression lens is given by

$$f = \frac{R \sin\theta}{2(1 - \cos\theta)} , \qquad (7.20)$$

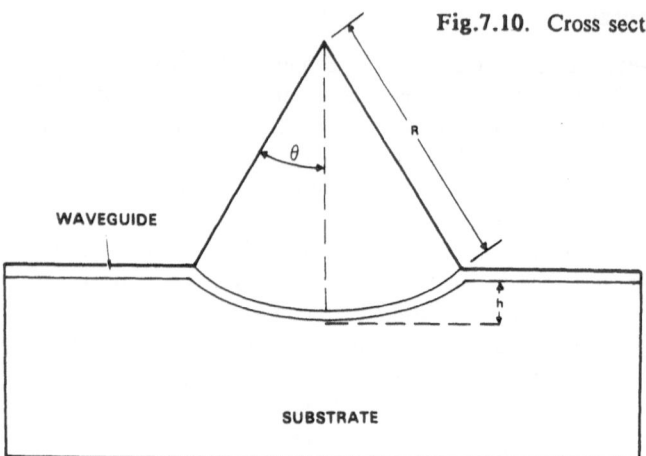

Fig.7.10. Cross section of a geodesic lens

where R is the radius of curvature of the surface and 2θ is the vertex angle subtended by the arc of the depression.

Just as do "bulk" optical elements with spherical surfaces, spherical geodesic lenses have spherical aberrations which prevent diffraction limited performance for low F/numbers. One method for removing the aberrations from these lenses is to introduce a refractive element with aberrations of the opposite sense. *Spiller* and *Harper* [7.53], and *Vahey* [7.54] were successful in achieving reductions in aberrations using this method. The disadvantage of this general approach is that a high quality, low F/number lens requires the precision fabrication of two lenses, a spherical depression lens and a refractive lens, rather than one. An alternate method for correction is to introduce an appropriate asphericity in the depression such that spherical aberration is removed. That this is possible follows from the knowledge that inhomogeneous planar refractors of the Luneburg type are aberration free and that such a planar lens can be mapped into an equivalent curved surface lens of constant refractive index, as shown by *Kunz* [7.55].

A procedure has been developed by *Toraldo* [7.56] giving a complete solution for a family of perfect aspheric lenses with the property that the focal length is equal to the radius of the depression. *Sottini* et al. [7.57] have found a general analytic solution for the profile of aspheric geodesic lenses able to form perfect geometrical images of the points of two given concentric circles on each other. Appoximate methods for determining the generating curve for rotationally symmetric lenses with arbitrary focal length have been developed by *Betts* et al. [7.58], and *Kassai* and *Marom* [7.59]. In the former case, the generating curve was expressed as a power series expansion in terms of the distance from the symmetry axis with coefficients chosen such that parallel rays are brought to a common focus. In the latter case, the equalivalent generalized Luneburg lens was determined numerically and then mapped into the required goedesic lens profile using the method of *Kunz* described ealier. Both methods incorporated rounding of the lens rim to prevent excessive radiation from occurring.

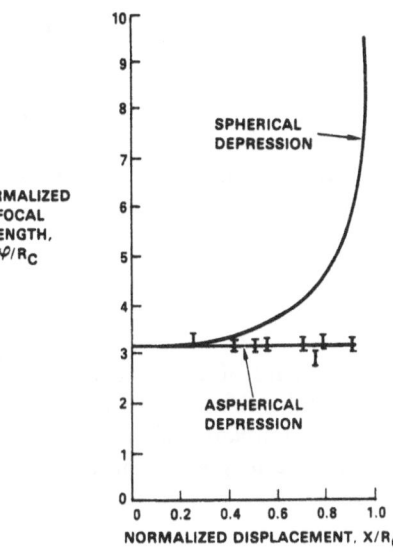

Fig.7.11. Measured focal distance versus lateral displacement of incident beam for geodesic lens fabricated on glass substrate. R is the radius of the lens depression. Reprinted with permission from [7.62]

Two methods for fabrication of aspheric lenses have been demonstrated. *Mergerian* et al. [7.60] used a numerically controlled, percision lathe for diamond turning the required surface directly in a y-cut LiNbO$_3$ substrate. After light polishing, titanium was subsequently diffused in to form a waveguide. The resulting 2.0 cm focal length lenses had low scattering but were not quite diffraction limited because of inaccurate edge rounding.

An alternative discussed by *Chen* and *Ramer* [7.61] is to precision machine a tool which could be used for ultrasonic grinding of a number of lenses. Using this technique, *Chen* et al. [7.62] made a 10 mm focal length lens with a 6.36 mm aperture in a glass substrate with a polyurethane waveguide. The lens had an aspheric shape calculated according to the procedure given in [7.59]. The lens quality is presented in Fig.7.11. The solid curve gives the calculated variation of the normalized focal length of a spherical lens for parallel rays of different normalized displacement from the optic axis. The data points show that the lens was corrected for spherical aberration over 90% of the aperture. The ultrasonic grinding technique has also been used to make geodesic lenses in LiNbO$_3$ substrates [7.63].

As was noted earlier in this chapter, low-cost production of an IOSA depends in part on the ability to repeatedly make lenses with specified parameters. An appreciation for the fabrication tolerances required can be obtained in the following way. Imagine that the depth h of the depression in Fig. 7.10 changes slightly such that R remains the same, but θ varies slightly. The change in focal length is given by

$$\delta f = \frac{4F(4F - 1)}{\sqrt{8F - 1}} \, \delta h \, , \tag{7.21}$$

255

where F is the lens F/number and δh is the "error" in depth. The depth of field of the lens is approximately

$$D \simeq 2\lambda F^2 \, , \tag{7.22}$$

where λ is the wavelength in the waveguide. If we require that the depth of field be at least as large as the focal length variation, it follows that

$$\delta h < \frac{\lambda F \sqrt{8F - 1}}{2(4F - 1)} \, . \tag{7.23}$$

This is a slow varying function of F. For values of F between 4 and 20, δh varies from 0.74λ to 1.6λ. These are tough tolerances to meet. Fortunately, the modern precision machining equipment referred to in this chapter is capable of submicrometer accuracies in positioning and surface contour.

7.3.4 Wideband Waveguide Bragg Cells

The principle requirement of the waveguide Bragg cell is to achieve sufficient light diffraction over a large bandwidth for adequate dynamic range and sensitivity. The need to limit third-order intermodulation products dictates that the diffraction efficiency for signals at detector saturation be no greater than about 5%. Amplifier and transducer power handling constraints limit power available for a strong signal to about 100 mW if multiple, simultaneous signals are to be coped with. Thus, desirable performance for the Bragg cell is of the order of 50% diffraction efficiency per Watt of RF input power. The bandwidth must be less than one octave in order to eliminate 2nd order intermodulation products.

a) Frequency Dependence of the Diffraction Efficiency

The expression for the diffraction efficiency of a waveguide Bragg cell is similar to that for bulk cells, except for the inclusion of a factor referred to as the coupling function which measures the relative overlap of the optical and acoustic fields. Since the extent of penetration of the acoustic fields below the surface is a function of the SAW wavelength, the coupling function varies across the RF frequency band of the Bragg cell. This variation, along with the frequency dependence of the phase matching condition and the transducer conversion efficiency, determines the overall frequency response. A detailed treatment on the overall frequency response has been presented in Chap.5.

The inherent efficiency of guided wave acousto-optic devices derives from the concentration of the SAW power within about one acoustic wavelength of the surface. This effect is a mixed blessing for wideband devices for if the SAW frequency is too high, the acoustic power will propagate closer to the surface than the optical power, resulting in inefficient diffraction. *Tsai* [7.64] has extensively studied the frequency de-

Table 7.1. Wide-band transducers

Transducer configuration	SAW direction	λ_0 [μm]	Bandwidth B [MHz]	Efficiency η [%/W(RF)]	Bη [GHz·%]
Multiple, tilted, frequency staggered [7.65]	z	0.633	320-1000	8	5.4
Phased-array [7.66]	21.8° from z	0.633	270-382	735	82.3
Chirped, tilted fingers [7.67]	z	0.633	380-850	80	37.3
Chirped, tilted fingers [7.68]	z	0.82	400-800	13	5.2
Multiple, tilted finger/ phased array [7.69]	z	0.82	780-1480	2.5	1.75

pendence of the coupling function for y-cut Ti:LiNbO$_3$ waveguides. He found that for these waveguides, with a typical optical mode depth of 2-3 μm, the coupling function peaks at about 500 MHz and falls by a factor of about 5 at 1400 MHz. These results are for an optical wavelength λ of 0.633 μm. Since the optical field depth is greater for longer wavelengths, it is expected that the mismatch would be even greater for a semiconductor laser ($\lambda \simeq 0.85 \mu$m). This suggests the desirability of a shallower diffusant than titanium. For high index film waveguides which are capable of guiding in much thinner layers, the coupling function effect is not as significant.

b) Wideband Transducer Configurations

Since the approaches to wideband transduction for waveguide Bragg cells have been covered in detail in Chap. 5, we will only summarize the published results in terms of efficiency and bandwidth. This is done in Table 7.1 for the most promising configurations. All of the devices that could reasonably be described as wideband have utilized y-cut, LiNbO$_3$ crystals, and this is the case for the example given.

It would appear that no single transducer configuration on Ti:LiNbO$_3$ will be capable of 500 MHz bandwidth and 50%/W(RF) diffraction efficiency. However, an optimum combination of different approaches such as a multiple phased-array of transducers may be capable of this level of performance. An array of multiple transducers has the advantage of making possible leveling of the overall frequency response by external electrical adjustment. If the device performance becomes sufficiently predictable and repeatable, then adjustments may be incorporated into the transducer design.

Table 7.2. SAW attenuation

Material	Ref.	f [MHz]	V_R [m/s]	d [μm]	d/Λ	α [db/μs]	$\bar{\alpha}$ [dg/μs]
$Ta_2O_5/SiO_2/Si$	7.70	200	$\simeq$3600	1.1	0.061	2.7	68
$Si_3N_4/SiO_2/Si$	7.70	200	$\simeq$5100	0.25	0.010	0.9	23
$ZnO/SiO_2/Si$	7.70	200	$\simeq$3900	0.65	0.033	0.5	13
$7059/SiO_2/Si$	7.70	200	$\simeq$4000	0.65	0.033	0.25	6.2
$Nb_2O_5/LiNbO_3$	7.71	320	$\simeq$3500	0.5	0.046	0.7	6.8
$As_2S_3/LiNbO_3$	7.72	200	$\simeq$3600	0.46	0.026	0.6	15
$As_2S_3/LiNbO_3$	7.72	200	$\simeq$3600	0.71	0.039	1.3	33
$As_2S_3/LiNbO_3$	7.72	200	$\simeq$3600	1.16	0.064	3.9	98
$Ti:LiNbO_3$	7.73	1470	$\simeq$3500	$\simeq$2	$\simeq$0.84	2.28	1.05

The data on the two chirped transducers in Table 7.1 provide a valuable comparison of wavelength dependence of the diffraction efficiency. The explicit λ^{-2} dependence of the diffraction efficiency accounts for a factor of 1.8 of the nearly 4 times difference in the results. It would appear that the remaining difference is attributable to the fall-off of the coupling function with increased optical wavelength.

c) SAW Attenuation

Acoustic attenuation limits the efficiency and frequency resolution attainable with a Bragg cell. Being a nonlinear function of frequency, SAW attenuation becomes a major concern at about 1 GHz and higher. Table 7.2 lists measured values of attenuation α for a variety of waveguide structures. Also given are the frequency f, the velocity V_R, the optical waveguide thickness d, and the ratio d/Λ of the waveguide thickness to the acoustic wavelength. This last quantity is a coarse measure of the relative overlap of the optical and acoustic fields. The quantity $\bar{\alpha}$ is the expected attenuation at f = 1.0 GHz assuming an f^2 dependence for α. This value is probably too small for all but the $Ti:LiNbO_3$ waveguide, since at higher frequencies a greater percentage of the SAW power is in the waveguide thin films, which are usually inherently lossier than the indicated substrates. Note that the silicon substates have a 1.0 μm layer of SiO_2 between the silicon and waveguide. The results for the $As_2S_3/LiNbO_3$ waveguides illustrate the importance of keeping d small for wideband devices where the waveguide material is inherently lossy. The attractiveness of $Ti:LiNbO_3$, from the standpoint of SAW propagation loss, should be apparent.

7.3.5 Detection and Signal Processing

The detector array is the optical processor signal output transducer. It provides the interface between the IOSA focal plane and electronic processing. A more detailed discussion of photodetector array technology than is possible here has been given by *Borsuk* [7.74]. Primary optical sensor considerations include implementation architectures, performance requirements and sensor coupling. Each is discussed below.

a) Detector Array Architecture

The simplest optical detector is a linear array of photo sensors with an electronic signal output for each. A diagram of such an array is indicated as the first block in Fig.7.12. This device is the simplest to fabricate. Dynamic range and speed are undegraded by the noise and circuit capacitance of additional monolithic circuitry. On the other hand, input/output limitations preclude large parallel output detector arrays. These arrays are best suited for optical processor applications requiring sensitive signal detection, operation over a large dynamic range and fine temporal resolution. The photosensor sensitivity is limited only by quantum noise, thermal noise and output amplifier noise. Dynamic range is determined by device sensitivity and the maximum input signal level. Response time is governed by circuit and diode capacitances, and drift velocity.

The detector array elements are usually PIN, Schottky, or avalanche photodiodes. The PIN and Schottky photodiodes provide a large linear dynamic range, whereas avalanche photodiodes offer increased sensitivity, but suffer in linearity. Efforts are currently being made to improve detector performance by incorporating photo-feedback techniques to enhance dynamic range, and floating gate techniques to enhance sensitivity.

The parallel detector array pinout makes this approach unattractive for high-resolution applications. Integrated circuit packages with more than twenty four electrical connections are awkward to use and the number of diodes that can be physically accommodated in a convenient package is limited to several tens. Thus, parallel detector arrays provide only coarse resolution and do not take full advantage of the IOSA optical resolution.

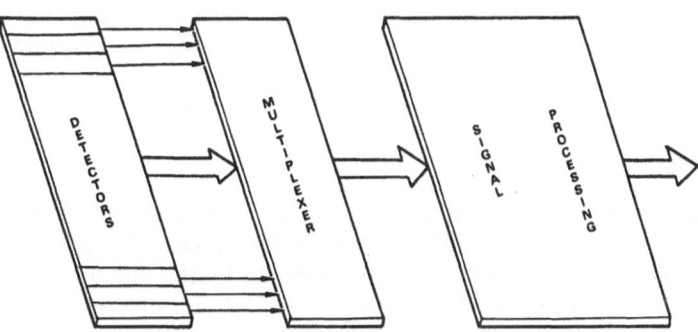

Fig.7.12. Detector array architecture

The self-scanned array is a parallel detector array alternative taking greater advantage of optical processor resolving capability. The self-scanned array architecture is represented as the detector and multiplexer combination in Fig.7.12. Charge excited by incident photons is stored. Upon command, the stored charges are transferred into an analog shift register and clocked from the circuit in a serial data stream. Mulitplexing all the sensor elements onto one terminal limits the revisit rate to the transfer rate divided by the number of sensors. A 1000 element array clocked at 1 MHz provides a revisit rate of 1 KHz.

Many detectors can be incorporated into the self-scanned array, taking full advantage of the IOSA resolving power. The array length is limited by photolithographic micro-electronics fabrication constraints. Linear CCD and photodetector arrays are available with up to 4000 elements. Overall, self-scanned arrays exhibit a lower dynamic range than a parallel output array. A typical value is in the 30-40 db range.

A combination of self-scanned arrays with parallel outputs can overcome some speed limitations of the self-scanned array. This can be accomplished by using parallel output from the multiplexer. A large self-scanned array may be segmented into a number of smaller arrays. All segments are clocked in parallel to enhance throughput rate. The series parallel array revisit rate is increased by the number of shift register outputs. For example, a 1000 element array with ten shift register outputs clocked at a 1 MHz rate results in a revisit rate of 10 kHz, thus providing a speed advantage of ten over the equivalent self-scanned array. However, pinout limitations are again incurred as speed is increased. Greater temporal measurement precision requires additional circuit capability that may be provided by increased clocking speeds, decreased resolution or more complex focal plane processing.

b) Post-Detection Processing

Array architectures may be adapted to specific functions such as thresholding, centroiding, blanking, siedelobe suppresion and A/D conversion. Monolithic, microelectronic signal processing technology, in conjuction with high performance sensor arrays, can enhance the sensor throughput rate and simplify post-sensor signal processing. Throughput rate is enhanced by transferring only signal information that fulfills predetermined criteria. Post-detection signal proessing is simplified because the centroiding algorithm has already filtered the detected signal and applied recognition criteria.

c) Detector Sensitivity

Sensitivity is a measure of the minimum detectable optical energy. It is a key detector characteristic that determines optical power budget and dynamic range. Detector array sensitivity is governed by the photodetector response and the effective noise background of the detector and associated multiplexing circuitry.

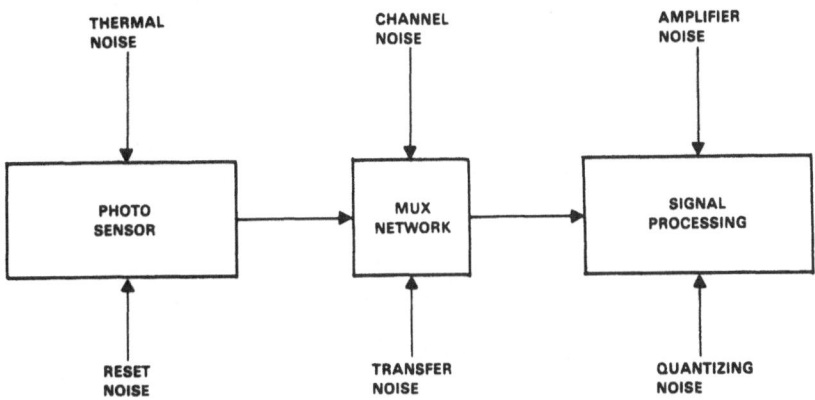

Fig.7.13. Detector array and signal processing noise sources

Photodetector array noise sources are indicated in Fig.7.13. They are associated with each stage of signal reporting, including photodetection, multiplexing and signal processing. Noise sources are a function of integration time, throughput rate and dimensional implementation constraints that also impact dynamic range. For conditions expected in spectrum analyzer applications, photodetector diode array noise levels of several thousand electrons have been measured while some charge-coupled detectors have shown only 200 electron noise levels at room temperature. At reduced temperatures noise levels of the order of ten electrons have been measured.

Noise, however, is only one aspect of sensitivity. Photodetector responsivity must also be considered. Responsivity is the ratio of signal current to the incident signal radiant power. It is determined by the transmissivity of the photodetector - optical medium interface, the internal quantum efficiency, and the collection efficiency of photo-generated electrons.

d) Dynamic Range

Photodetector dynamic range is defined as the range of input photon flux between the sensitivity level and saturation. Sensitivity may be defined by the noise equivalent signal level or by a threshold false alarm rate which relates the threshold signal level to the noise level.

For scanned arrays, dynamic range is established by photosensor sensitivity level, the transfer characteristic and the electron storage capacity. Charge storage capacity is a function of geometry in both photodiode or CCD photodetecting devices. The saturation charge (Q_{sat}) that can be stored by a detector element is given by

$$Q_{sat} = CV_R , \qquad (7.24)$$

where V_R is the reset voltage and C is the detector storage capacitance. CCD photosensor storage capacitance is typically 0.05 pf with a saturation

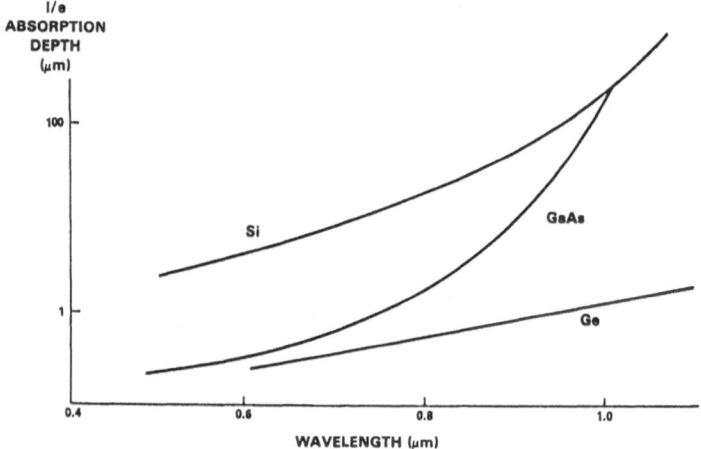

Fig.7.14. Wavelength dependence of 1/e absorption depth

charge of $0.25 \cdot 10^6$ electrons. Photodiode storage-well capacitance is typically larger (0.5 pf) with a 10^7 electron storage; but, storage-well capacitances of 5.0 pf have been fabricated with electron capacity in excess of 10^8. Speed and noise performance suffer considerably at the high storage levels.

Dynamic range and resolution are also affected by cross-talk between photodetectors. Photons entering one photodetector element result in some electrons that are collected by adjacent photodetectors. The degree of cross-talk is dependent upon the photon absorption depth. Small absorption depths minimize the probability of electron migration into adjacent photodetector sites. The half power absorption depth for various materials systems is indicated in Fig.7.14. For closely spaced photosensors (less than $15\,\mu$m) at $0.83\ \mu$m wavelength, pixel isolation requires photosensors in an absorbing material such as germanium or gallium arsenide, or interelement physical isolation in silicon.

e) Speed

Applications demanding accurate time determination impose severe speed requirements on self-scanned detector arrays. Time resolution required may be in the sub-microsecond range. Even for moderate array lengths the clocking speed to provide this revisit time is beyond the capability of currently available commercial devices. The upper limit on clock rate is now about 10 MHz. For an array of 100 elements, the revisit time would be 10 μs.

In addition to the signal charge transfer rate, sensor interrogation also affects speed. It is essential to quickly and efficiently address a photodetector element and remove the stored charge for subsequent serial readout. Rapid photodetector interrogation is required to minimize the inactive period and maximize the sensor duty factor. All the charge resident in a detector must be transferred into the serial readout circuit to enhance the signal-to-noise ratio and prevent charge from one frame being held over to the succeeding frame. This "holdover" cross-talk is especially difficult to suppress when high resolution time of arrival measurement are to be made.

A final consideration is the output data rate of a high speed sensor array. An array of 100 elements would provide 7-bit frequency quantization. Suppose that the amplitude of each channel's output is quantized to 6 bits. If the revisit time for the array were 1 μs, the output data rate would be $1.3 \cdot 10^9$ bits per second. Digital processing at these rates is presently impractical for applications where system cost and size are constrained. Thus, it is important that algorithms be implemented for filtering out all signals but those of interest.

f) Photodetector Pitch

The photodetector array dimensional constraints strongly impact IOSA frequency resolution. Photosensor pitch is related to frequency resolution through (7.12). Detector arrays are commerially available with pitch as small as 12 μm. Developmental arrays have been made with a pitch of 8 μm [7.77]. Following the discussion of Sect.7.2.2, this would permit a single frequency resolution of 1.5 MHz for a 5 cm focal length lens.

g) Photodetector Coupling

Photodetector circuit-mating to the IOSA presents problems not encountered in the corresponding bulk acousto-optic technology. Since the IOSA optical beam is confined to the thin optical waveguide, the photodetector must be placed close to the waveguide output edge to capture the rapidly diverging light. Reflection of light incident on the photodetectors back into the optical circuit substrate is of particular concern. The high dielectric constant substrate tends to confine reflected light, where it is scattered and reflected internally until it reappears as optical noise at the detector array. The background light level is increased and the resulting IOSA dynamic range decreased. It is imperative that photodetector reflections be minimized to prevent loss of sensitivity and dynamic range. This can be achieved by mounting the detector array at a tilt with respect to the substrate and applying antireflection coatings to the detectors and waveguide edge.

Ultimately, photodetectors might be a part of a monolithic IOSA circuit. The entire circuit could be fabricated in a microelectronics compatible material such as silicon or gallium arsenide. This would greatly ease the coupling problem. One approach to this has been developed by *Boyd* and *Chen* [7.76] and is illustrated in Fig.7.15. Here the IOSA with photodetector circuit is fabricated on a silicon substate. The coupling problem is simpler since photodetector/optical waveguide alignment depends only upon mask alignment. Coupling is effected by guiding the deflected light directly to the photosensor as shown.

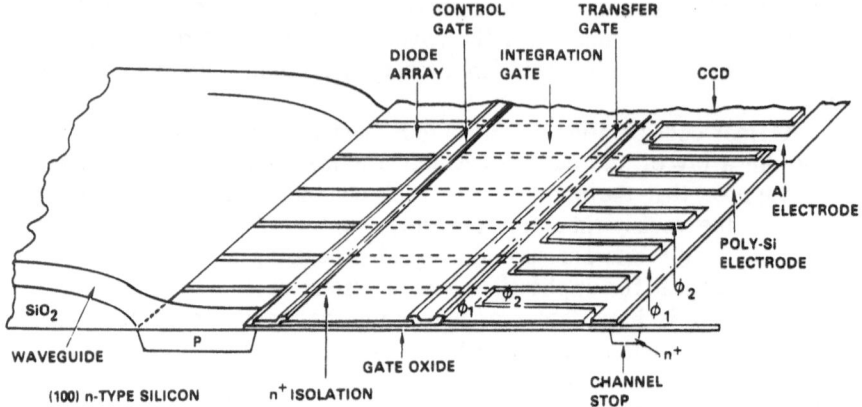

Fig.7.15. Integrated optical waveguide and CCD detector array. Reprinted with permission from [7.76]

7.4 IOSA Demonstrations

The first demonstration of a partially assembled IOSA was reported by *Mergerian* et al. [7.77]. The word "partially" is used because the laser source was a helium-neon gas laser that was focused with external lenses onto the input edge of the waveguide. The waveguide was formed by titanium diffusion into a $LiNbO_3$ substrate of 7 cm length. The crystal was x-cut so that the SAW power flowed along the c-axis and optical power along the y-axis. Two single-point diamond-turned geodesic lenses of nominal 2.5 cm focal length were used for collimation and focussing. A two element SAW transducer array was used to cover the 400 MHz bandwidth centered at 600 MHz. A separate measurement for the transducer performance yielded 5% diffraction efficiency for 60 mW of RF drive power.

The detector array was a 140 element self-scanned array, read out in groups of 10 at a 5 MHz clock rate. The resulting access time to any pixel was 2 μs. The single frequency resolution was limited by the array pitch of 12 μm to about 5.3 MHz. The actual optical spot size was 9-10 μm in diameter, or about 1.2 times diffraction-limited. The dynamic range for a cw RF input was limited to about 20 db because of the necessity to restrict the input optical power at 0.63 μm in order to avoid optical damage to the waveguide. A later version of this device, which used a diode-laser source, achieved 40 db dynamic range [7.78].

Ranganath et al. [7.79] have reported operation of the first fully assembled IOSA, which included a butt-coupled semiconductor laser. The waveguide was formed by titanium diffusion into a 5.5 cm long y-cut $LiNbO_3$ substrate. The optical beam propagated in the x-direction and the SAW beam along the c-axis. The 2.0 cm focal length geodesic lenses were formed by ultrasonic grinding. A single, chirped-SAW transducer with continuously varied finger tilt was used for light diffraction. The acousto-optic bandwidth was nearly 400 MHz, centered at 600 MHz. Using an optical wavelength of 0.82 μm, separate measurements with identical transducers showed 7% diffraction efficiency for 500 mW of RF power across the full 400 MHz band.

The detector array consisted of 100 silicon MOS photodetectors which transferred signal charge into a CCD array for serial readout in two data streams. At a clock rate of 2 MHz, this provided an access time to each detector of 25 μs. The 8 μm pitch of the array corresponded to an RF frequency increment between detectors of 4 MHz. However, the single-frequency resolution of the IOSA was limited to about 16 MHz by the diameter of the focused spot at the detector array. This was attributed to the detector being positioned somewhat out of the small depth-of-field of the low F/number optical train. The observed dynamic range of 20 db for a cw input signal was limited by detector output noise, as the charge storage capacity could be saturated at a laser output power of about 1 mW and an RF input power of 200 mW.

Since these milestone demonstrations of functioning IOSAs, a number of variants have been reported which have chiefly involved alternate lens structures and waveguide materials [7.80]. Among LiNbO$_3$-based circuits, *Davis* and *Hickernell* [7.81] have achieved an IOSA that utilized sputtered Ta$_2$O$_5$ Luneburg lenses. *Suhara* et al. [7.48] obtained 4 MHz resolution with 1 GHz bandwidth in a folded path arrangement using chirped, reflection-grating lenses that were butt-coupled to the substrate. Folding permitted reduction of the overall length to 15 mm. Also in LiNbO$_3$, an IOSA with 1 GHz bandwidth and 3 MHz resolution was developed by *Suhara* et al. [7.82], which employed Fresnel-type diffraction lenses formed by proton-exchange. *Li* et al. [7.83] have demonstrated a very noval approach in which the IOSA is fabricated on a spherical LiNbO$_3$ substrate. The substrate acts as a single geodesic lens, performing both collimation of the input light beam and focussing of the diffracted beam. The first device of this type has shown 75 MHz bandwidth and 1.5 MHz resolution. A subsequent device that has a bandwidth of 250 MHz centered at 500 MHz is described in Chap.8.

Two silicon-based IOSAs have been reported, both using Fresnel-type diffraction lenses. *Suhara* et al. [7.84] used a sputtered As$_2$O$_3$ waveguide for high-AO efficiency (1.2%/mW) and over 80 MHz of bandwidth. *Valette* et al. [7.85] used a Si$_3$N$_4$ waveguide and achieved a 250 MHz bandwidth with 6 MHz resolution. Both of these silicon-based circuits used ZnO transducers.

7.5 Related Applications

The IOSA circuit discussed to this point in this chapter represents the simplest form of an acousto-optic processor. There are, of course, numerous other applications for acousto-optic processors which, in principle, could be implemented with integrated optics technology. In this section, we will suggest and briefly discuss a few of these.

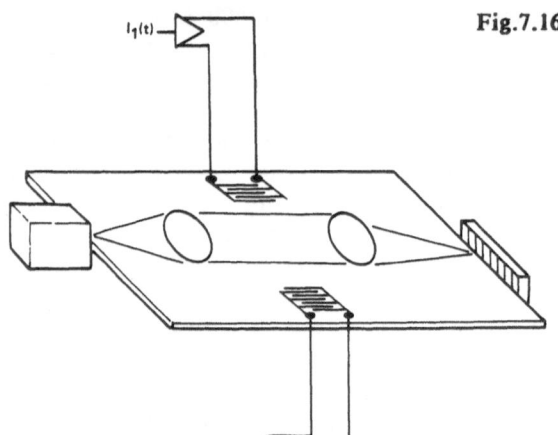

Fig.7.16. Space-integrating correlator

7.5.1 Acousto-Optic Correlation

Acousto-optic correlators may be categorized as space-integrating, time-integrating, or a hybrid combination of the two. A simple one-dimensional correlator performs the operation

$$R(x') = \int I_1(x)I_2(x-x')dx , \qquad (7.25)$$

where $I_1(x)$ and $I_2(x-x')$ are two functions of x and $R(x')$ is their correlation function. In a space-integrating correlator, the variable x is a spatial coordinate; in a time-integrating correlator, x represents time.

An extensive review of bulk-wave acousto-optic correlators has been given by *Rhodes* [7.86] and a brief review on their guided-wave counterparts by *Tsai* [7.64]. Examples of space- and time-integrating correlators are given in Figs.7.16 and 17, respectively. The intent here is only to make observations regarding the prospects for their implementation in integrated optical form from the standpoint of the components required and their layout.

It is apparent from an inspection of the figures that correlators have a great deal in common with spectrum analyzers. The componentry is essentially the same, except that the time-integrating correlator requires a zero-order beam stop. This could be accomplished by terminating the waveguide at the desired location so that light is dumped into the substrate. It is to be expected that the specifications for all the components would be similar for both correlators and spectrum analyzers. It is possible, though, that some applications might not require the large bandwidth normally desired for spectrum analysis. In that case a silicon based device with ZnO transducers might be more attractive than a LiNbO$_3$ based system. In some configurations, such as in Fig.7.17, more lenses are required than for the IOSA; consequently, the circuit length will be greater, leading to increased cost, more scattering and greater throughput loss. *Tsai* et al. [7.87-89] have reported on the construction of

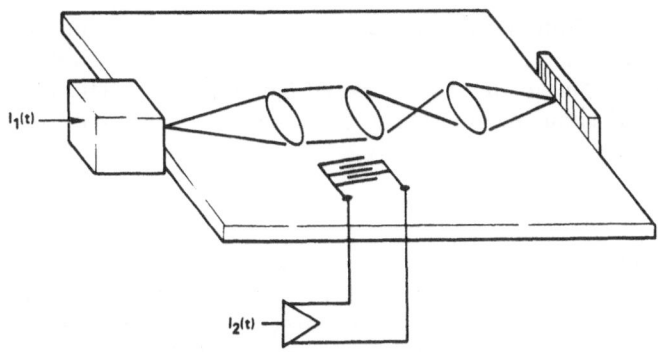

Fig.7.17. Time-integrating correlator

LiNbO$_3$-based correlators that are based on isotropic and anisotropic acousto-optic Bragg diffractions. A detailed description of such correlators is give in Chap.8.

7.5.2 Dual Bragg Cell Configurations

A natural extension of the simple IOSA is to enhance the cost and space savings of the technology by adding a second Bragg cell and detector which share the laser source and lenses, as illustrated in Fig.7.18. When the two transducers (or transducer arrays) are properly oriented to satisfy the Bragg condition, the upper transducer will diffract light to the upper detector array and the lower transducer to the lower detector array. The undiffracted beam will fall in between.

The frequency bands of the two transducers could be the same or different. An application where they are the same will be discussed in the next subsection. If they are different, then the scheme represents a way to double the bandwidth of the processor with a modest additional investment in the optical circuit. Of course, multiplexing the total RF band into

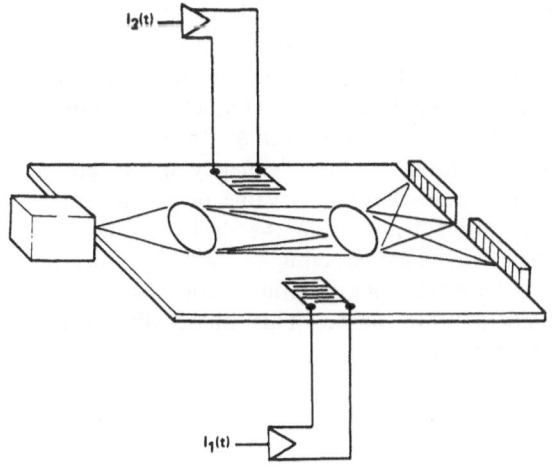

Fig.7.18. Dual channel IOSA

two channels requires additional RF hardware. In arranging the transducers, care must be taken to avoid spurious responses due to acoustic reflections. It should be recognized that the increased bandwidth may mean increased probability of simultaneous signals and resulting cross-modulation effects.

7.5.3 Other Processing Options

By employing appropriate signal conditioning prior to input to the SAW transducers, it should be possible to enhance the utility of the IOSA. For example, the RF input could be stored in a charge-coupled device which is then read into the Bragg cell at a slower or faster rate. This, in effect, performs a frequency multiplication which shifts the signal into the IOSA processing band and changes the frequency scale, providing a variable zoom capability. At present the input/output rates of CCDs are limited to the sub-gigahertz range, but development efforts are underway which may extend performance into the 1-5 GHz range.

Another interesting variation on the use of the IOSA involves the combination of a microwave hybrid circuit with the dual Bragg cell concept depicted in Fig.7.18. The microwave hybrid is a device with two inputs, S_1 and S_2, and two outputs, S_3 and S_4, which are related through

$$S_3 = S_1 + S_2 \cos\theta ,$$

$$S_4 = S_1 - S_2 \cos\theta , \qquad (7.26)$$

where θ is the relative phase of S_1 and S_2. If S_1 and S_2 are equal in amplitude and frequency, then it is found that

$$\theta = \cos^{-1}\left[\frac{S_3 - S_4}{S_3 + S_4}\right] . \qquad (7.27)$$

Thus we can derive the relative phase of two signals by performing an amplitude measurement on the outputs of the microwave hybrid.

Each output of the hybrid would be fed to one or the other of the transducers in Fig.7.18. The IOSAs would perform frequency channelization in the usual way so that outputs would appear at the detector locations of the twin arrays corresponding to the frequency of S_1 and S_2. Subsequent electronic processing could derive the value of θ. The interest in performing this sort of operation arises from the desire to determine the angle of arrival of signals. A signal arriving at two elements of a phased array antenna would generate equal outputs from each, at the same frequency, that differed only in phase. This phase difference is directly related to the angle of arrival.

7.6 Summary

The basic technology of the IOSA has been seen to be quite versatile in its applicability to a broad range of signal processing functions. In addition to the attractive performance characteristics that are inherent to acousto-optic signal processors, the IOSA offers the potential advantages of smaller size, lower cost and greater reliability. Numerous technology challenges must be addressed before the IOSA concept can realize its potential. Semiconductor lasers with greater single-mode output power and less spontaneous emission are needed. Scattering levels must be reduced in optical waveguides. Although geodesic lenses have good quality, reproducibility needs to be improved. Diffraction-grating lenses are highly reproducible but need to be improved in quality. Greater dynamic range requires more efficient Bragg cells and input/output coupling, and more sensitive detector arrays. To make use of the large effective processing speed of acousto-optic systems, electronic interface circuitry is needed to efficiently and quickly filter out redundant or uninteresting information and digitize useful information for subsequent data-processing. If these things come to pass, then the technology of the IOSA may indeed realize its promising potential.

References

7.1   R. Schubert, J.H. Harris: IEEE Trans. MTT-16, 1048 (1968)

7.2   D.L. Hecht: Opt. Engineering 16, 461 (1977)

7.3   M.C. Hamilton, D.A. Wille: Topical Meeting on Integrated Optics, New Orleans, LA (1974) Digest Paper WA8

7.4   M.C. Hamilton, D.A. Wille, W.J. Miceli: Opt. Engg 16, 475 (1977)

7.5   D.B. Anderson, T.J. Boyd, M.C. Hamilton, R.R. August: IEEE J. QE-13, 268 (1977)

7.6   M.K. Barnoski, B.U. Chen, T.R. Joseph, J.Y.M. Lee, O.G. Ramer: IEEE Trans. CAS-26, 1113 (1979)

7.7. A. Spezio, R. Orsino: 1984 IEEE Int'l Workshop on Integrated Optical and Related Technologies of Signal Processing, Florence, Italy, Techn. Digest

7.8   J.T. Boyd, D.B. Anderson: IEEE J. QE-14, 437 (1978)

7.9   D.W. Vahey: Proc. SPIE 126, 62 (1979)

7.10 G.B. Brandt: Proc. SPIE 176, 70 (1979)

7.11 D.W. Vahey, N.F. Hartmann, R.C. Sherman: Contract F33615-79-C-1852, Final Technical Report, AFWAL-TR-80-1186 (May 1981)

7.12 H.C. Casey Jr., M.B. Panish: *Heterostructure Lasers* (Academic, New York 1978)

7.13 H. Kressel, J.K. Buttler: *Semiconductor Lasers and Heterojunction LEDs* (Academic, New York 1977)
       H. Kressel (ed.): *Semiconductor Devices for Optical Communication*, 2nd. ed., Topics Appl. Phys., Vol.39 (Springer, Berlin, Heidelberg 1982)

7.14 M. Nakamura: IEEE Trans. CAS-26, 1055 (1979)

7.15 W.K. Burns, R.P. Moeller: Appl. Opt. 20, 913 (1981)

7.16 C.T. Müller, C.T. Sullivan, W.S.C.Chang, D.G.Hall, J.D.Zino, R.R.Rice: IEEE J. QE-16, 363 (1980)

7.17 D.G. Hall, R.R. Rice, J.D. Zino: Opt. Lett. 4, 292 (1979)

7.18 T. Tamir (ed.): *Integrated Optics*, 2nd ed., Topics Appl. Phys., Vol.7 (Springer Berlin, Heidelberg 1979)
T. Tamir (ed.): *Guided-Wave Optoelectronics*, Springer Ser. Electron. Photonics, Vol.26 (Springer, Berlin, Heidelberg 1988)
7.19 J.E. Goell, R.D. Standley: Proc. IEEE **58**, 1504 (1970)
7.20 J.E. Goell: Appl. Opt. **12**, 737 (1973)
7.21 Y.C. Cheng, W.D. Westwood: J. Electron. Mat. **3**, 36 (1974)
7.22 R.L. Aagard: Appl. Phys. Lett. **27**, 605 (1975)
7.23 W. Stutius, W. Streifer: Appl. Opt. **16**, 3218 (1977)
7.24 S. Dutta, H.E. Jackson, J.T. Boyd: 3rd Int'l Conf. on Integrated Optics and Optical Fiber Communications, San Francisco, CA (1981) Techn. Digest, Paper WB5
7.25 R.V. Schmidt, I.P. Kaminow: Appl. Phys. Lett **25**, 458 (1974)
7.26 G.B. Hocker, W.K. Burns: IEEE J. QE-11, 270 (1975)
7.27 H. Naitoh, M. Nunoshita, T. Nakayama: Appl. Opt. **16**, 2546 (1977)
7.28 W.K. Burns, G.B. Hocker: Appl. Opt. **16**, 2048 (1977)
7.29 S.K. Sheem, W.K. Burns, A.F. Milton: Opt. Lett. **3**, 76 (1978)
7.30 D.W. Vahey, R.P. Kenan, W.K. Burns: Appl. Opt. **19**, 270 (1980)
7.31 A.M. Glass, I.P. Kaminow, A.A. Ballman, D.H. Olsen: Appl. Opt. **19**, 276 (1980)
7.32 W.K. Burns, P.H. Klein, E.J. West, L.E. Plew: J. Appl. Phys. **50**, 6175 (1979)
7.33 K. Sugii, M. Fukuma, H. Iwasaki: J. Mater. Sci. **13**, 523 (1978)
7.34 R. Schubert, J.H. Harris: J. Opt. Soc. Am. **61**, 154 (1971)
7.35 R. Ulrich, R.J. Martin: Appl. Opt. **10**, 2077 (1971)
7.36 D. Zang, C. Tsai: Appl. Phys. Lett. **46**, 703 (1985)
7.37 F. Zernike: Opt. Commun. **12**, 379 (1977)
7.38 G.C. Righini, V. Russo, S. Sottini, G. Toraldo di Francia: Appl. Opt. **12**, 1477 (1973)
7.39 W.H. Southwell: J. Opt. Soc. Am. **67**, 1010 (1977)
7.40 J. Sochacki: J. Lightwave Technol. LT-3, 684 (1986)
7.41 S.K. Yao, D.B. Anderson: Appl. Phys. Lett. **33**, 307 (1978)
7.42 S.K. Yao, D.E. Thompson: Appl. Phys. Lett. **33**, 635 (1978)
7.43 S. Forouhar, W.S.C. Chang, S.K. Yao: IEEE J. LT-2, 503 (1984)
7.44 P.R. Ashley, W.S.C. Chang: Appl. Phys. Lett. **33**, 490 (1978)
7.45 G. Hatakoshi, S. Tanaka: Opt. Lett. **2**, 142 (1978)
7.46 P.K. Tien: Opt. Lett. **1**, 64 (1977)
7.47 T. Fujita, H. Nishihara, J. Koyama: Opt. Lett. **7**, 578 (1982)
T. Suhara, H. Nishihara, J. Koyama: IEEE J. QE-18, 1057 (1982)
T. Suhara, K. Kobayashi. H. Nishihara. J. Koyama: Appl. Opt. **21**, 1966 (1982)
7.48 W.S.C. Chang, P.R. Ashley: IEEE J. QE-16, 744 (1980)
7.49 S. Valette, A. Morque, P. Mottier: Electron. Lett. **18**, 13 (1982)
7.50 T.Q. Vu, J.A. Norris, C.S. Tsai: Opt. Lett. **13**, 1141 (1988)
7.51 T.Q. Vu, J.A. Norris, C.S. Tsai: Appl. Phys. Lett. **54**, 1098-1100 (1989)
7.52 T. Van Duzer: Proc. IEEE **58**, 1230 (1970)
7.53 E. Spiller, J.S. Harper: Appl. Opt. **13**, 2105 (1974)
7.54 D.W. Vahey: Final Report, Contract F33615-77-C-1153, "Optical Waveguide Geodesic Lenses", AFWAL/TR-78-85 (June 1978)
7.55 K.S. Kunz: J. Appl. Phys. **25**, 642 (1954)
7.56 G. Toraldo di Francia: Opt. Acta. **1**, 157 (1955)
7.57 S. Sottini, V. Russo, G. Righini: J. Opt. Soc. Am. **69**, 1248 (1979)
7.58 G.E. Betts, J.C. Bradley, G.E. Marx, D.C. Schubert, H.A. Trenchard: Appl. Opt. **17**, 2346 (1978)
7.59 D. Kasai, E. Marom: J. Opt. Soc. Am. **69** (1979)
7.60 D. Mergerian, E.C. Malarkey, R.P. Pautienus, J.C. Bradley: SPIE **176**, 85 (1979)

7.61　B. Chen, O.G. Ramer: IEEE J. QE-15, 853 (1979)

7.62　B. Chen, E. Marom, R.J. Morrison: Appl. Phys. Lett. 33, 511 (1978)

7.63　B. Chen, E. Marom, A. Lee: Appl. Phys. Lett. 31, 263 (1977)

7.64　C.S. Tsai: IEEE Trans. CAS-26, 1072 (1979)

7.65　C.S. Tsai, M.A. Alhaider, Le T. Nguyen, B. Kim: Proc. IEEE 64, 318 (1976)

7.66　Le T. Nguyen, C.S. Tsai: Appl. Opt. 16, 1297 (1977)

7.67　C.C. Lee, K.Y. Liao, C.L. Chang, C.S. Tsai: IEEE J. QE-15, 1166 (1979)

7.68　T.R. Joseph, B. Chen: 1979 IEEE Ultrasonics Symp., New Orleans, LA (1979)

7.69　R.L. Davis, F.S. Hickernell, F.V. Richard, R. Ward: Contract F33615-80-1048, Final Technical Report, AFWAL-TR-81-1289, February (1982)

7.70　F.S. Hickernell, R.L. Davis, F.V. Richard: 1978 IEEE Ultrasonics Symp. (Proc.: IEEE Cat. No. 78CH1344-1SU, Cherry Hill, NJ)

7.71　M.C. Hamilton: Unpublished

7.72　Y. Ohmachi: J. Appl. Phys. 44, 3928 (1973)

7.73　R.L. Davis, F.S. Hickernell: Am. Phys. Soc. Meeting, Phoenix, AZ (March 1981)

7.74　G.M. Bursuk: Proc. IEEE 69, 100 (1981)

7.75　J.Y.M. Lee, B. Chen: Abstracts, Spring Meeting, Electrochemical Society, St. Louis, MO, 80-1, 416 (1980)

7.76　J.T. Boyd, C.L. Chen: IEEE J. QE-13, 282 (1977)

7.77　D. Mergerian, E.C. Malarkey, R.P. Pautienus, J.C. Bradley, G.E. Marx, L.D. Hutcheson, A.L. Kellner: Appl. Opt. 19, 3033 (1980)

7.78　D. Mergerian, E.C. Malarkey, R.P. Pautienus: Int'l Conf. Integrated Opt. and Optical Fiber Commun., Tokyo, Japan (1983) Paper 30B3-5

7.79　T.R. Ranaganath, T.R. Joseph, J.Y. Lee: 3rd Int'l Conf. on Integrated Optics and Optical Fiber Communications, San Francisco, CA (1981) Techn. Digest, Paper WH3

7.80　T. Suhara, H. Nishihara: IEEE J. QE-22, 845 (1986)

7.81　R.L. Davis, F.S. Hickernell: Int'l Conf. Integrated Opt. and Optical Fiber Commun., San Francisco, CA (1981) Paper We-6

7.82　T. Suhara, S. Fujiwara, H. Nishihara: Topical Meet. Gradient Index. Opt. Imaging Syst., Palermo, Italy (1985) Paper A3

7.83　Q. Li, C.S. Tsai, S. Sottini, C. Lee: Appl. Phys. Lett. 46, 707 (1985)

7.84　T. Suhara, T. Shiono, H. Nishihara, J. Koyama: IEEE J. LT-1, 624 (1983)

7.85　S. Valette, J. Lizet, P. Mottier, J. Jadot, S. Renard, A. Fournier, A Grouillet, P. Gidon, H. Denis: Electron. Lett. 9, 883 (1983)

7.86　W.T. Rhodes: Proc. IEEE 69, 65 (1981)

7.87　I.W. Yao, C.S. Tsai: 1978 IEEE Ultrasonics Symp. Proc., IEEE Cat. No.78, CH 1344-154, pp.87-90

7.88　C.S. Tsai, J.K. Wang, K.Y. Liao: Proc. SPIE 180, 160-162 (1979)

7.89　K.Y. Liao, C.C. Lee, C.S. Tsai: 1982 Topical Meeting on Integrated and Guided-Wave Optics, Pacific Grove, CA, Techn. Digest, WA4-1 to 4, IEEE Cat. No.82 CH 1719-4

8. Integrated Acousto-Optic Device Modules and Applications

Chen S. Tsai

With 20 Figures

As discussed in Chaps. 5-7, a great deal of advancements have been made in planar guided-wave acoustooptics. These advancements include the analytical treatment of complex interaction geometry, preparation of waveguide materials, design and fabrication of wide-band Bragg modulators and deflectors (Bragg cells), and the demonstration of a number of simple applications. Continuing progress in the fabrication and performance of other components including optical waveguides, waveguide lenses, diode laser sources and photodetector arrays, and their integration have significantly advanced the prospects for realization of a variety of integrated acousto-optic (AO) device modules and circuits. The most notable example of such modules is the integrated optic RF spectrum analyzer with $LiNbO_3$ that has been presented in detail in Chap. 7. This chapter provides a review on the spectrum analyzer modules on nonpiezoelectric substrate materials and a number of other planar AO device modules with $LiNbO_3$ and GaAs as well as a spherical waveguide AO device module in $LiNbO_3$ that are being developed. The results obtained thus far have shown that such AO device modules will have small substrate dimensions along the optical path and will also be inherently of high modularity and versatility. Consequently, such integrated AO device modules should find novel applications in wide band multichannel integrated- and fiber-optic communication, signal processing, and computing systems. Some of these potential applications are also described and discussed.

8.1 RF Spectrum Analyzer Modules in Nonpiezoelectric Substrates

The integrated AO device module that has received the most attention is the so-called integrated optic RF spectrum analyzer for real-time processing of wide-band RF signals. A fully-integrated or monolithic version is depicted in Fig. 5.43. As described in Sect. 5.5, the AO spectral analysis of RF signal is based on the properties that the deflection angle and the intensity of the Bragg-diffracted light are, respecitively, proportional to the frequency and the power of the RF signal being applied to the surface acoustic wave (SAW) transducer. As presented in Chaps. 5 and 7, a $LiNbO_3$ substrate for the AO Bragg cell and the waveguide lenses together with hybrid integration of both the light source and the photodetector array constitutes the most common approach at present. In this hybrid approach

Springer Series in Electronics and Photonics, Vol. 23
Guided-Wave Acoustooptics   Editor: C. S. Tsai
© Springer-Verlag Berlin, Heidelberg 1990

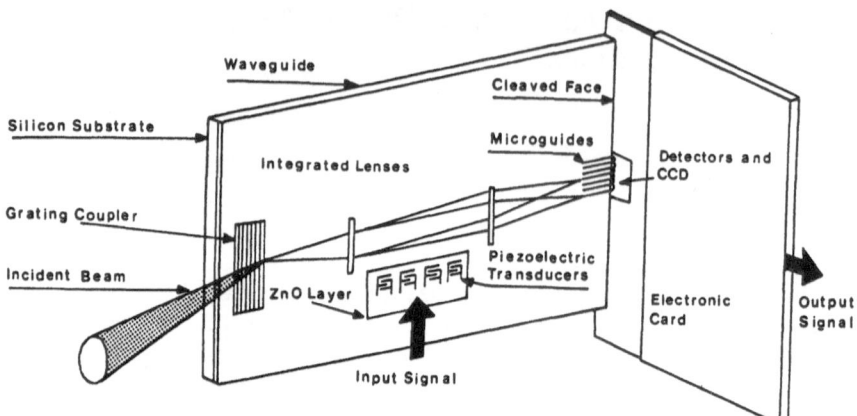

Fig.8.1. Schematic of experimental setup for integrated optical spectrum analyzer in oxidized silicon substrate [8.9]

the GaAs-based laser diode and the photodetector array consisting of either PIN, Schottky, or avalanche photodiodes are butt-coupled to the input and the output end faces of the LiNbO$_3$ substrate [8.1]. Since a detailed treatment of this particular module in the LiNbO$_3$ piezoelectric substrate has been given in the preceding chapters, this section will focus on the modules that were realized in the nonpiezoelectric substrates.

To facilitate a total or monolithic integration, GaAs will be the ideal substrate to use because both the diode laser and the photodetector array as well as the associated electronic devices can be fabricated in the resulting GaAs-GaAlAs planar waveguide [8.2]. On this substrate the light source takes the convenient form of either a distributed feedback [8.3] or a Bragg-reflector laser [8.4], and the photodetector array [8.5] can be fabricated in the same substrate. As discussed in Sect.8.7, very significant progress has been made on the realization of efficient and wideband AO Bragg cells on this substrate material [8.6,7]. Most recently, waveguide lenses have also been fabricated successfully in GaAs waveguides by using ion-milling [8.8]. Thus, monolithic integrated AO RF spectrum analyzers and other monolithic integrated AO circuits are on their way to be realized. Other substrate materials that are capable of integrating the AO Bragg cell and the photodetector array on the same substrate are thermally oxidized silicon (SiO$_2$/Si) [8.9] and arsenic trisulphide on thermally oxidized silicon (As$_2$S$_3$/SiO$_2$/Si) [8.10]. Figures 8.1 and 2 show, respectively, spectrum analyzer modules using these nonpiezoelectric substrate materials.

In the first spectrum analyzer module shown in Fig.8.1 [8.9], the AO Bragg cell utilized a four-element tilted-array transducer (Chap.5) that was fabricated via a silicon nitride/zinc oxide (Si$_3$N$_4$/ZnO) overlay. The SAW generated by the ZnO overlay transducer [8.11] was centered at 300 MHz with a bandwidth of about 100 MHz. Both the collimation and the Fourier- transform lenses were of the Fresnel type referred to in Chap.7. The incident He-Ne laser light at 6328 Å wavelength was coupled into the waveguide through a grating coupler [8.12] and the spectrum of the Bragg-

274

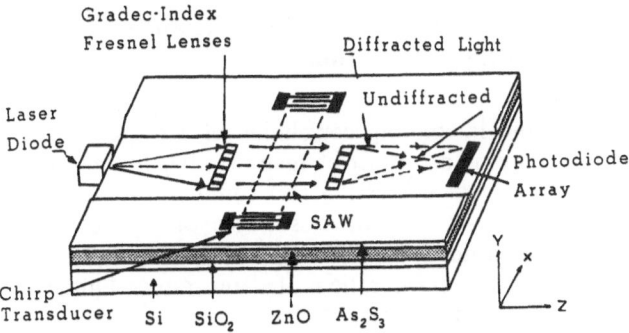

Fig.8.2. An integrated-optic Fourier processor consisting of an acousto-optic deflector and graded-index Fresnel lenses in an As_2S_3 waveguide on a SiO_2/Si substrate [8.10]

diffracted light was relayed through an array of microguides in the same substrate prior to detection by a photodetector array-CCD register combination. A frequency resolution of 10 MHz, a dynamic range of 15 to 20 dB, and a crosstalk of -7 db between adjacent channels were measured in this preliminary module. As a result of relatively low AO figure of merit with the SiO_2/Si substrate and low conversion efficiency of the ZnO transducer, an RF drive power of 2 W was required to diffract 2% of the incident light power. An advanced version that incorporates an edge-coupled diode laser and a SAW transducer array of higher center frequency and larger bandwidth was also developed subsequently.

In the second spectrum analyzer module shown in Fig.8.2 [8.10], a ZnO film was first deposited on a portion of the SiO_2/Si substrate to serve as the piezoelectric film for SAW transduction. Both the tilted-array transducer and the tilted-finger chirp transducer (Chap.5) covering the frequency range from 90 to 170 MHz were fabricated on the ZnO film [8.11]. An amorphous chalcogenide As_2S_3 film waveguide was subsequently grown on both the region with ZnO film and the region without it. The As_2S_3 film was used to facilitate efficient AO Bragg diffraction because of its high AO figure of merit. Both the collimation and the Fourier transform lenses were of the graded-index Fresnel type fabricated by electron-beam direct-writing technique (Chap.7). Spectral analysis measurement on the prototype just described was carried out using a Nd:YAG laser at 1.064 μm wavelength that was coupled into the optical waveguide via a grating coupler. The diffracted-light spots were coupled via the cleaved waveguide edge and projected upon a CCD camera through an image lens. As a result of a much degraded focal spot size due to in-plan scattering of the waveguide, namely a measured focal spot width of 9.3 μm defined at -3 db intensity points, a relatively low frequency resolution of 19 MHz was measured. A dynamic range of approximately 20 dB was also demonstrated.

8.2 Acousto-Optic Time-Integrating Correlator Module Using Anisotropic Bragg Diffraction

As indicated in Sect.5.5, time-integrating correlation of RF signals using AO Bragg diffraction has become a subject of many studies because of its unique applications in radar signal processing and communications. While most of these studies have utilized bulk-wave isotropic Bragg diffraction, some studies have also utilized guided-wave isotropic and anisotropic Bragg diffractions. In the conventional configurations that use either bulk-wave [8.13] or guided-wave [8.13] *isotropic* Bragg diffraction, a pair of imaging lenses and a spatial filter are required to separate the diffracted light beam from the undiffracted light beam. Some results of the experiments that utilize guided-wave isotropic Bragg diffraction have already been presented in Chap.5. Hybrid and monolithic structures for integrated optic implementations have also been suggested [8.14]. This section describes a novel structure which utilizes guided-wave *anisotropic* Bragg diffraction and hybrid integration [8.15,16], as depicted in Fig.8.3a. Unlike the conventional configurations, this particular structure can conveniently incorporate a thin-film polarizer to separate the diffracted light from the undiffracted light prior to detection and, thus, eliminates the need for imaging lenses and the spatial filter. The resulting AO time-integrating correlator is much smaller in the dimension along the optical path and capable of providing a larger time window and lower optical insertion loss. Furthermore, it is easier to implement in the integrated optic format. For example, in the correlator module [8.15], as shown in Fig.8.3b, a laser diode and a thin-film polarizer/photodetector array (CCPD) composite were, respectively, butt-coupled to the input and output end faces of a Y-cut LiNbO$_3$ plate $2 \times 12 \times 15.4$ mm^3 in size. A single geodesic lens (with an 8 mm focal length) was used to collimate the input light beam prior to interaction with the SAW. The SAW propagates at 5° from the X-axis of the LiNbO$_3$ plate to facilitate anisotropic Bragg diffraction between the TE$_0$- and the

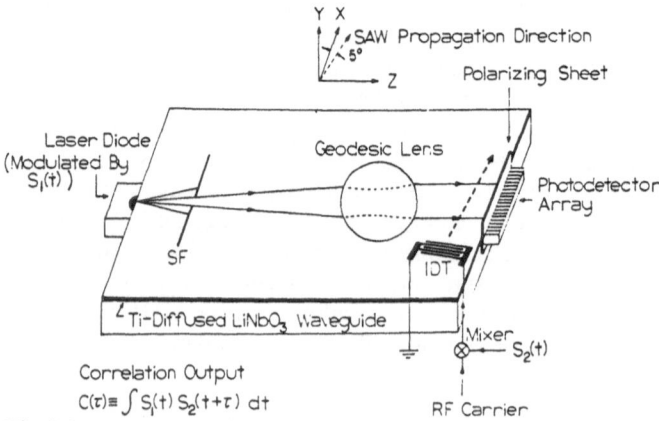

Fig.8.3a. Acousto-optic time-integrating correlator module using anisotropic Bragg diffraction

Fig.8.3b. Hybrid integrated acoustooptic time-integrating correlation of RF signals

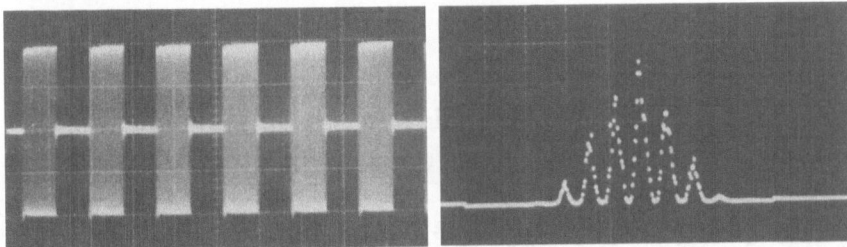

Fig.8.3c. Auto-correlation waveform (*right*) for a 10 MHz square-wave modulated signal (*left*)

TM_0-modes [8.17]. The correlation between the two signals $S_1(t)$ and $S_2(t)$ was performed by modulating the laser diode with $S_1(t)$ and the RF carrier applied to the SAW transducer with $S_2(t)$. The time-integrating correlation waveform was read out from the photodetector array.

Experiments carried out with this correlator of an incomplete hybrid integration at 0.6328 μm optical wavelength and 391 MHz acoustic center frequency, demonstrated a bandwidth of 60 MHz, a time-bandwidth product of $4.2 \cdot 10^5$, and a dynamic range of 27 dB [8.15]. Figure 8.3c shows the autocorrelation waveform obtained with a 10 MHz square-wave modulation signal. Note that in the correlator module of Fig.8.3b the geodesic lens is located at the center and the SAW transducer at the right end of the $LiNbO_3$ substrate. A complete hybrid integration using a diode laser at 0.78 μm wavelength and the SAW at 314 MHz center frequency was also realized subsequently [8.18]. A considerably larger bandwidth should be obtainable by using the wideband Bragg cells presented in Chap.5 and direct modulation of the diode laser at GHz rates [8.19]. Also, the geodesic lens may be conveniently replaced by the titanium-indiffused proton-exchanged (TIPE) lens that was developed recently [8.20]. A brief discussion on the TIPE lens will be given in Sect.8.4.

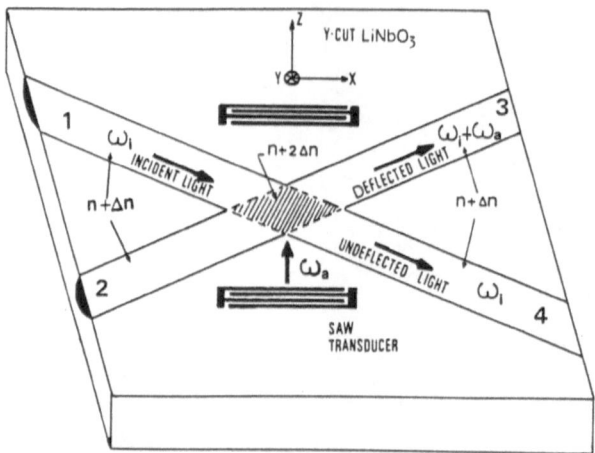

Fig.8.4a. Acousto-optic diffraction from surface acoustic wave in a LiNbO$_3$ crossed-channel waveguide

8.3 Crossed-Channel Waveguide Acousto-Optic Modulator/Deflector and Frequency-Shifter Modules

While *planar* waveguide AO Bragg devices have already reached some degree of sophistication and found immediate applications, *channel* waveguide AO devices that result from AO deflection in the channel waveguides [8.21] have only started to receive interest and attention. This interest was motivated by the fact that comparable cross sections of the channel waveguide and the single-mode optical fiber would greatly facilitate interfacing of the resulting AO channel devices with the fiber optic systems. One interaction configuration of particular interest is shown in Fig.8.4a [8.21]. Two identical channel waveguides in an Y-cut LiNbO$_3$ substrate are crossed at an angle ψ to form a 2Δn straight intersection [8.22]. Unlike the conventional Δn intersection, the refractive index change in the crossover region is twice that in the remaining parts of the channel waveguide. As a result, the light wave is better guided in the crossover region. This LiNbO$_3$ crossed-channel waveguide was first used to modulate and switch the light wave by the electrooptic effect [8.22]. For the similar application that utilizes the AO effect, a pair of interdigital SAW transducers was symmetrically positioned, as shown in Fig.8.4a, so that the SAW generated would propagate in the crossover region of the crossed-channel waveguide. The center frequency of the SAW was chosen such that the corresponding Bragg angle was equal to one half of the intersection angle. The working principle of the resulting device is now briefly described. An optical wave incident at guide #1 is diffracted by the moving optical grating induced in the crossover region by the SAW. Consequently, a portion of the incident light is deflected into guide #3, and the frequency of the deflected light is upshifted by an amount equal to the acoustic frequency. Similarly, a por-

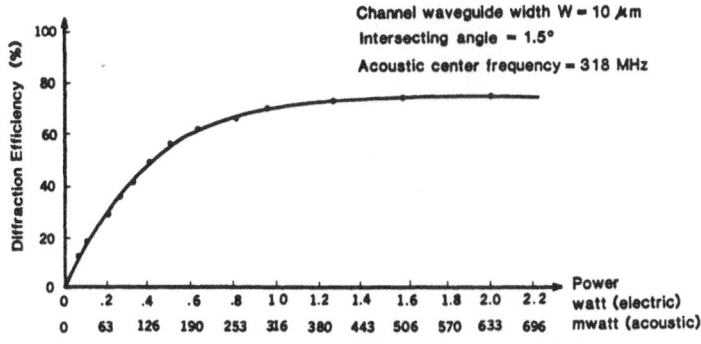

Fig.8.4b. Acoustooptic diffraction efficiency versus drive powers

tion of an optical wave incident at guide #2 will be deflected into guide #4 and the frequency of the deflected light will be downshifted by the same amount. Such a device module may find applications in future integrated and fiber optic systems. In the application for heterodyning detection the frequency-shifted light can be conveniently used as a reference signal (local oscillator) in connection with optical communications and fiber-optic sensing.

High diffraction efficiency was first demonstrated in a preliminary experiment with *multimode* crossed-channel waveguides of 30 μm channel width on a Y-cut LiNbO$_3$ substrate with a SAW operating at 634 MHz center frequency [8.21]. Subsequently, the experiment was extended to *single-mode* crossed-channel waveguides of 10 μm channel width, with similar results [8.23]. Figure 8.4b shows the measured diffraction efficiency versus both the RF and the acoustic drive powers. Specifically, a 50% diffraction efficiency was measured with 0.13W of acoustic power centering at 318 MHz. A bandwidth of 13.4 MHz was also measured. This result clearly indicates the feasibility for realization of an active integrated optic module with a 50-50 power split and a tunable frequency offset. Figure 8.4c is a photograph of the complete module. Located in the center of the device holder is the LiNbO$_3$ plate 0.2×1.0×1.4 cm^3 in size. A pair of RF connectors for excitation and detection of the SAW are also shown. The same device configuration was subsequently utilized to perform frequency shifting [8.24] and side-band modulation [8.25]. While both prism and edge couplings of the light beam have been used successfully, a more rigid and robust coupling should utilize single-mode fibers together with any of the existing approaches such as the flip-chip technique [8.26].

8.4 Channel-Planar Composite Waveguide Acousto-Optic Bragg Modulator Modules

The first successful experiment on AO Bragg diffraction in a LiNbO$_3$ single-mode channel-planar composite waveguide that incorporates a TIPE linear microlens array was carried out using the arrangement shown in

Fig.8.4c. Single-mode crossed-channel waveguide acousto-optic modulator module in LiNbO₃ substrate

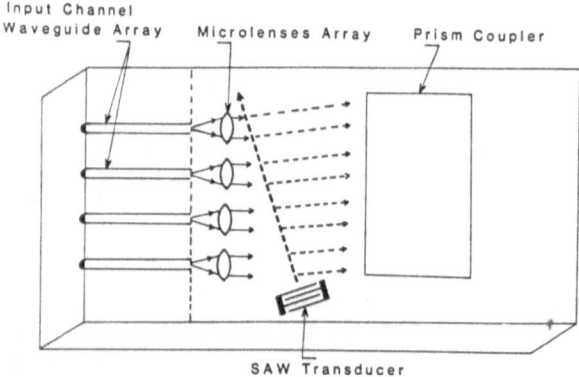

Fig.8.5. Acousto-optic Bragg diffraction in a LiNbO₃ channel-planar composite waveguide

Fig.8.5 [8.27]. Subsequently, a compact integrated modulator module was realized by further incorporating a large-aperture TIPE integrating lens with the channel-waveguide array, the planar waveguide, the linear microlens array, and the SAW transducer in a substrate $0.2 \times 1.0 \times 2.0$ cm³ in size (Fig.8.6a) [8.28]. A photograph of this integrated AO Bragg modulator module is shown in Fig.8.6b. The extension of the channel-waveguide array *directly* into the planar waveguide is of particular interest because collimation of the multiple incident light beams, AO Bragg diffractions, and the subsequent focusing can readily take place in the planar waveguide.

Both the single TE_0- mode channel-waveguide array and the single TE_0-mode planar waveguide (having the same Δn) were fabricated using the TI process while the linear microlens array was formed using the TIPE technique [8.20]. The linear microlens array (each with a 0.12 mm aperture and 0.6 mm focal length) and the integrating lens (with a 4.0 mm aperture and 10.0 mm focal length) were formed using the TIPE method [8.29]

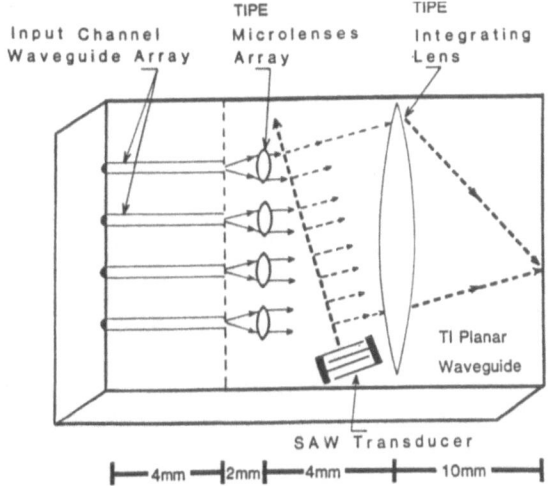

TIPE Microlenses Array

TIPE Integrating Lens

Input Channel Waveguide Array

TI Planar Waveguide

SAW Transducer

├──4mm──┤2mm├──4mm──┤──10mm──┤

Fig.8.6a. An integrated acoustooptic Bragg modulator module in a LiNbO₃ channel-planar composite waveguide

Fig.8.6b. Photograph of an integrated acousto-optic Bragg modulator module

which combined the titanium-indiffusion (TI) [8.30] and the Proton-Exchange (PE) [8.31] processes. The TIPE technique was developed originally for fabrication of planar waveguides [8.29]. This technique has demonstrated several capabilities which are not possible with either the TI process or the PE process alone. For fabrication of the single-mode microlenses and microlens arrays [8.20] the well-established TI process was first applied in a Y-cut LiNbO₃ substrate to form a planar waveguide that supports a single TE mode of the lowest order. Subsequently, a masking material such as Si_3N_4 or Cr/Au with a designed lens contour was deposited on the TI waveguide (Fig.8.7a). The sample was then immersed in molten benzoic acid at 230°C for six hours. As a result of the selective proton exchange, the region (the shaded area in Fig.8.7a) without the masking material had its extraordinary refractive index increased by as much as 0.11 in comparison to the remaining TI region. Consequently, this PE region of appropriate contour will function as a planar waveguide lens. For example, using

281

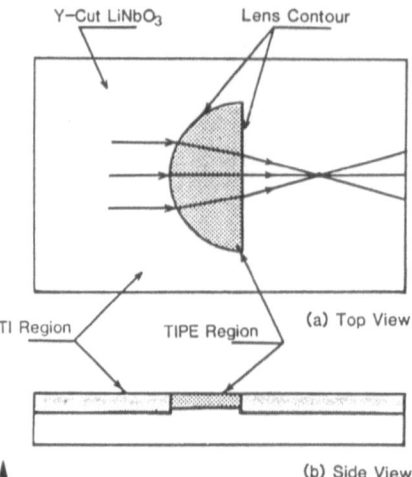

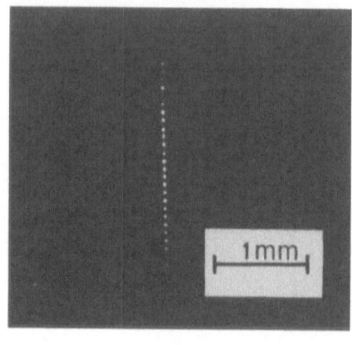

Fig.8.7a. Planar waveguide lens in LiNbO$_3$ formed by titanium-indiffusion proton exchange (TIPE) technique

Fig.8.7b. Multiple focused light spots from a TIPE 60-element linear microlens array in a LiNbO$_3$ waveguide (60μm aperture and 200μm focal length for each element lens)

the Fermat principle the contour for a plano-convex lens has been shown to be an ellipse [8.20].

A variety of basic (single) lenses with plano-convex and double-convex contours of various apertures and focal lengths have been fabricated and tested. The measured half-power (3 db) width of the focal spot in light intensity was typically 2.0 μm. The strength of the highest sidelobe was typically 12 to 16 db lower than that of the main lobe. The measured focal length of the lens agrees well with the design value. The average insertion loss of the lens was measured to be 1.5 db which corresponds to a throughput efficiency of 71%. An angular field of view of 10° has been measured with the plano-convex lenses. In the case of double-convex lenses an angular field of view as large as 25° has also been measured.

A large number of the basic single-mode microlenses, as described above, but of much smaller dimensions in aperture and focal length have also been configured into a linear array in the LiNbO$_3$ substrate. For example, a 60-element linear microlens array with each lens element having a 60 μm aperture and 200 μm focal length has been successfully fabricated. Figure 8.7b shows the multiple focused light beam spots that result from the illumination of a portion of the lens array with a collimated light beam, using the configuration depicted in Fig.4 of [8.20], again using a He-Ne laser at 6328 Å wavelength. Measurement of the intensity profile of the multiple-focused light spots has indicated a high degree of uniformity among the elements in the lens array. The TIPE microlenses and linear lens arrays fabricated thus far have demonstrated a combination of desirable properties including short focal length, large numerical aperture (low f-number), small focal spot size, large angular field of view, and low optical insertion loss at both the 6328 Å and the 7920 Å laser wavelengths. Based on the results obtained so far, we expect that the aperture of element microlenses can be reduced considerably from 60 μm and the number of microlenses in the linear array can be much higher than 60. Accordingly, the number of element microlens in the array can substantially exceed 166 per cm. Thus, this TIPE fabrication method has demonstrated its potential capability of forming simultaneously various types of planar waveguide microlenses using a simple masking technique, and thus facilitate implementation of a variety of multichannel integrated-optic modules.

While the linear microlens array was used to collimate the multiple light beams from the channel-waveguide array before incidence upon the SAW for Bragg diffraction, the large-aperture integrating lens served to collect and focus the multiple Bragg-diffracted light beams. In a preliminary experiment, the AO Bragg diffraction was measured by placing a prism coupler in the planar waveguide [8.27]. Efficient and wideband Bragg diffraction from the SAW centering at 500 MHz was measured at 6328 Å optical wavelength. In addition, the beam profiles of both the diffracted and the undiffracted light were found to be highly uniform.

Detailed measurements were carried out using the module of Fig.8.6. Both end faces of the LiNbO$_3$ substrate were polished to optical quality in order to permit edge coupling, and the output end face was cut to coincide with the focal plane of the integrating lens. As a preparatory testing prior to the actual Bragg diffraction experiment, a light beam at 6328 Å wavelength was edge-coupled to the channel waveguides at the input end face, one at a time. The focused light beam at the output end face was then examined in detail. The half-power width of the focal spot at the output end face was measured to be 16 μm and the optical through-put was measured to -18 db. Subsequently, efficient and wide-band Bragg diffraction was again measured at the same optical wavelength using the SAW centered at 500 MHz. Specifically, an AO Bragg bandwidth of 90 MHz and a 50% diffraction at an RF drive power of 470 mW were obtained. The AO bandwidth for this preliminary device was limited by the bandwidth of the SAW transducer that utilized six pairs of interdigital finger electrodes with an aperture of 1.0 mm.

Figure 8.8a shows the oscilloscope traces of the envelope of the pulse-modulated SAW at 500 MHz and the resulting modulations of approxi-

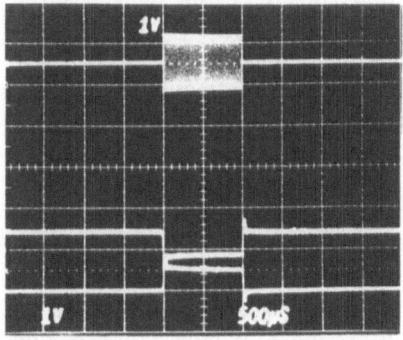

Fig.8.8a. Waveforms of acousto-optic Bragg diffraction in a LiNbO$_3$ channel-planar composite waveguide: pulse-modulated SAW at 500 MHz (*top trace*), undiffracted light (*middle trace*), and diffracted light (*bottom trace*)

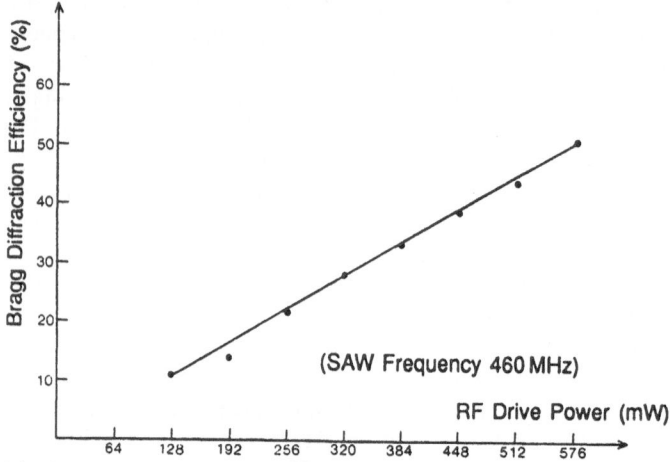

Fig.8.8b. Acousto-optic Bragg diffraction efficiency versus RF drive power in a LiNbO₃ composite waveguide

mately 40% in the diffracted and undiffracted light at an RF peak power of 400 mW. As shown in Fig.8.8b, a good linearity between the diffracted light power and the RF drive power was measured up to a diffraction efficiency of at least 55%. The measured conversion efficiency of -16 db for the simple SAW transducer used in this device module was considerably lower than the state-of-the-art. Consequently, even higher diffraction efficiency should be achievable at a lower RF drive power when a transducer of higher efficiency is used. Also, a considerably higher center frequency and a much larger AO bandwidth can be obtained using a transducer of sophisticated design, for example, the multiple tilted transducer or the tilted-finger chirp transducer described in Chap.5. Finally, by measuring the rise and the decay times of modulation in the diffracted light caused by a pulse-modulated SAW, the aperture of each individual light beam from the linear microlens array was determined to be 110 μm. This measured light beam aperture is consistent with the design aperture of the element microlens. Since the velocity for the SAW propagating in the Z-axis of the Y-cut LiNbO₃ substrate is $3.488 \cdot 10^5$ cm/s, this light beam aperture corresponds to 16 acoustic wavelengths at 500 MHz and thus ensures the common situation in which the diffraction spread of the light beam is much smaller than that of the acoustic beam.

The compact integrated AO Bragg modulator module described above may find a variety of applications in RF signal processing (correlation, spectral analysis, etc.) [8.32], digital transversal filtering, and optical systolic array processing [8.33,34], in addition to optical switching and routing. For example, the module has been utilized to perform matrix-vector multiplication [8.28]. Specifically, a simple experiment on matrix-vector multiplication involving a 2×2 matrix and a two-dimensional vector using two channels of the channel-waveguide array and the SAW, as shown in Fig.8.9a, has yielded some encouraging results. The light beams from two descrite diode lasers at the wavelength of 0.792 μm were edge-coupled, one

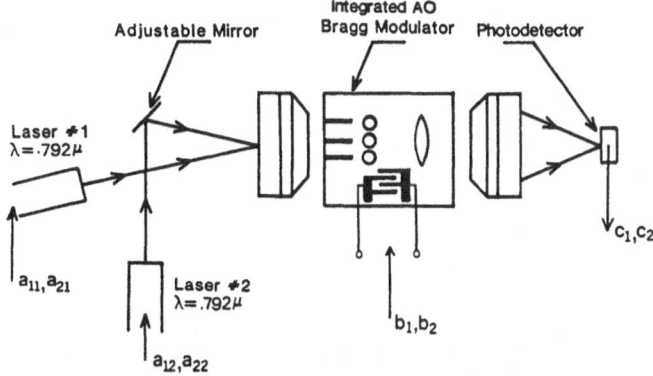

Fig.8.9a. Block diagram for matrix-vector multiplication using acoustooptic Bragg diffraction in a LiNbO$_3$ channel-planar-composite waveguide together with a TIPE linear microlens array

by one, into two of the channel waveguide array. The matrix elements of the two columns, $\begin{bmatrix} a_{11} \\ a_{12} \end{bmatrix}$ and $\begin{bmatrix} a_{21} \\ a_{22} \end{bmatrix}$, were used to pulse-modulate separately the output intensities of the two diode lasers prior to coupling to the waveguide. Similarly, the components of the vector $\begin{bmatrix} b_1 \\ b_2 \end{bmatrix}$ were used to pulse-modulate the 500 MHz SAW. A master pulse generator was employed to facilitate the required synchronization among the matrix elements and the vector components. Finally, the components of the matrix-vector product $\begin{bmatrix} c_1 \\ c_2 \end{bmatrix}$ were obtained from the output of the photodetector which was located at the focal point of the Bragg-diffracted light. The waveforms as shown in Fig.8.9b correctly represent the product of the multiplication $\begin{bmatrix} 5 \\ 4 \end{bmatrix}$ between the matrix $\begin{bmatrix} 1 & 2 \\ 2 & 1 \end{bmatrix}$ and the vector $\begin{bmatrix} 1 \\ 2 \end{bmatrix}$. A dynamic range of 27 to 29 dB was measured in this particular experiment.

The simple experiment just described has clearly demonstrated the capability of the integrated AO Bragg modulator module to simultaneously perform "multiplication" and "addition" on analog and digital data. A number of alternate architectures are possible. Thus the TIPE microlens-based integrated AO Bragg modulator modules should facilitate realization of some forms of integrated optical computers [8.28, 34]. The resulting integrated optical computers may possess some advantages over their bulk counterparts in terms of size, speed, drive power, hardware complexity, and ultimate cost.

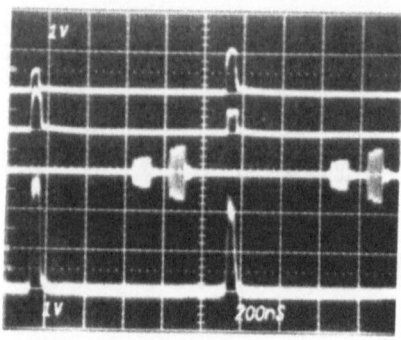

Fig.8.9b. Oscilloscope traces for a matrix-vector multiplication involving a 2x2 matrix and a two-dimensional vector. Photodetector output representing the first column $\begin{bmatrix} 1 \\ 2 \end{bmatrix}$ of the matrix (*top trace*); Photodetector output representing the second column $\begin{bmatrix} 2 \\ 1 \end{bmatrix}$ of the matrix (*second trace*); Modulated SAW representing the vector $\begin{bmatrix} 1 \\ 2 \end{bmatrix}$ (*third trace*); and photodetector output representing the matrix-vector product $\begin{bmatrix} 5 \\ 4 \end{bmatrix}$ (*bottom trace*)

8.5 Multichannel RF Correlator Modules Using Acousto-Optic and Electro-Optic Bragg Diffractions

The capability of the TIPE technique introduced in Sect.8.4 for simultaneous fabrication of linear microlens arrays and large-aperture lenses has also facilitated realization of multichannel integrated-optic modules that simultaneously utilize AO Bragg diffraction and an array of electrooptic (EO) Bragg diffractions [8.35,36] in a common LiNbO₃ substrate [8.37]. A version of such ingrated optic correlators that have been fabricated and tested recently [8.38] is shown in Fig.8.10a. In this 10-channel correlator module

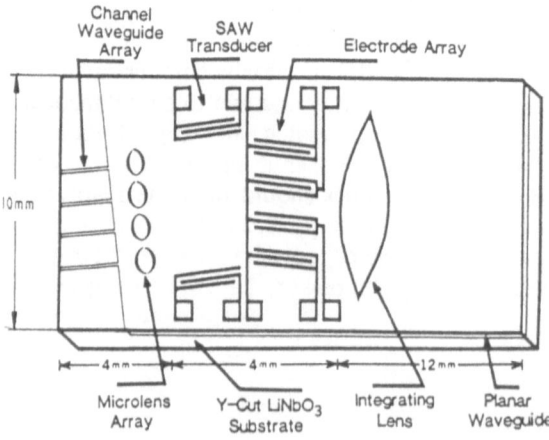

Fig.8.10a. Multichannel RF correlator module in LiNbO₃ waveguide

286

Table 8.1. Design parameters of integrated optic correlator module

Element	Design Values
TIPE Integrating Lens:	
Focal length	10 mm
Typical focal spot size	7 μm[a]
E-O Bragg grating array:	
Finger length	1900 μm
Finger width	3.3 μm
Periodicity	13.2 μm
Number of periods	15
Q factor	19.7
Bragg angle θ_B ($\lambda_0 = 0.633 \mu$m)	0.624°
Interval between gratings	10 μm
Number of gratings	10
AO SAW Transducer:	
Finger length	1900 μm
Finger width	3.3 μm
Number of finger pairs	5
Acoustic center frequency	260 MHz
Acoustic bandwidth	50 MHz
Bragg angle θ_B ($\lambda_0 = 0.633 \mu$m)	0.624°

[a] Obtained with 2mm beam width

each channel waveguide was followed by a TIPE microlens that served to enlarge and collimate the output light beam. The SAW transducer pair was followed immediately by a 10-element EO Bragg diffraction grating array that had a separate and independent terminal electrode for each element, and, subsequently, by a large-aperture TIPE lens. The width and length of each EO Bragg grating were 0.2 and 2.0 mm, respectively. Other device parameters are given in Table 8.1. Excellent performance characteristics had been obtained with such EO Bragg diffraction array [8.37, 38]. In one of the experiments that have been carried out, a sequence of binary data word was used to amplitude-modulate the SAW. This amplitude-modulated SAW then caused Bragg diffraction of a large-aperture guided-light beam incident from the left. Since the EO diffraction grating array was oriented such that the AO Bragg diffracted light would incident upon it at Bragg angle, each of the resulting multiple AO Bragg diffracted light beams would subsequently encounter a second Bragg diffraction when the elements of the EO diffraction grating array were activated by another sequence of binary data word. A multiple of these doubly-diffracted light beams were then collected by the large-aperture integrating lens and focussed upon a photodetector. Thus, the "Multiplication" and "Addition" required in performing correlation between the two sequences of data word

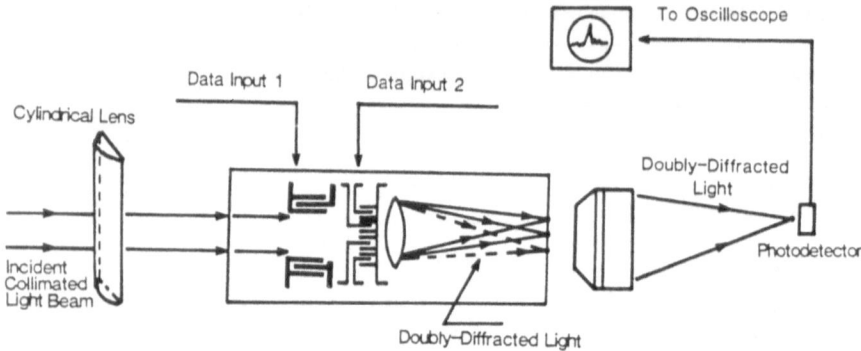

Fig.8.10b. Arrangement for correlation experiment

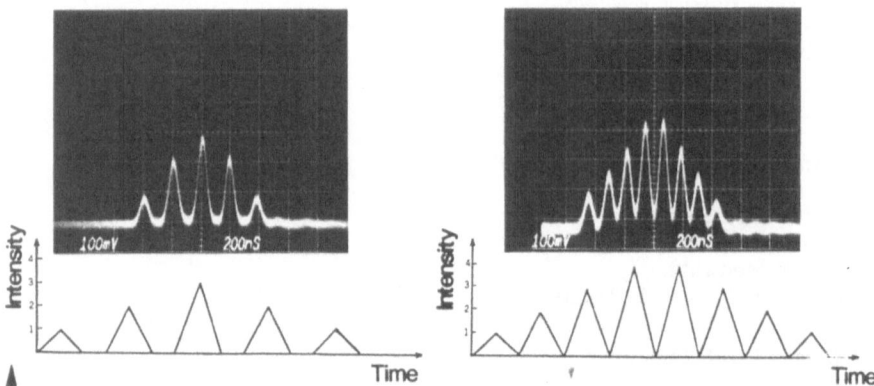

Fig.8.10c. Auto-correlation waveform of a nine-bit binary sequence obtained using a multi-channel integrated optic module (binary sequence: 100100100)

Fig.8.10d. Cross-correlation waveform of a nine-bit binary sequence obtained using a multi-channel integrated optic module (binary sequences: 101010101 and 010101010)

were readily accomplished by the combination of the AO and the EO Bragg diffraction gratings and the large-aperture TIPE lens, respectivély.

Figure 8.10b depicts the arrangement used to perform a variety of correlation experiments. Note that in this arrangement a linear array of an ion-milled planar microlens array was used at the input plane to convert a collimated incident light beam into a linear array of focused light spots for simultaneous and efficient excitation of the channel-waveguide array. Examples of the auto-correlation waveform and the cross-correlation waveform obtained using a He-Ne laser at 0.6238 μm wavelength and the amplitude-modulated SAW at 17.5 Mbits/s centered at 260 MHz are shown in Figs.8.10c and d, respectively, together with the calculated wave forms. The agreement between the calculated correlation waveforms and the measured correlation waveforms is seen to be very satisfactory. The overall dynamic range of the correlator was measured to be 29dB.

In summary, since the width (or aperture) of each elementary EO Bragg diffraction grating can be as small as 40 μm, a bit capacity (word) as large as 200 bits and a data rate as high as 88 Mbits/s can be obtained in a correlator which employs an incident light of 1.0 cm aperture. Thus, it is concluded that programmable integrated optic digital correlator or filter modules of very large bit word and very high data rate can be realized.

8.6 Planar-Waveguide Acousto-Optic Frequency-Shifter Module

As mentioned in Sect.5.5, guided-light sources with electronically tunable frequency shift are needed as local oscillators in heterodyned integrated- and fiber-optic communication and sensor systems [8.39,40]. A number of schemes for producing such frequency-shifted light sources have been reported [8.41-49]. Also, as mentioned in Sect.5.1, the frequency of the Bragg-diffracted light is shifted from that of the incident light by the SAW frequency. A unique feature of the resulting AO frequency shifting scheme is its inherent single-sideband suppressed-carrier property. Earlier AO fre-quency-shifting experiments were performed using the SAW in planar-, crossed channel- [8.50,51], and spherical- [8.52] waveguides in LiNbO$_3$. However, these experiments all utilized a single Bragg diffraction and, therefore, all suffered from an undesirable characteristic in that the propa-gation direction of the frequency-shifted light varies with the acoustic fre-quency. This undesirable characteristic will result in a severe limitation in the useful bandwidth of frequency shift which is especially acute in a single-mode fiber optic system in which the very small cross-section of the fiber core is fixed in space.

Most recently a novel device configuration that is free from the afore-mentioned undesirable characteristic was reported [8.53]. Figure 8.11 shows

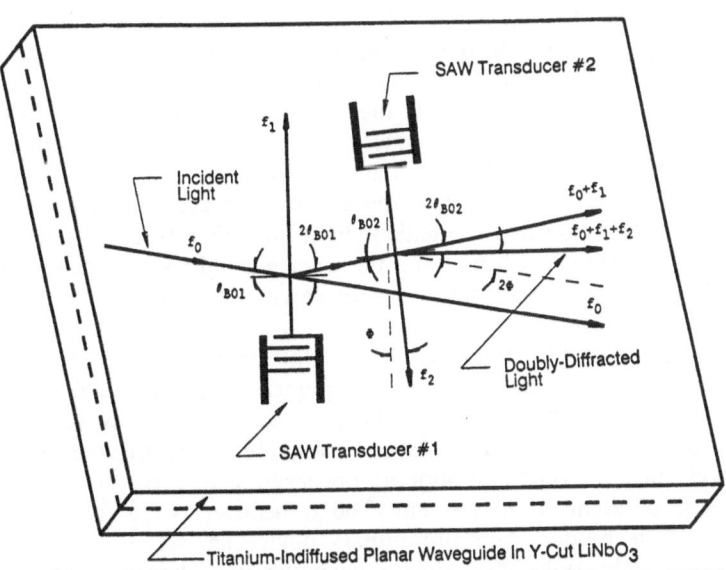

Fig.8.11. Optical frequency shifting scheme using guided-wave acousto-optic Bragg diffractions in cascade

the device configuration that consists of two SAW transducers fabricated in a Y-cut LiNbO$_3$ planar waveguide. The pair of transducers are used to generate two tilted- but counter-propagating SAWs side by side to facilitate efficient Bragg diffractions in cascade. The center frequencies of the transducers, f_{01} and f_{02}, are chosen judiciously to differ by a suitable amount Δf_{012}, i.e. $\Delta f_{012} = f_{01} - f_{02}$. The frequency differential Δf_{012} is determined by the requirement that the doubly (or cascaded) Bragg-diffracted light be resolved angularly (or spatially) from the incident light. The orientations of the transducers and, thus, the propagation directions of the SAWs are tilted by an angle ϕ that is identical to the difference between the two Bragg angles ϕ_{BO1} and ϕ_{BO2} at the two center frequencies. Now, a channel waveguide guided-light beam, expanded and collimated by the first TIPE lens, incident from the left upon the first SAW at Bragg angle ϕ_{BO1} will encounter efficient Bragg diffraction in tandem as the diffracted light from the first SAW will be incident upon the second SAW at Bragg angle ϕ_{BO2}. It has been shown that the doubly-diffracted light will propagate at an angle which deviates from the incident light (undiffracted light from the first SAW) by 2ϕ [8.53] and carries a frequency shift of $(f_{01} + f_{02})$. Also, as the frequencies of the driving signals to the two transducers are tuned simultaneously by an equal amount of δf, the doubly-diffracted and frequency-shifted light will maintain its propagation direction, regardless of the magnitude of frequency tuning. Obviously, the corresponding frequency shift in the doubly-diffracted light equals $(f_{01} + f_{02} + 2\delta f)$. Now, a second TIPE lens is used to bring the doubly-diffracted light to a focus at the input cross section of the output channel waveguide. The resulting light source of large tunable frequency shift may be interfaced with a single-mode optical fiber jointed permanently to the output channel waveguide. Note that the frequency-shifting scheme just described can readily be carried over to bulk-wave AO Bragg devices. It should also be noted that as the frequency differential between the two SAWS approaches zero- the tilt angle between the two transducers approches zero likewise. In other words, the two transducers will then be excited at the same frequency and the resulting SAWs propagate at 180 degrees with each other. The frequency shift in the doubly-diffracted light is then twice the SAW frequency. Unfortunately, in this case the doubly-diffracted light propagates in the same direction as the incident light and, therefore, cannot be separated spartially from it. Clearly, the undiffracted light which does not carry any frequency shift will cause severe carrier leakages in the frequency-shifted light desired.

A frequency shifter that employed neither the channel waveguides nor the lenses but two transducers at the center frequencies of $f_{01} = 559$ MHz and $f_{02} = 459$ MHz was fabricated in a Y-cut LiNbO$_3$ waveguide $0.2 \times 1.0 \times 2.0$ cm^3 in size [8.53]. The planar waveguide was fabricated using the TI method and was found to support propagation of TE$_0$-mode at a propagation loss of about 1 dB per cm. The transducers were fabricated using the lift-off technique. This preliminary device has served to verify the working principle and the desirable characteristics described above. Experiments were carried out using a He-Ne laser at an optical wavelength of 0.6328 μm. The efficiency of the Bragg diffractions in cascade was as high as 76% at a RF drive power of 100 mW per transducer.

An optical heterodyning setup that consisted of a 5 milliwatt 0.63 μm He-Ne laser and a Mach-Zehnder interferometer was employed to measure the device performances including frequency shift, tunable bandwidth, dynamic range, crosstalks, optical carrier leakages, image sidebands, and any higher-order nonlinear distortions. As expected, the frequency shift in the doubly-diffracted light was found to be equal to the sum of the center frequencies of the two SAWs, i.e., 1018 Mhz, and a tunable bandwidth of 165 Mhz was obtained. The signal-to-noise ratio (SNR) of the frequency-shifted component was measured to be 51 dB at an RF drive power of 100 mW per transducer. Neither optical carrier leakage nor image sideband was detected in the present setup. Furthermore, no frequency-shifted components resulting from acoustic harmonics, e.g., the second- or third-harmonics, was detected. Thus, a suppression of unwanted components by more than 51 dB was demonstrated in this preliminary frequency shifter at a light power of 5 mw. Finally, the throughput of the frequency shifter was measured to be -10 dB. Most recently, a frequency shifter module that incorporates a channel waveguide and a TIPE collimating lens at the input, a TIPE focusing lens at the output has been realized in a Y-cut LiNbO$_3$ substrate 0.1×1.0×1.5 cm^3 in size. Measured performance figures comparable to those described above have been obtained.

In summary, an integrated AO frequency shifter module that has its frequency-shifted light propagating in a fixed direction but resolved angularly from the incident light, irrespective of the magnitude of frequency turning has been constructed in a Y-cut LiNbO$_3$ waveguide 0.1×1.0×1.5 cm^3 in size. Since, as shown in Sect.5.4, guided-wave AO Bragg diffraction of high efficiency with GHz center frequency and GHz bandwidth is obtainable, such frequency shifter modules should be capable of operating at multi-gigahertz center frequency with multi-gigahertz electronically tunable bandwidth.

8.7 GaAs Acousto-Optic Bragg Cell and RF Spectrum Analyzer Modules

Despite the various successes of the LiNbO$_3$-based AO Bragg device modules referred to in Chaps.5 and 7 and the preceding sections of this Chapter, they have one inherent disadvantage: only a partial or hybrid integration can be realized in such a substrate. In contrast, GaAs-based device modules are potentially capable of total or monolithic integration because both the laser sources and the photodetector arrays as well as the associated electronic circuits can be integrated in the same substrate. Clearly, one of the key components in such future GaAs-based monolithic integrated optic device modules or circuits is an efficient wide-band AO Bragg cell. In this section we review the realization of miniaturized GaAs-GaAlAs planar waveguide AO Bragg cells in a substrate size of 0.035×0.7

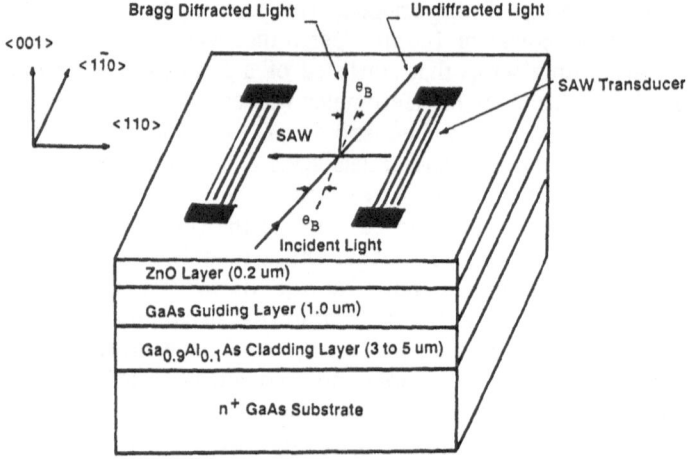

Fig.8.12a. Guided-wave acousto-optic Bragg diffraction from surface acoustic waves in a ZnO-GaAs-GaAlAs composite waveguide

×0.8 cm³. High diffraction efficiency and large bandwidth have been accomplished at 1.15 μm optical wavelength using a SAW with a center frequency ranging from 190 to 1,150 MHz in a Z-cut GaAs substrate using the interaction configuration depicted in Fig.8.12a [8.7]. Formation of waveguide lenses [8.8] and their integration with the AO Bragg cell in a common GaAs substrate have also been made most recently.

8.7.1 Geometry, Design and Fabrication of Acousto-Optic Bragg Cells in GaAs Waveguides

A previous theoretical study [8.6] by us has predicted, as shown in Fig.8.12b, an AO Bragg bandwidth as large as 1.6 and 1.4 GHz for the $\langle 100 \rangle$- and $\langle 110 \rangle$-propagation SAW, respectively, in a Z-cut GaAs-GaAlAs waveguide that supports TE_0-mode at 1.15 and 1.30 μm wavelength. The AO Bragg cells to be described here were designed using the particular interaction configuration of Fig.8.12a. Although some AO Bragg diffraction experiments were previously carried out using 200 MHz SAW in both GaAs-GaAsP waveguides [8.6] and GaAs-GaAlAs waveguides [8.6], lack of the technologies for growth of high-quality, large-size GaAs planar waveguides and fabrication of efficient zinc oxide (ZnO) transducers has heretofore prevented experiments at a higher acoustic frequency. Recent advances in these two fabrication technologies have facilitated further experimental study at significantly higher SAW frequency [8.7]. First, a high-quality GaAs-$Ga_{0.85}Al_{0.15}$As planar optical waveguide of large size (typically 2.2×2.7cm²) was prepared using an in-house liquid-phase epitaxy (LPE) system on a 0.35 mm thick silicon-doped GaAs substrate. The cross section of the optical waveguide together with the ZnO overlay and the transducer geometry are again shown in Fig.8.12a. The waveguide specimen was then cut into a number of samples each with the size of 0.7x0.8 cm² prior to optical propagation measurement. Measurement of the

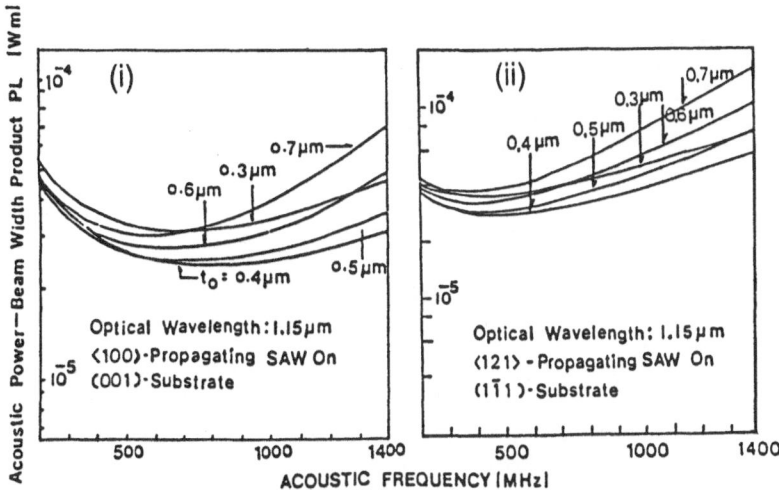

Fig.8.12b. SAW power density required vs. acoustic frequency for 100% Bragg diffraction (TE_0-TE_0) in a single-mode GaAs waveguide at 1.15 μm optical wavelenght: (i) $\langle 100 \rangle$-propagating SAW on (001) substrate, (ii) $\langle 121 \rangle$-propagating SAW on (111) substrate

mode structure and the propagation loss of the waveguide were facilitated using edge-coupling through the (110) cleavage plane. A typical propagation loss that has been measured for the TE_0-mode waveguides is 1.5 to 2.0 db/cm. Also the best throughput coupling efficiency was measured to be -1.6 db for an optical path of 0.8 cm.

Subsequently, a ZnO overlay film and an IDT were fabricated on top of the optical waveguide for efficient excitation of the SAW. The number of finger-electrode pairs and the electrode aperture used were typically 9 and 1.0 mm. The finished device was first attached to a commercial 16-pin DIP (dual in-line package) IC holder which greatly facilitated subsequent SAW propagation and Bragg diffraction measurements. Figure 8.12c shows the complete package of the resulting Bragg cell module.

8.7.2 Measured Performances of AO Bragg Cells

The Bragg diffraction measurements for the various bands of acoustic center frequency, namely 200, 400, 600, 800, 960, and 1,150 MHz were all performed using the 1.15 μm line of a He-Ne laser and the 1.30 μm wavelength of a diode laser. High diffraction efficiency was measured in each band. For example, a diffraction efficiency of 25% was measured with an acoustic drive power of 14.3 mW at 400 MHz. This measured figure is only slightly lower than the theoretical value. It is to be noted that this acoustic drive power requirement is comparable to an equivalent Bragg cell that utilizes the Z-propagation SAW in a Y-cut $LiNbO_3$ substrate. However, it should also be noted that as a result of the lower conversion efficiency of the ZnO transducer the corresponding RF drive power (1.0W) required for this diffraction efficiency is higher than its $LiNbO_3$ counterpart. Figure

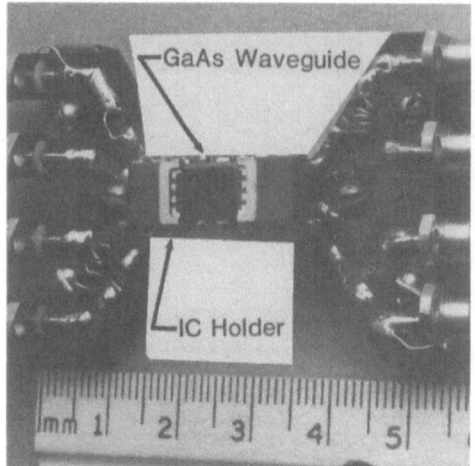

Fig.8.12c. Photograph of GaAs acousto-optic Bragg cell module

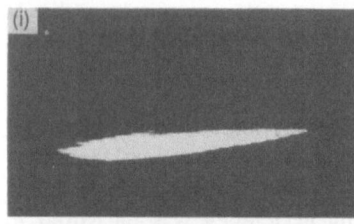

Undiffracted Light

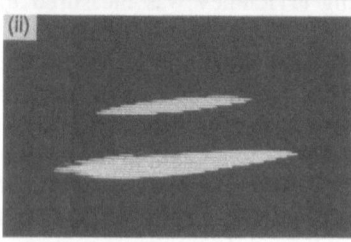

Diffracted Light

Undiffracted Light

Fig.8.12d. Photograph of diffracted and undiffracted light beams: (i) RF power off, (ii) RF power on

8.12d shows the diffracted and the undiffracted light spots obtained using an ir vidicon camera. Note that the fine lines in the light spots were caused by the raster scan of the vidicon camera.

The measured performance figures of the Bragg cells operating at various frequency bands are summarized in Table 8.2. As an example, consider the Bragg cell that utilized a tilted-finger chirp transducer centered at 485 MHz. Figures 8.13a and b show, respectively, the frequency responses of the tilted-finger chirp transducer and the resulting AO Bragg cell. Clearly, a -3 db acoustic bandwidth of 250 MHz and a -3dB AO bandwidth of 245 MHz have been obtained. Note that Fig.8.13b also shows the AO frequency response measured at the other incident angle indicated in the inset. As expected, due to lack of Bragg condition tracking, the result-

Table 8.2. Measured performance on acoustooptic Bragg diffraction

Center Frequency of SAW	200	360	400	485	600	800	960	1115
Transducer types [MHz]								
I.Parallel-finger synchronous II.Tilted-finger chirp	I	II	I	II	I	I	I	I
Diffraction efficiency (per mW acoustic power per mm acoustic aperture) [%]	1.0	3.2	1.9		2.3			15.0
-3 db AO device bandwidth [MHz]	10	201	37	240	45	35	40	80
Maximum measured AO diffraction efficiency [%]	65	15	28		9	9.6	18	38
Measured AO diffraction efficiency at 1.0 W RF power	40	6.5	23	5.0	7.3	15.0	6.6	14.0
RF drive power/MHz bandwidth at 1% diffraction efficiency [mW/MHz]	2.5	0.8	1.2		3.0	5.7	3.7	0.9
SAW propagation loss [db/μs]	0.32	1.4	1.5	2.0	2.8			7.7

ing AO Bragg bandwidth of 55 MHz was considerably smaller than that measured for the preferred incident angle. The diffraction efficiency was measured to be 5.0% at 1.0 W RF drive power. Next, the frequency resolution δf_R and the number of resolvable spots or channels N_R in connection with the application in light beam scanning and RF spectral analysis, were measured using two RF signals of varying frequency separation. Figure 8.13c shows the photographs of the two diffracted light spots obtained at the far-field with the frequency separation as a parameter. Obviously, the Bragg cell provided diffracted light spots of very fine quality. A uniform light beam with an aperture of 1.2 mm at 1.15 μm wavelength was used in this experiment, giving a measured frequency resolution of approximately 3 MHz as compared to a calculated value of 2.4 MHz based on Rayleigh criteria. The measured bandwidth of 245 MHz therefore provided approximately 82 resolvable spots or channels.

Next, consider the Bragg cell that utilized the parallel-finger synchronous transducer centered at 800 MHz. A maximum diffraction efficiency, namely, 9.6% at 1.9 Watt RF drive power (as limited by the available RF drive power) was also measured in this case. In order to alleviate the resolution limitations with our existing fabrication facility, a parallel-double-electrode synchronous transducer [8.54] operating at its third harmonics was employed for this purpose. An AO bandwidth of 35 MHz was measured and a diffraction efficiency of 5.0% at 1.0 Watt RF drive power could be projected. Finally, the last column in the Table shows the measured data for the Bragg cell that centered at 1,115 MHz. To the best of our knowledge this represents the highest acoustic frequency that has been employed for AO Bragg diffraction in a GaAs planar waveguide.

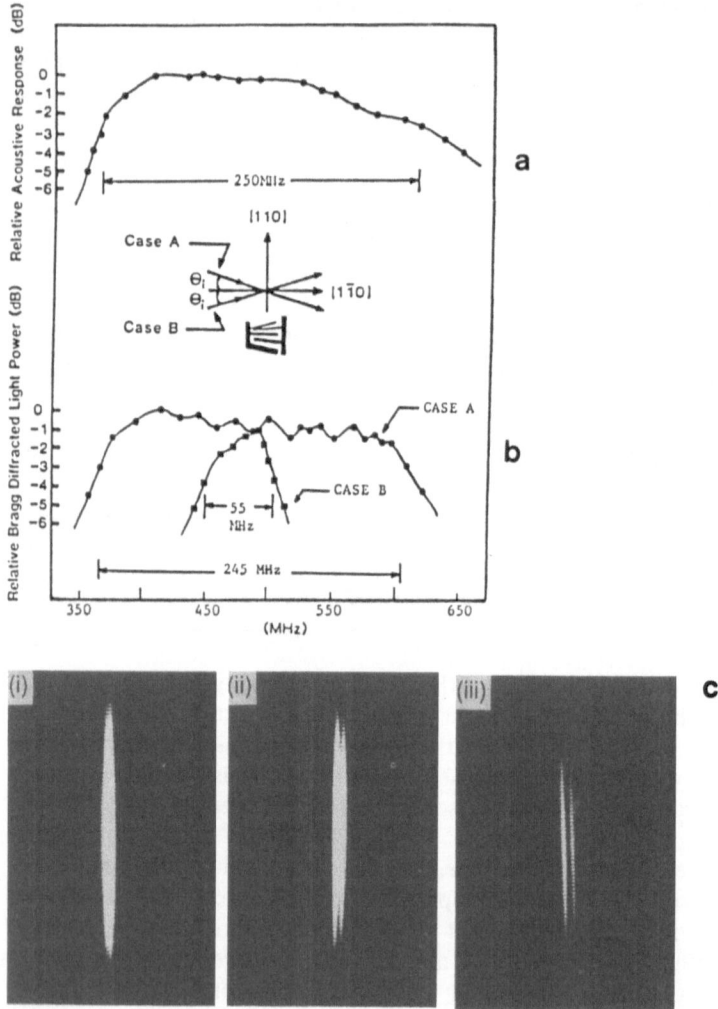

Fig.8.13a-c. Measured RF response of the tilted-finger chirp transducer (a), measured frequency responses of the resulting acousto-optic Bragg cell (b), (c) diffracted light spots resulting from two RF signals of varying frequency separation and drive power: (i) δf = 2 MHz, (ii) δf = 3 MHz, and (iii) δf = 5 MHz

There exists an increasing interest in development and realization of GHz SAW devices in waveguide-free GaAs substrates [8.55, 56]. Consequently, it is important to measure the propagation losses of the SAW in waveguide-free GaAs substrates at GHz range [8.57-59]. It is also of great importance to measure the propagation loss of the SAW in the GaAs-GaAlAs waveguides as it is one of the major parameters that will ultimately determine the highest center frequency and the maximum bandwidth achievable with the resulting AO Bragg cells. The propagation loss of the GaAs-GaAlAs waveguide was measured by translating the incident light beam long the SAW propagation path and recording the Bragg diffracted light power while keeping both the frequency and the power of the SAW fixed. The propagation losses measured at the various center frequencies are also presented in Table 8.2. These measured data indicate that the SAW propagation losses in the GaAs-GaAlAs waveguides do not differ drastically from those measured in the waveguide-free GaAs substrates.

In summary, design, fabrication, and performance characteristics measurement of compact miniaturized GaAs waveguide AO Bragg cells that operate at acoustic center frequencies from 190 to 1,115 MHz have been carried out. A high diffraction efficiency was measured at each frequency band. A 245 MHz bandwidth has been achieved with the Bragg cell that employed a single tilted-finger chirp transducer centered at 485 MHz. The SAW propagation losses in the GaAs waveguides were also measured at various center frequencies using AO Bragg diffraction as probe. The measured performance figures serve to project realization of Bragg cells with a center frequency of 1.0 GHz or higher and an octave bandwidth and, thus, suggest further component integration in a common GaAs waveguide toward a monolithic integrated AO module.

8.7.3 GHz Acousto-Optic Bragg Cells

As mentioned in the last subsection, AO Bragg cells at gigahertz frequencies have been realized most recently [8.7]. These latest Bragg cells utilize the GaAs-Ga$_{0.85}$Al$_{0.15}$As waveguides in the same configuration as that shown in Fig.8.12a and the SAW transducers consisting of 33 double-electrode fingers and an acoustic aperture of 0.76 mm. A He-Ne laser of 1.15 μm wavelength and 5 mW output power was used as the optical source in the AO diffraction experiment. The light beam was focused and edge-coupled into the cleaved edge of the Bragg cell sample. At the acoustic center frequency of 1.115 GHz the Bragg angle in air was 14.4°. The undiffracted and diffracted light spots were imaged upon an InGaAs PIN photodetector at the far-field. For the case involving an incident light at TE$_0$-mode and the resulting diffracted light at TE$_0$-mode a diffraction efficiency as high as 38% was achieved at an acoustic drive power of 2.72 mW or an RF drive power of 2.72 W. Figures 8.14a and 8.14b show, respectively, the

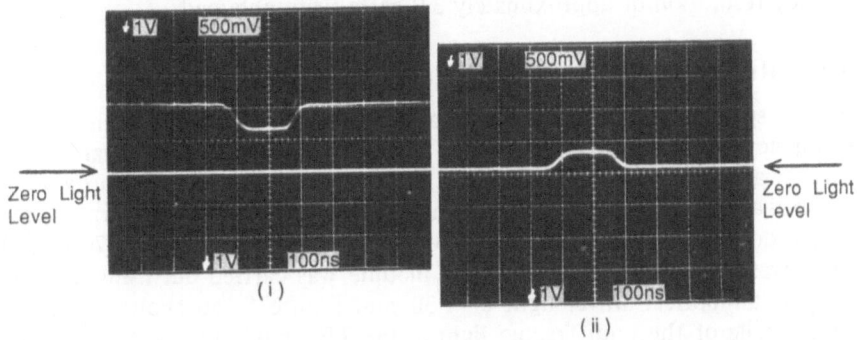

Fig.8.14a. Waveforms of undiffracted and diffracted light by pulse-modulated SAW at 1.115 GHz: (i) Depletion in undiffracted light, (ii) increase in diffracted light

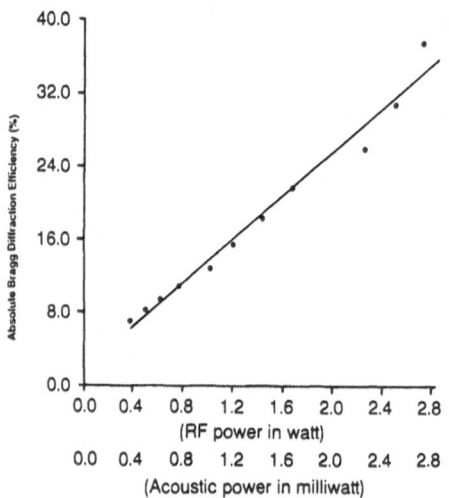

Fig.8.14b. Bragg diffraction efficiency versus the drive power at the SAW frequency of 1.115 GHz for a TE_0-mode incident light

pulse-modulated waveforms of the diffracted and undiffracted light, and the diffraction efficiency versus drive powers. For the case involving an incident light at TM_0-mode and the resulting diffracted light at TM_0-mode the measured diffraction efficiency was 32% under the same drive powers.

It is to be noted that a linear relationship between the diffracted light intensity and the drive powers was maintained for the entire range of drive powers employed. Higher diffraction efficiency is yet to be accomplished due to lack of signal generator of higher power at the frequency used. Mode-conversion between the incident and the diffracted light, if any, was not observed. The measured -3 dB AO bandwidth was nearly the same as the bandwidth of the SAW transducer, namely, 80 MHz.

Finally, the GHz Bragg cell described above has been used to perform light beam deflection/switching and RF spectral analysis. For example, in the experiment with RF spectral analysis a total of 27 channels at a frequency resolution of approximately 3.0 MHz were obtained.

8.7.4 RF Spectrum Analyzer Module

An RF spectrum analyzer module that consists of a Bragg cell at the center frequency of 500 MHz and a pair of ion-milled collimation-Fourier-transform waveguide lenses [8.8] (Fig.8.15a) was realized most recently in a GaAs waveguide $0.1 \times 0.7 \times 2.3$ cm³ in size. A tilted-finger chirp transducer of the double-electrode type that covers the frequency range of 370 to 630 MHz was used. Measurement of the module was carried out using a diode laser at an optical wavelength of 1.30 μm. Figure 8.15b shows the focal spot profile of the undiffracted light at the TE_0-mode that was focused at the output end face of the $GaAs-Ga_{0.85}A\ell_{0.15}As$ waveguide, indicating that a -3dB focal spot size of 6.0 μm was achieved. For the Bragg cell a

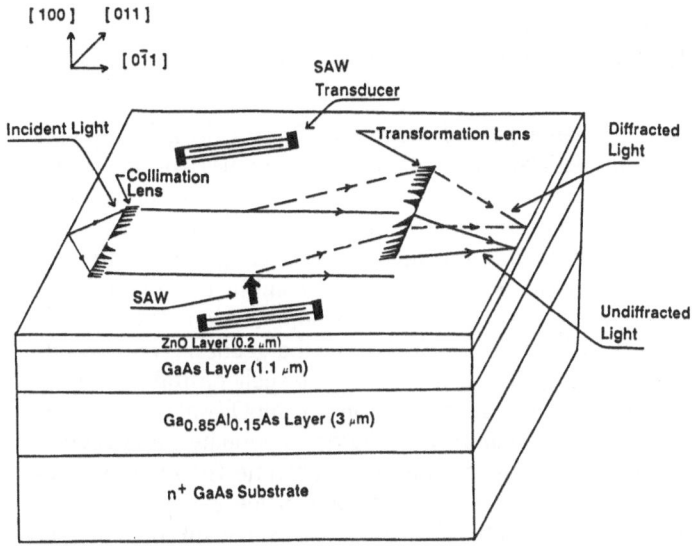

Fig.8.15a. Acousto-optic RF spectrum analyzer module in a ZnO-GaAs GaAlAs composite waveguide

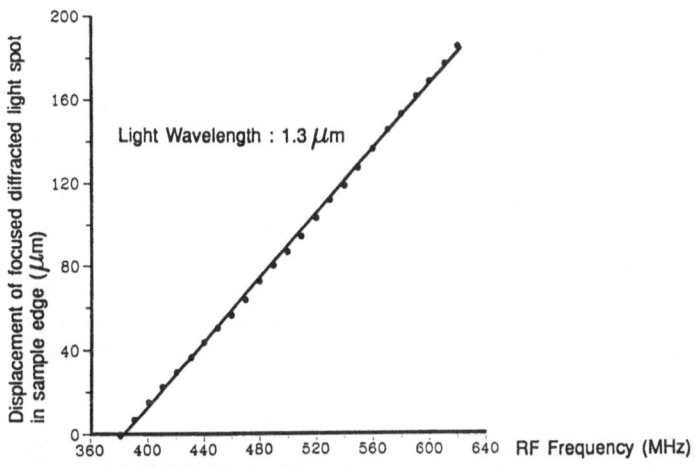

Fig.8.15b. Frequency scanning of focused diffracted light spot from integrated AO Bragg cell module

diffraction efficiency of 8% per watt of RF drive power and a -3dB bandwidth of 240 MHz were measured. By varying the frequency of the RF drive signal the focal spot of the diffracted light was seen to scan at the rate of 0.98 μm per MHz. Thus a frequency resolution of approximately 7.0 MHz and a total of 34 channels were accomplished with this GaAs AO spectrum analyzer module.

8.8 Spherical Waveguide Acousto-Optic Bragg Modulator/Deflector and Frequency-Shifter Modules

As shown in Chaps.5-7, planar guided-wave AO Bragg cells are now used in the development and realization of microoptic modules for RF signal processing such as spectral analysis and correlation. In such applications, waveguide lenses are required to collimate and focus the light beams in the substrate. These lenses, formed and positioned separately and accurately on the same substrate, must provide simultaneously a combination of desirable characteristics. Consequently, research and development on a variety of planar waveguide lenses [8.8, 20, 60-75] including the geodesic, chirp-grating, Fresnel, Luneburg, and TIPE types aimed at such desirable characteristics are being carried out. In this subsection a LiNbO$_3$ *spherical* waveguide substrate [8.76, 77] that *simultaneously* guides, collimates, and focuses the light beams, and thus serves as an alternate substrate for realization of integrated AO Bragg device modules is described. Some experimental results on optical guiding, collimating, focusing, AO Bragg diffraction, and applications in multiport light beam switching, RF signal processing, and frequency shifting are also presented.

8.8.1 Optical Guiding and Propagation in a Spherical Waveguide and Acousto-Optic Interaction Configuration

The LiNbO$_3$ spherical waveguide and the AO interaction configuration are depicted in Figs.8.16a and b, respectively. A focused light beam entering at a point P on the rim of the base plane of the hemisphere block that has a titanium-diffused layer right beneath the spherical surface is guided, collimated at the top of the hemisphere, and refocused at a point P' on the opposite side of the rim. It is clear that the spherical waveguide just described functions like a double-diffraction system in bulk optics. The aberration of this spherical waveguide structure is potentially much smaller than that in a planar waveguide of identical aperture and field angles [8.

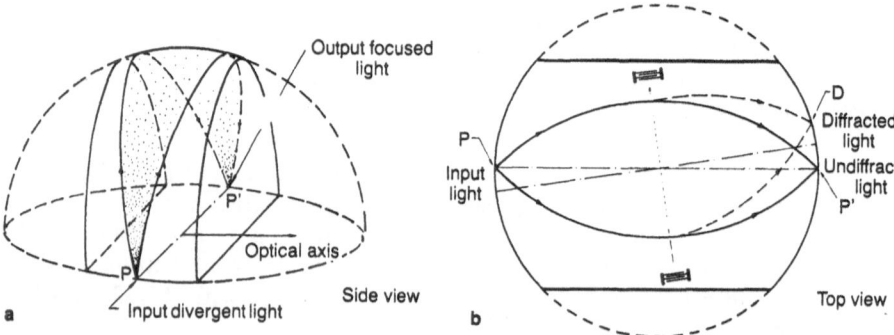

Fig.8.16. (a) A LiNbO$_3$ spherical waveguide. (b) Acousto-optic Bragg diffraction from surface acoustic waves in a LiNbO$_3$ spherical waveguide

75]. Consequently, the diffraction-limited aperture of the light beam can be made considerably larger than that provided by a planar waveguide lens. Accordingly, a larger processing gain or time-bandwidth product can be realized with the resulting spherical waveguide AO signal processors. Now, if an interdigital transducer of proper orientation and center frequency is deposited near the top of the hemisphere, the SAW generated will interact with the incident guided light wave in a manner similar to the AO interaction in the planar waveguides. In the case of Bragg diffraction, the diffracted light propagates at twice the Bragg angle from the incident light, and is also focused at a point D on the rim. As the frequency of the SAW is varied, the focal spot of the Bragg-diffracted light is scanned along the rim. Thus, the spherical waveguide substrate serves simultaneously as the optical waveguide and the collimating Fourier-transform lens pair when it is used to implement an integrated-optic RF spectrum analyzer.

The spherical waveguide described above was formed in a slice of the central portion of the LiNbO$_3$ hemisphere having 31 mm diameter and crystal orientation, as shown in Fig.8.16a. In order to minimize variation in the thickness of the titanium film prior to indiffusion, three successive depositions on adjacent areas of the spherical surface were made by rotating the hemisphere block with respect to the titanium source. After the characteristics on guiding and focusing of the light wave at 6328 Å He-Ne laser wavelength had been measured, a pair of the SAW transducers having the design center frequency of 250 MHz and aperture of 1.0 mm was fabricated near the top of the hemisphere, as shown in Fig.8.16b, using the well-established lift-off technique [8.78].

Excitation of guided-light waves in the spherical waveguide and subsequent measurements of the guiding and focusing characteristics were carried out using a standard arrangement that utilized spherical lenses and edge-coupling. First, a spherical lens was used to focus the He-Ne laser light beam into a spot size of 2.20 μm. The focused light beam was then edge-coupled to the rim of the hemisphere block. Figure 8.16c shows a photograph of the guided-light-beam propagating in the TE$_0$ mode with a 8.4° divergence angle. The bright lines in the photograph were caused by

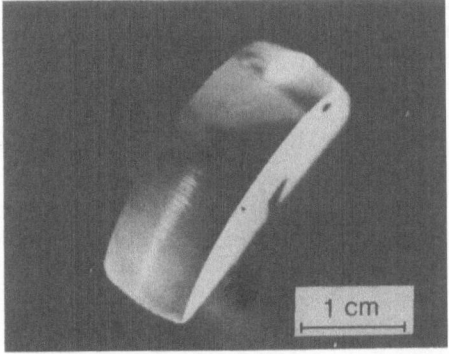

Fig.8.16c. Photograph showing guiding, collimation, and focusing of a 6328 Å He-Ne laser light beam in a LiNbO$_3$ spherical waveguide

the interference between the unguided incident light and the reflected light from the spherical surface. In fact, in a separate study guiding of the light beam via total internal reflection from the spherical surface, which is analogous to the so-called whistling gallery mode, has also been observed [8.79]. The unguided incident light was in turn due to some mismatch between the profile of the input light and that of the waveguide mode. From the intensity distribution of the focused light beam at the output port as imaged through a relay lens and displayed by a charge-coupled photodetector (CCPD) array, a focal plane spot size of 2.33 μm (in the dimension along the rim) was measured. Note that this measured focal spot size at the output port is practically identical to that at the input port. The optical throughput of the spherical waveguide was measured to be -12 dB. Significant increase in the throughput can be expected if antireflection coatings are provided at both input and output edges and/or if matching of the profile of the focal spot at the input edge with that of the waveguide mode is improved.

8.8.2 Acousto-Optic Diffraction Experiments

Subsequent measurement on AO diffraction was also carried out at the 6328 Å wavelength. The measured AO frequency response of the device is shown in Fig.8.16d. It is to be noted that for this particular device the -3 dB bandwidth of 75 MHz centering at 245 MHz was limited by the bandwidth of the SAW transducer. At the center frequency 25% diffraction was obtained at 124 mW SAW power or 1.8 W RF drive power. Note that combination of the center frequency and the aperture of the SAW for this preliminary device was such that the diffraction process lay between the Raman-Nath and the Bragg regimes. Note also that the penetration depth of the SAW at 245 MHz was considerably larger than that of the guided optical wave. Thus a significant improvement in diffraction efficiency should be possible with a SAW of higher frequency, which would render the two penetration depths more comparable. Since the measured conversion efficiency of the SAW transducer was only -11.7 dB, it is also reasonable to project some reduction in the RF drive power requirement. Figure 8.16e shows the intensity profiles of both undiffracted and diffracted light beams at the focal plane, as scanned by a CCPD array. These profiles clearly indicate that the focusing characteristic of the incident light is pres-

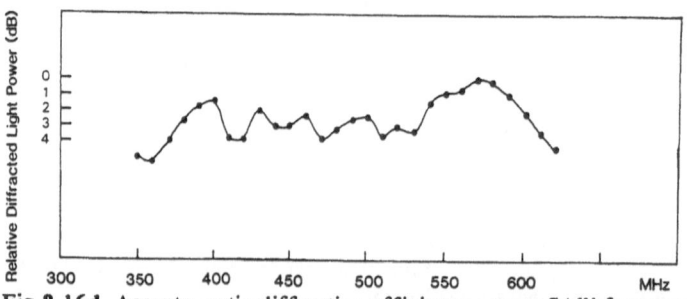

Fig.8.16d. Acousto-optic diffraction efficiency versus SAW frequency

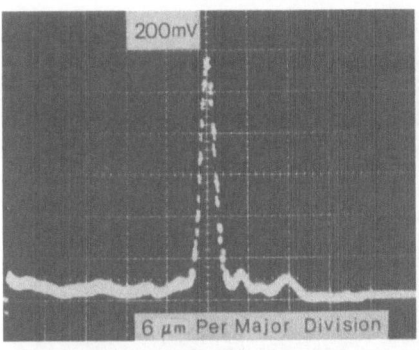 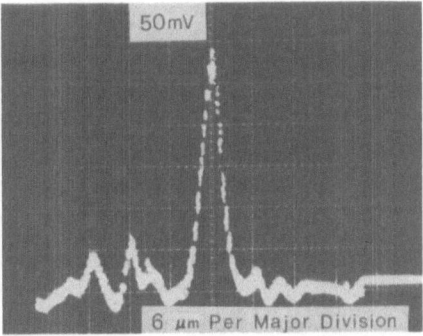

Fig.8.16e. Profiles of undiffracted (*left*) and diffracted (*right*) light beams measured using charge-coupled photodetector array

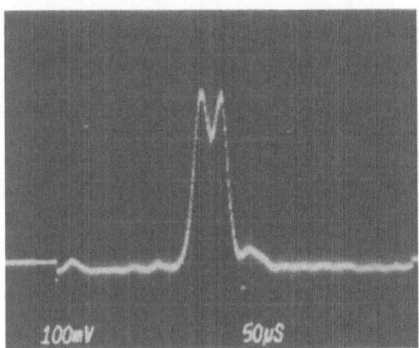 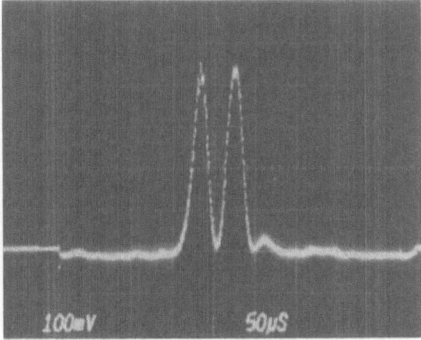

Fig.8.17. Spectral analysis of RF signals using acoustooptic diffraction in a LiNbO₃ spherical waveguide (center frequency: 240MHz). $\Delta f = 2$ MHz (*left*) and $\Delta f = 4$ MHz (*right*)

erved in the diffracted light. In a subsequent experiment with a second device which employed a tilted-finger chirp transducer operating at 500 MHz center frequency, a device bandwidth as large as 250 MHz was obtained [8.80].

8.8.3 Applications to Communications and RF Signal Processing

The various applications that had been demonstrated using the planar guided-wave AO devices can also be served by the spherical guided-wave AO device just described. For example, the applications to optical beam scanning and switching as well as RF spectral analysis were readily demonstrated by varying the frequency of the SAW. A 1.5 MHz frequency resolution (based on Rayleigh criterion) yielding a total of 40 frequency channels within a 60 MHz band, was measured in a standard RF spectral analysis experiment (Fig.8.17). The measured frequency resolution was found to be in good agreement with the calculated value. Other RF signal processing applications such as convolution and spatial-integrating correlation have also been successfully demonstrated. Figures 8.18 and 19 show, respectively, the waveforms of the convolution and the space-integrating

303

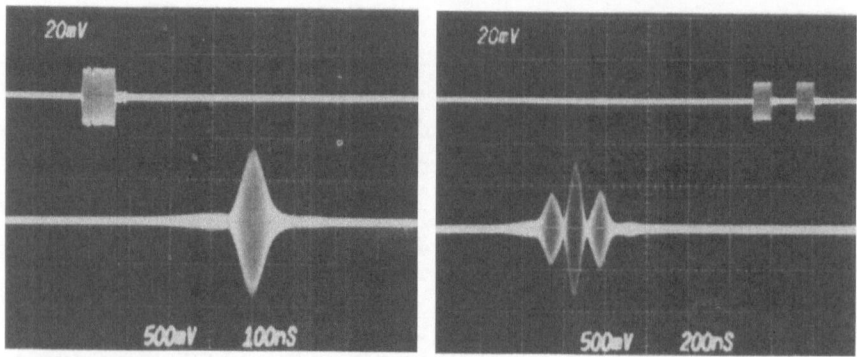

Fig.8.18. Convolution of RF signals using acousto-optic diffraction in a LiNbO$_3$ spherical waveguide (center frequency: 240MHz)

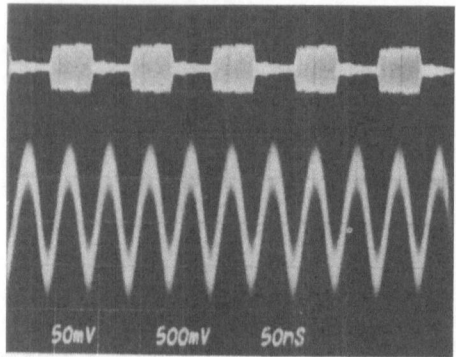

Fig.8.19. Auto-correlation of square-wave modulated RF signal using acousto-optic diffraction in a LiNbO$_3$ spherical waveguide (RF carrier frequency: 240MHz, modulation frequency: 10MHz)

correlation outputs. It suffices to emphasize that since the spherical waveguide also provides the function of a collimating-Fourier-transform lens pair, the undesirable characteristics such as critical tolerance in lens placement and separation, severe lens aberration, long focal length, and high optical losses, which are concomitant with the present planar waveguide signal processor modules, can be significantly reduced.

8.8.4 Spherical Waveguide Acousto-Optic Frequency Shifter Module

In regard to the application to optical frequency shifting [8.41-53], we first observe the following unique features of AO Bragg diffraction in spherical waveguide that have been demonstrated experimentally thus far: (1) a focused light beam edge-coupled at the input port is simultaneously guided, collimated, and refocused into a spot size of a few microns at the output port, (2) the focal spot size of the light beam at the input port is preserved in both diffracted and undiffracted light beams at the output port, and (3) the desirable focusing characteristics of the diffracted light beam are also preserved for the entire AO Bragg bandwidth. These experimental obser-

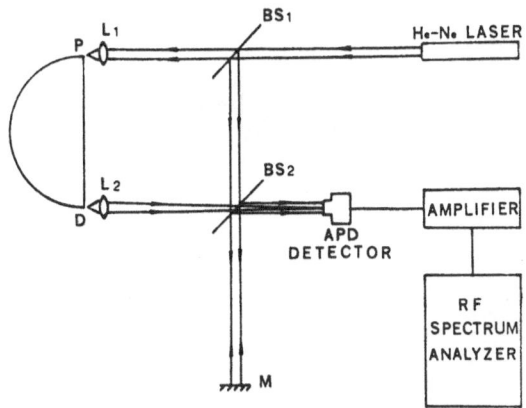

Fig.8.20a. Heterodyning system for measurement of acousto-optic frequency shifting in a spherical waveguide

vations have clearly suggested the potential advantage of the spherical waveguide over the planar waveguide in regard to wide-band AO frequency shifting, namely the diffracted light is automatically focused into a convenient spot size for efficient coupling with single-mode optical fibers. We have successfully carried out broadband AO frequency shifting experiments in two $LiNbO_3$ spherical waveguide Bragg cells using SAW centered at 250 and 500 MHz, respectively, and a He-Ne laser at 6328 Å wavelength [8.52].

Figure 8.20a shows the optical heterodyning system used to measure the single sideband frequency shift and related performance parameters such as conversion efficiency, tuning bandwidth, dynamic range, strength of unwanted sidebands, etc. The He-Ne laser beam is first divided by a beam splitter BS_1. One light beam is edge-coupled into the spherical waveguide and propagates in a TE_0-mode with an angular aperture of approximately 10°, and then partially Bragg-diffracted by the SAW. The frequency-shifted diffracted light which serves as the signal beam is then refocused upon a high-speed avalanche photodetector (APD) through a second beam splitter BS_2. The other light beam from BS_1 propagates through BS_2 and is subsequently reflected by the mirror M. A portion of the reflected light provided by BS_2 is used as a reference beam and is thus impinged upon the same APD to undergo mixing with the frequency-shifted signal beam. The RF spectra of the output from the APD is then displayed using a Tektronics spectrum analyzer.

The first frequency-shifting experiment was performed using the first spherical waveguide AO Bragg modulator with the SAW passband centered at 250 MHz. The spectrum of diffracted light was obtained at the various frequencies within the bassband. The amount of frequency up-shift was correctly measured with a maximum signal-to-noise ratio of 40 dB. A tunable bandwidth of 75 MHz, as defined by 3 dB reduction in the intensity of the diffracted light from that at the center frequency, was obtained with this particular device. Although measurement for the maximum linear dynamic range has not been carried out for fear of possible damage to the

305

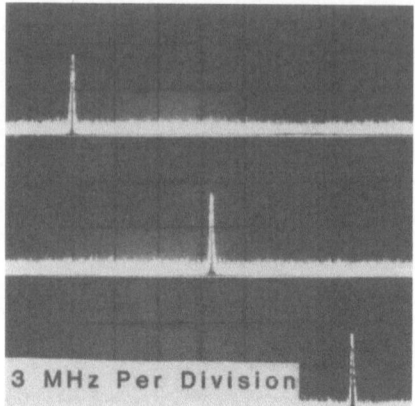

Fig.8.20b. Single-side-band RF spectra obtained using acousto-optic Bragg diffraction in a LiNbO$_3$ spherical waveguide: SAW frequency: 484.5 MHz (*upper trace*), 494.5 MHz (*middle trace*), 504.4 MHz (*bottom trace*)

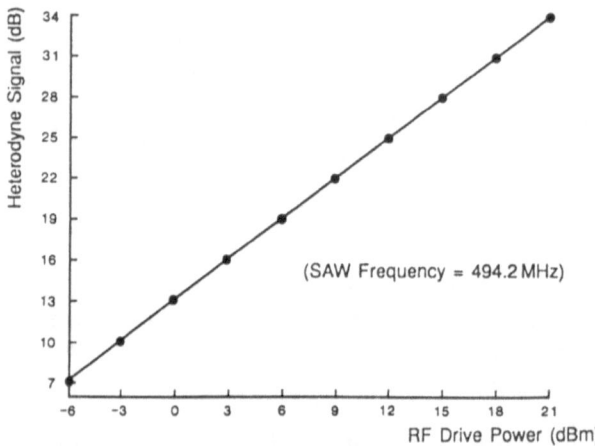

Fig.8.20c. Heterodyned signal of frequency-shifted light from a LiNbO$_3$ spherical waveguide acoustooptic Bragg cell as a function of RF drive power

transducer at the corresponding maximum RF drive power, a linear dynamic range of 34 dB was measured at an RF drive power of 500 mW. At this particular drive power the measured diffraction efficiency was 25%. Herewith linear dynamic range is defined as the range [dB] in which the intensity of the diffracted light remains directly proportional to the RF drive power. It is limited by the nonlinear behavior at the upper end and by noise at the lower end.

The second frequency-shifting experiment was carried out using the second spherical waveguide AO Bragg modulator that was excited with the SAW centered at 500 MHz. Figure 8.20b shows the RF spectra obtained as the frequency of the RF driver signal was tuned from the center frequency. It is seen that, as expected in AO Bragg diffraction from a traveling acoustic wave, neither the optical carrier nor the image sideband spectra appeared in Fig.8.20b. At the center frequency (494.2 MHz) the linear dynamic range of this particular single-sideband frequency shifter was measured to be approximately 27 dB at an RF drive power of 150 mW (Fig.8.20c). The corresponding conversion efficiency was measured to be 7%. Again, no attempt was made to obtain a higher linear dynamic range or a conversion efficiency for fear of possible damage to the SAW transducer at higher RF drive power. However, since the measured insertion loss of the SAW transducer used in this second device was as high as -16 dB, it should be possible to achieve a considerably higher conversion efficiency and thus linear dynamic range at even lower RF drive power. Similar experiments performed at 375 and 625 MHz have clearly demonstrated a tunable bandwidth of 250 MHz, as defined at 3 dB reduction in light intensity. The optical throughput between the input and the output ports was measured to be -10 to -12 dB. Finally, in order to provide a realistic estimate of the frequency tuning range for applications in fiber-optical systems we have also measured the maximum frequency deviation at 3 dB reduction in the strength of the single-sideband spectrum without any realignment of the reference beam. The measured frequency deviation of 11 MHz is in agreement with the calculated frequency increment required for deflecting the diffracted beam by one resolvable spot diameter. Since the measured -3 dB focal spot size of the diffracted light at the output edge (point p' in Fig.8.20a) was 2.5 μm, a tunable range of 22 MHz from the center frequency can be expected when the frequency shifter is permanently attached to a single-mode optical fiber with a 5.0 μm core diameter.

We have also measured the frequency spectrum of the diffracted light beams that result from simultaneous application of multiple RF signals to the SAW transducer. For example, Fig.8.20d shows the spectrum obtained when two RF signals of different frequencies were simultaneously applied. No unwanted spectra such as that due to intermodulation could be detected in the present setup, thus indicating the suppression of unwanted side bands by a minimum of 27 dB. This estimate on unwanted sideband suppression was made by measuring the total attenuation required to reduce the strength of the desired frequency-shifted spectra to 3 dB above the noise level. This observation clearly suggests another unique capability of this spherical waveguide AO device: to provide an array of simultaneous and independent frequency-shifted light sources that are spatially separated. Frequency shift and tuning bandwidth considerably larger than 500 and 250 MHz, respectively, can be accomplished using multiple tilted synchronous transducers and/or tilted-finger chirp transducers. It should be emphasized that together with the unshifted light beam, such an array of light beams of independent frequency shifts are all focused and separated spatially, and can therefore be directly coupled to an optical fiber array.

In summary, wide-band single-sideband suppressed-carrier frequency shifting has been accomplished using AO Bragg diffraction from traveling surface acoustic waves in a LiNbO$_3$ spherical waveguide. Tunable bandwidth of 75 and 250 MHz centering at 250 and 500 MHz, respectively, and the corresponding conversion efficiency of 25 and 7% have been achieved at 6328 Å optical wavelength. Linear dynamic range of 34 and 27 dB and suppression of unwanted side bands by more than 27 dB have been measured. A considerably larger frequency shift and tuning bandwidth as well

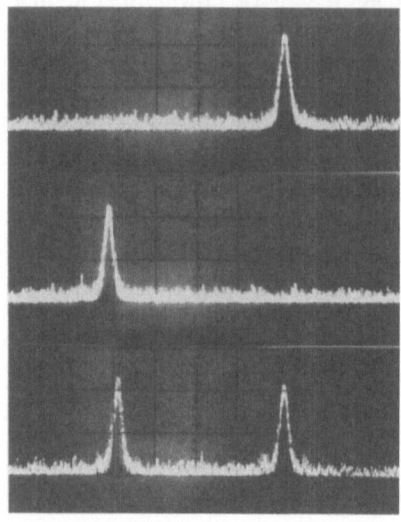

Fig.8.20d. Multiple frequency-shifting using a LiNbO$_3$ spherical waveguide acoustooptic Bragg cell. Single-sideband RF spectra obtained using: (*upper trace*) a single SAW at 500.67 MHz, (*middle trace*) a second SAW at 500.27 MHz, (*bottom trace*) two SAWs simultaneously at 500.67 and 500.27 MHz (horizontal scale: 0.1 MHz/cm)

as higher dynamic range and conversion efficiency should be achievable using transducers of more sophisticated design and lower insertion loss. Unlike its counterparts that utilize AO or EO diffraction in planar waveguides, this spherical waveguide AO device can be directly and rigidly coupled with single-mode optical fibers to provide simultaneously unshifted light and an array of frequency-shifted light sources, and thus can be developed into a compact integrated-optic frequency-shifter module.

8.8.5 Summary

In summary, guiding, collimation, and focusing of a light beam at 6328 Å wavelength in a LiNbO$_3$ spherical waveguide as well as AO diffraction in such a waveguide using the SAW centering at 245 and 500 MHz have been achieved. The experiments carried out thus far have demonstrated a number of unique features of AO Bragg diffraction in a spherical waveguide. Near diffraction-limited characteristics have been measured in the incident light, and the diffracted light has shown to preserve this desirable characteristic. Light beam scanning and switching and various RF signal processing experiments including spectral analysis, convolution, correlation, and frequency shifting have been carried out with encouraging results. Thus, a variety of compact AO device modules, including RF spectrum analyzers, correlators, convolvers, deflectors, switches, frequency shifters, and time- and wavelength-division demultiplexers may be realized. Similarly, a variety of compact EO device modules such as the deflectors and switches, the time- and wavelength-division demultiplexers and multiplexers can also be realized by depositing electrodes or gratings on top of the spherical waveguide. Again, no separate lenses are required in coupling the resulting devices to the single-mode optical fibers or the photodetector arrays [8.5].

8.9 Conclusion

The recent development of Titanium-Indiffusion Proton-Exchange (TIPE) microlenses and lens arrays has made possible the construction of a variety of single- and multi-channel integrated Acousto-Optic (AO) and acousto-optic Electro-Optic (EO) circuits in LiNbO$_3$ channel-planar waveguides $0.1 \times 1.0 \times 2.0$ cm^3 in size. These hybrid AO and AO-EO circuits can be fabricated through compatible and well-established techniques. The recent realization of ion-milled microlenses and lens arrays together with the recent development of GHz AO Bragg modulators and EO Bragg modulator arrays have also paved the way for construction of similar, but monolithic AO and AO-EO GaAs/GaAlAs waveguides of comparable size. Both types of integrated AO and AO-EO circuits suggest versatile applications in communications, signal processing, and computing.

Acknowledgement. The author wishes to express his deep gratitude to Dr. Robert Adler of Zenith Corporation for his thorough review of the chapter and for his many valuable suggestions.

References

8.1   R.G. Hunsperger, A. Yariv, A. Lee: Parallel end-butt coupling for optical integrated circuits. Appl. Opt. 16, 1026 (1977)

8.2   E. Garmire: "Semiconductor Components for Monolithic Applications", Chap.6, in *Integrated Optics*, ed. by T. Tamir, Topics Appl. Phys., Vol.7 (Springer, Berlin, Heidelberg 1975) pp.243-302;
N. Bar-Chaim, S. Margalit, A. Yariv, I. Ury: GaAs Integrated Optoelectronics. IEEE Trans. ED-29, 1372-1381 (1982)
U. Koren, S. Margalit, T.R. Chen, K.L. Yu, A. Yariv, N. Bar-Chaim, K.Y. Lau, I. Ury: Recent developments in monolithic integration of InGaAsP/InP optoelectronic devices. IEEE J. QE-18, 1653-1662 (1982)
A. Yariv: The beginning of integrated optoelectronic circuits. IEEE Trans. ED-3, 1656 (1984)
I. Hayashi: Research aiming for future optoelectronic integration. The Optoelectronics Joint Research Laboratory, IEE Proc.133, Pt.j, 237 (1986)
O. Wada, T. Sakurai, T. Nakagami: Recent progress in optoelectronic integrated circuits. IEEE J. QE-22, 805 (1986)

8.3   I. Hayashi, M.B. Panish, P.W. Foy, S. Sumski: Junction lasers which operate continuously at room temperature. Appl. Phys. Lett. 17, 109-111 (1970)
H. Kogelnik, C.V. Shank: Stimulated emission in a periodic structure. Appl. Phys. Lett. 18, 152-154 (1971)
M. Nakamura, A. Yariv, H.W. Yen, S. Somekh, H.L. Garvin: Optically pumped GaAs surface laser with corrugation feedback. Appl. Phys. Lett. 22, 515 (1973)
H. Yonezu, I. Sakuma, K. Kobayashi, T. Kamejima, M. Ueno, Y. Nannichi: A GaAs-Al$_x$Ga$_{1-x}$As double heterostructure planar strip laser. Jpn. J. Appl. Phys. 12, 1485-1492 (1973);
Y. Suematsu, M. Yamada, K. Hayashi: Integrated twin-guide AlGaAs laser with multiheterostructure. IEEE J. QE-11, 457-460 (1975);
M. Nakamura, K. Aiki, J. Umeida: CW operation of distributed-feedback

GaAs-GaAlAs diode lasers at temperatures up to 300 K. Appl. Phys. Lett. 27, 403-405 (1975);

A. Yariv: *Introduction to Optical Electronics*, 2nd ed. (Holt/Rinehart/Winston, New York 1976)

C.C. Ghizoni, J.M. Ballantyne, C.L. Tang: Theory of optical-waveguide distributed feedback. - A Green's function approach. IEEE J. QE-13, 843 (1977)

L.A. Coldren, K. Iga, B.I. Miller, J.A. Rentschler: GaInAsP/InP stripe-geometry laser with a reactive-ion-etched facet. Appl. Phys. 37, 681 (1980)

Y. Suematsu: Advances in semiconductor lasers", Physics Today 32, 32-39 (May 1985)

8.4   S. Wang: Principles of distributed feedback and distributed Bragg reflector lasers. IEEE J. QE-10, 413-427 (1974)

F.K. Reinhart, R.A. Logan, C.V. Shank: GaAsAl$_x$Ga$_{1-x}$As injection lasers with distributed Bragg reflectors. Appl. Phys. Lett. 27, 45-58 (1975)

H. Kawanishi, Y. Suematsu, K. Kishino: GaAsAl$_x$Ga$_{1-x}$As integrated twin-guide lasers with distributed Bragg reflector. IEEE J. QE-13, 64 (1977)

W. Streifer, D.R. Scifres, R.D. Burnham: Coupled wave analysis of DFB and DBR lasers. IEEE J. QE-13, 134 (1977)

H.W. Yen, W. Ng, I. Samid, A. Yariv: GaAs distributed Bragg reflector lasers. Opt. Commun. 17, 213 (1976)

C.C. Tseng, D. Botez, S. Wang: Optically pumped epitaxial GaAs waveguide lasers with distributed Bragg reflectors. IEEE J QE-12, 549 (1976)

8.5   D.B. Ostrovsky, R. Poirier, L.M. Reibor, C. Deverdun: Integrated optical photodetector. Appl. Phys. Lett. 22, 263-464 (1973);

H. Stoll, A. Yariv, R.G. Hunsperger, G.L. Tangonan: Proton-implanted optical waveguide detectors in GaAs. Appl. Phys. Lett. 23, 664 (1973)

G.E. Stillman, C.M. Wolfe, I. Melngailis: Monolithic integrated In$_x$Ga$_{1-x}$As Schottky-barrier waveguide detector. Appl. Phys. Lett. 25, 36 (1974)

C.C. Tseng, S. Wang: Integrated grating-type Schottky-barrier photodetector with optical channel waveguide. Appl. Phys. Lett. 26, 632 (1975)

J.T. Boyd, C.L. Chen: Integrated optical solicon photodiode array. Appl. Opt. 15, 1389 (1976)

J.L. Merz, R.A. Logan: Integrated GaAs-Al$_x$Ga$_{1-x}$ injection lasers and detectors with etched reflectors. Appl. Phys. Lett. 30, 530 (1977)

G.M. Borsuk, A. Turley, G.E. Marx, E.C. Malarkey: Photosenser array for integrated optical spectrum analyzer systems. Proc. SPIE 176, 109 (1979)

J.C. Grammel, J.M. Ballantyne: A high speed photoconductive detector and waveguide structure. Appl. Phys. Lett. 36, 149-152 (1980)

G.W. Anderson, A.E. Spezio: Photodetector approaches for acousto-optics spectrum analysis: Proc. SPIE 477, 161-164 (1984)

I. Melngailis: Laser sources and detectors for guided-wave optic signal processsing. Proc. SPIE 185 (1979)

N. Bar-Chaim, K. Y. Lau, I.Ury, A. Yariv: Gallium Aluminum Arsenide/Gallium Arsenide Integrated Optical Repeater. Seventh Topical Meeting on Integrated and Guidesd-Wave Optics, Tech. Digest., pp.TuD1-4, IEEE Cat.#84CH1997-6, Kissimmeee, Florida, April 24-26, 1984;

G. Eisenstein, T.P. Lee, R.S.Tucker, C.A. Burrus, W.B. Sessa, P. Besomi: InGaAsP semiconductor lasers with integrated semiconductor amplifier-modulators at 1.3 μm wavelength. Ibid, pp.TuD2-4

8.6   K.W. Loh, W.S.C. Chang, W.R. Smith, T. Grudkowski: Bragg coupling efficiency for guided acoustooptic interaction in GaAs. Appl. Opt. 15, 156-166 (Jan. 1976)

O. Yamazaki, C.S. Tsai, M. Umeda, L.S. Yap, C.J. Lii, K. Wasa, J. Merz: Guided-wave acousto-optic interactions in GaAs-ZnO composite structure, in

1982 Ultrasonics Symposium Proc., 418-420, IEEE Cat. No. 82CH1823-4

C.J. Lii, C.C. Lee. O. Yamazaki, L.S. Yap, K. Wasa, J. Merz, C.S. Tsai: Efficient Wideband Acoustooptic Bragg Diffraction in GaAs - GaAlAs Waveguide Structure. Proc. of 1983 International Conf. on Integrated Optics and Optical Fiber Communications, pp. 30B3-2 to -4 (Aug. 1983)

A.A. Ilyich, S.M. Kikkarin, D.V. Petrov, A.V. Tsarev, I.B. Yakovkin: A comparison of acoustooptic interaction in $LiNbO_3$ and GaAs waveguides. Opt. Commun. 56, 161-166 (1985)

8.7 C.J. Lii, C.S. Tsai, C.C. Lee: Wideband guided-wave acousto-optic Bragg cells in GaAs-GaAs-GaAlAs waveguide. IEEE J. QE-22, 868-872 (1986);

C.J. Lii, C.S. Tsai, C.C. Lee, Y.A. Abdelrazek: Wideband Acoustooptic Bragg Cells in GaAs Waveguide. Proc. 1986 IEEE Ultrasonics Symp. IEEE Cat. No. 86CH2375-4, pp.429-433

Y. Abdelrazek, C.S.Tsai: High-performance acoustooptic Bragg cells in ZnO-GaAs waveguide at GHz frequencies. Optoelectronics - Device and Technologies 4, 33-37 (1989)

8.8 T.Q. Vu, J.A. Norris, C.S. Tsai: Planar waveguide lenses in GaAs by using ion milling. Appl. Phys. Lett. 54, 1098-1100 (1989)

8.9 S. Valette, J. Lizet, P. Mottier, J.P. Jadot, S. Renard, A. Fournier, A.M. Grouillet, P. Gidons, H. Denis: Integrated optical spectrum analyzer using planar technology on oxidized silicon substrate. Electron. Lett. 19, 883-885 (1983); and Integrated-optical circuits achieved by planar technology on silicon substrates: application to the optical spectrum analyzer. IEEE Proc. 131, Pt.H, 325-332 (1984)

8.10 T. Suhara, T. Shiono, H. Nishihara, J. Koyama: An integrated-optic Fourier processor using an acousto-optic deflector and Fresnel lenses in As_2S_3 waveguide. J. Lightwave Techn. LT-1, 624-630 (1983)

8.11 F.S. Hickernell: "Zink-oxide thin-film surface wave transducers. Proc. IEEE 64, 631 (1976);

G.S. Kino, R.S. Wagers: Theory of interdigital couplers on nonpiezoelectric substrates. J. Apppl. Phys. 44, 1480 (1973)

8.12 See, for example, M.L. Dakss, L. Kuhn, P.F. Heidrich, B.A. Sott: Appl. Phys. Lett. 16, 523 (1970)

H. Kogelnik, T.P. Sosnowski: Bett System Tech. J. 49, 1602 (1970)

T. Tamir, H.L. Bertoni: J. Opt. Soc. Am. 61, 1397 (1971)

C.C. Ghizoni, B.U. Chen, C.L. Tang: IEEE J. QE-12, 69 (1976)

Many others cited in *Integrated Optics*, 2nd edn., ed. by T. Tamir, Topics Appl. Phys., Vol.7 (Springer, Berlin, Heidelberg 1979) Chap.3

8.13 R.A. Sprague, K.L. Koliopaulos: Time-integrating acousto-optic correlator. Appl. Opt. 15, 89 (1976);

R.A. Sprague: A review of acoustooptic signal correlations. Opt. Eng. 16, 467 (1977);

T.M. Turpin: Time-integrating optical processor. SPIE 154, 196 (1978)

I.W. Yao, C.S. Tsai: A time-integrating correlator using guided-wave acousto-optic interactions, in *1978 IEEE Ultrasonics Symposium Proc.*, IEEE Cat. No. 78CH1344-1SU, pp.87-90;

N.J. Berg, I.J. Abramovitz, J.N. Lee, M.W. Casseday: A niew surface-wave acousto-optic time integrating correlator. Appl. Phys. Lett. 36, 256-258 (1980)

8.14 C.S. Tsai, J.K. Wang, K.Y. Liao: Acousto-optic time-integrating correlators using integrated optics technology. Proc. SPIE 180, 160-162 (1979)

8.15 K.Y. Liao, C.C. Lee, C.S. Tsai: Time-integrating correlator using guided-wave anisotropic acousto-optic Bragg diffraction and hybrid integration. 1982 Topical Meeting on Integrated and Guided-Wave Optics, Pacific Grove, CA, *Technical Digest*, WA4-1 to 4, IEEE Cat. No. 82CH1719-4;

C.C. Lee, K.Y. Liao, C.S. Tsai: Acousto-optic time-integrating correlator using hybrid integrated optics. *1982 IEEE Ultrasonics Symposium Proc.*, 405-407, IEEE Cat. No. 82CH1823-4

8.16 E.T. Aksenov, A.V. Kukharev, A.A. Lipovskii, A.V. Pavlenko. Acoustooptic convolver using integrated-optic elements. Sov. Tech. Phys. Lett. 7, 513-514 (Oct.1981);
T. Kitano, H. Nioshimoto, N. Takado: Hybrid-Integrated Acoustooptic Time-Intergrating Correlator. IEEE Ultrasonics Symp., Nov. 14-16, 1984 Dallas Texas, Tech. Digest, p.14

8.17 C.S. Tsai, I.W. Yao, B. Kim, Le T. Nguyen: Wideband guided-wave anisotropic acousto-optic Bragg diffraction in LiNbO$_3$ waveguides. 1977 Int'l. Conf. on Integrated Optics and Fiber Communications, Tokyo, Japan, *Digest of Technical Papers*, pp.57-60

8.18 Q. Li, C.S. Tsai: An acousto-optic time-integrating correlator module with hybrid integration in a spherical waveguide. (unpublished)

8.19 C.S. Tsai, C.C. Lee, K.Y. Liao: RF correlation with integrated acousto-optic modules. 1982 WESCON, Anaheim, CA

8.20 D.Y. Zang, C.S. Tsai: Single-mode waveguide microlenses and microlens arrays fabrication in LiNbO$_3$ using titanium-indiffusion proton-exchange technique. Appl. Phys. Lett. 46, 703-705 (1985)

8.21 C.S. Tsai, C.L. Chang, C.C. Lee, K.Y. Liao: Acousto-optic Bragg deflection in channel optical waveguides. 1980 Topical Meeting on integrated and Guided-Wave Optics, in: *Technical Digest of Post-Deadline Papers*, PD7-1 to 4, IEEE Cat. No. 80CH1489-4QEA

8.22 C.L. Chang, C.S. Tsai: GHz bandwidth optical channel waveguide TIR switches and 4x4 switching networks. at 1982 Topical Meeting on integrated and Guided-Wave Optics, 6-8 Jan, Pacific Grove, CA, in: *Technical Digest*, ThD2-1 to 4, IEEE Cat. No. 83CH1719-4

8.23 C.S. Tsai, C.T. Lee, C.C. Lee: Efficient acousto-optic diffraction in crossed-channel waveguides and resultant integrated optic module. *1982 IEEE Ultrasonics Symposium Proc.*, 422-425, IEEE Cat. No. 82CH1823-4

8.24 C.S. Tsai, C.T. Lee: Optical frequency shifting using acousto-optic diffraction in crossed-channel waveguides in LiNbO$_3$ (unpublished)

8.25 C.H. von Helmolt, R.T. Kersten, W. Auch, W. Steudle: Bragg switch/SSB-modulator with integrated optic single-mode waveguides. in: *Proc. of Second European Conf. on integrated Optics*, IEEE Publ. No. 227, 132-135 (1983)

8.26 H.P. Hsu, A.F. Milton, W.K. Burns: Multiple fiber end fire coupling with single-mode channel waveguides. Appl. Phys. Lett. 33, 603-605 (1978)

8.27 C.S. Tsai, D.Y. Zang, P. Le: An integrated acousto-optic module for optical computing. postdeadline paper presented at Topical Meeting on Optical Computing, sponsored by OSA/IEEE, 15-18 March 1985, Incline Village, Nevada, in: *Technical Digest*, PD5-1 to 4

8.28 C.S. Tsai, D.Y. Zang, P. Le: Guided-wave acousto-optic Bragg diffraction in a LiNbO$_3$ channel-planar waveguide with application to optical computing. Appl. Phys. Lett. 47, 549-551 (1985)
C.S. Tsai: LiNbO$_3$-based integrated-optic device modules for communication, computing, and signal processing. CLEO'86, San Francisco, Calif., Techn. Digest 44-46 (IEEE Cat. No.86 CH2274-9)
C.S. Tsai: Titanium-indiffused proton-exchanged microlens-based integrated optic Bragg modulator modules for optical computing. *Optical and Hybrid Computing*, ed. by H.H. Szu, SPIE 634 409-421 (Jan. 1987)
C.S. Tsai: Integrated-optical device modules in LiNbO$_3$ for computing and signal processing. J. Mod. Opt. 35, 965-977 (1988)

8.29 M. DeMicheli, J. Botineau, P. Sibillot, D.B. Ostrowsky, M. Papuchon. Fabrication and characterization of titanium-indiffused proton-exchanged (TIPE) waveguide in LiNbO$_3$. Opt. Comm. **42**, 101-103 (1982)

8.30 R.V. Schmidt, I.P. Kaminow: Metal-diffused optical waveguides in LiNbO$_3$. Appl. Phys. Lett. **25**, 458 (1974)

8.31 J.L. Jackel, C.E. Rice, J.J. Veselka: Proton-exchange for high-index waveguide in LiNbO$_3$. Appl. Phys. Lett. **47**, 607-608 (1983)

8.32 D.Y. Zang, C.S. Tsai: Titanium-indiffused proton-exchanged waveguide lenses in LiNbO$_3$ for optical information processing. Appl. Opt. **25**, 2264-2271 (1986)

8.33 *Proc. IEEE, Special Issue on Optical Computing*, 758-965 (1984)

8.34 A. Vander Lugt: Signal detection by complex spatial filtering. IEEE Trans. Inform Theory IT-10, 139 (1964)

H.T. Kung: Why systolic architectures?. Computer **15**, 37 (1978);

J.W. Goodman, A.R. Dias, L.M. Woody: Fully parallel, high-speed incoherent optical method for performing discrete Fourier transforms. Opt. Lett. **2**, 1 (1978)

M. Tur, J.W. Goodman, B. Moslehi, J.E. Broon, J.H. Shaw: Fiber-optic signal processor with application to matrix-vector multiplication and lattice filtering. Opt. Lett. **7**, 463 (1982)

H.J. Caulfield, W.J. Rhodes, M.J. Foster, S. Horvitz: Optical implementation of systolic array processing. Opt. Comm. **40**, 86 (1981)

T.M. Turpin: Spectrum analysis using optical processing. Proc. IEEE **69**, 79 (1981)

D. Casasent: Acousto-optic transducers in iterative optical vector-matrix processors. Appl. Opt. **21**, 1958 (1982)

A.W. Lohmann: Chances for optical computing. Optik **65**, 9 (1983)

R.A. Athale, W.C. Collins, P.D. Stilwell: High accuracy matrix multiplication with outer product optical processor. Appl. Opt. **22**, 368 (1983)

H.J. Caulfield, J.A. Neff, W.T. Rhodes: Optical computing: the coming revolution in optical signal processing. Laser Focus **19**, 100 (1983)

R.P. Bocker, S.R. Clayton, K. Bromley: Electro-optical matrix multiplication using the twos complement arithmetic for improved accuracy. Appl. Opt. **22**, 2019 (1983)

R.P. Bocker, W.J. Miceli: Optical matrix-vector multiplications using floating point arithmetic. OSA Topical Meeting on Optical Computing, Incline Village, Nev. (1985), Techn. Digest TuD 3-1 to 4

C.M. Verber, R.P. Kennan, Integrated optical circuit for numerical computation. Proc. SPIE **408**, 57 (1983)

C.M. Verber: Integrated-optical approaches to numerical optical processing. Proc. IEEE **72**, 942-953 (1984)

A.A. Sawchuk, T.C. Strand: Digital optical computing. Proc. IEEE **72**, 758 (1984)

A. Huang: Architectural considerations involved in the design of an optical digital computer. Proc. IEEE **72**, 780 (1984)

P. Guilfoyle: Systolic acousto-optic binary convolver. Opt. Eng. **23**, 20 (1984)

J.L. Jewell, M.C. Rushford, H.M. Gibbs: Use of a single nonlinear Fabry-Perot etalon as optical logic gates. Appl. Phys. Lett. **44**, 172-174 (1984)

P.W. Smith: All-Optical Switching and Logic: Potential and Limitations. 1984 Conf. on Lasers and Electro-Optics, June 19-22, Anaheim, California, Tech. Digest, p.184, IEEE Cat.#CH1965-3

S.D. Smith: Optical bistability, photonic logic, and optical computation: Appl. Opt. **25**, 1550-1564 (1986)

D. Psaltis: Two-dimensional optical processing using one-dimensional input devices. Proc. IEEE **72**, 962 (1984)

T.J. Bicknell, D. Psaltis, A.R. Tanguay: Integrated-optical synthetic aperture radar processor. OSA Annual Meeting (1985) Paper TuE6

8.35 J.F. St. Ledger, E.A. Ash: Laser beam modulation using grating diffraction effects. Electron. Lett. **4**, 99-100 (1968)

M.A.R.P. De Barros: "High-speed electro-optic diffraction modulator for base-band operation". Proc. Inst. Elec. Eng. 119, 807-814 (1972)

J.N. Polky, J.H. Harris: Interdigital electro-optic thin-film modulator. Appl. Phys. Lett. 21, 307-309 (Oct. 1972)

J.M. Hammer, W. Phillips: Low-loss single-mode optical waveguides and efficient high-speed modulators of $LiNb_xTa_{1-x}O_3$. Appl. Phys. Lett. 24, 545-547 (1974)

Y.K. Lee, S. Wang: Electrooptic Bragg deflection modulators: Theoretical and experimental studies. Appl. Opt. 15, 1565 (1976)

X. Cheng, C.S. Tsai: Electrooptic Bragg diffraction modulators in GaAs/AℓGaAs heterostructure waveguides. J. Lightwave Tech. 6, 809-817 (1988)

8.36 E.N. Glytsis, T.K. Gaylord, M.G. Moharam: Electric Field, Permittivity, and Strain Distribution Induced by Interdigitated Electrodes on Electro-Optic Waveguides. J. Light Wave Tech. **LT-5**, 668-683 (1987)

8.37 D.Y. Zang, P. Le, C.S. Tsai: "Integrated Electrooptic Bragg Modulator Modules for Optical Computing", Tech. Digest, 2nd Topical Meeting on Optical Computing, Incline Village, Nev. (1987) pp.193-196

8.38 P. Le, D.Y. Zang, C.S. Tsai: Integrated electrooptic Bragg modules for matrix-vector and matrix-matrix multiplications. Appl. Opt. 27, 1780-1785 (1988)

C.S. Tsai, D.Y. Zang, P. Le: High-packing density integrated optic device modules in $LiNbO_3$ for programmable correlation of binary sequences. Opt. Lett. 14, 889-891 (1989)

8.39 S. Ezekiel, S.R. Balsamo: Passive ring resonator laser gyroscope. Appl. Phys. Lett. **30**, 478-480 (May 1977)

J.L. Davis, S. Ezekiel: Closed-loop, low-noise fiber-optic rotation sensor. Opt. Lett. 6, 505 (1981)

8.40 R.F. Cahill, E. Udd: Phase-nulling fiber-optic laser gyro. Opt. Lett. **4**, 93 (1979)

8.41 K.K. Wong, S. Wright: An optical serrodyne frequency translator. Proc. 1st Europ. Conf. on Integrated Optics. IEEE Conf. Publ. 201, 63 (1981)

8.42 F. Heismann, R. Ulrich: Integrated-optical single-sideband modulator and phase shifter. IEEE J. **QE-18**, 767 (1982)

8.43 B. Culshaw, M.G.F. Wilson: Integrated optic frequency shifter modulator. Electron. Lett. 17, 135 (1981)

8.44 F. Heismann, R. Ulrich: Integrated-optical frequency translator with strip waveguide. Appl. Phys. Lett. 45, 490 (1984)

8.45 M. Izutsu, S. Shikama, T. Sueta: Integrated optical SSB modulator and frequency shifter. IEEE J. QE-17, 2225 (1981)

8.46 K. Nosu, S.C. Rashleigh, H.F. Taylor, J.F. Weller: Acousto-optic frequency shifter for birefringent fiber. Electron. Lett. 19, 816 (1983)

8.47 W.P. Risk, R.C. Youngquist, G.S. Kino, H.J. Shaw: Acousto-optic frequency shifting in birefringent fiber. Opt. Lett. 9, 309 (1984)

8.48 R.H. Kingston, R.A. Becker, F.J. Leonberger: Broadband guided-wave optical frequency translator using an electro-optical Bragg array. Appl. Phys. Lett. 42, 759 (1983)

8.49 L.M. Johnson, R.A. Becker, R.H. Kingston: Integrated-optical channel-waveguide frequency shifter. at 1984 Topical Meeting on integrated and Guided-Wave Optics, in: *Technical Digest*, IEEE Cat. Not. S4CH1997-6, p.WD4-1

8.50 C.S. Tsai, C.L. Chang, C.C. Lee, K.Y. Liao: Acousto-optic Bragg deflection in channel optical waveguides. 1980 Topical Meeting on Integrated and Guided-Wave Optics, *Technical Digest of Post-Deadline Papers*, IEEE Cat. No. 80CH1489-4QEA, pp.PD7-1

8.51 C.S. Tsai, C.T. Lee, C.C. Lee: Efficient acousto-optic diffraction in crossed channel waveguides and resultant integrated optic module. *1982 IEEE Ultrasonics Symposium Proc.*, IEEE Cat. No. 82CH1823-4. pp.422-425

8.52 C.S. Tsai, Q. Li: Wideband optical frequency shifting using acousto-optic Bragg diffraction in a LiNbO$_3$ spherical waveguide. Proc. 5th Int'l Conf. on Integrated Optics and Optical Fiber Communications, Venezia, Italy (1985), Techn. Digest 129-132

8.53 C.S. Tsai, Z.Y. Cheng: Novel guided-wave acousto-optic frequency shifting scheme using Bragg diffractions in cascade. Appl. Phys. Lett. 54, 1616-1618 (1989)
Z.Y. Cheng, C.S. Tsai: A novel integrated acoustooptic frequency shifter. J. Lightwave Tech. LT-7, 1575-1580 (1989)

8.54 T.W. Bristol, W.R. Jones, P.B. Snow, W.R. Smith: "Applications of Double Electrodes in Acoustic Surface Wave Device Design", 1972 IEEE Ultrasonics Symp. Proc., IEEE Cat.#72 CH708-8SU, p.343

8.55 T.W. Grudkowdki, G.K. Montress, M. Gilden, J.B. Black: "GaAs Monolithic SAW Devices for Signal Processing and Frequency Control", 1981 Ultrasonics Symp. Proc., pp.88-97 (1980)

8.56 R.T. Webster, P.H. Carr: "Rayleigh Waves on Gallium Arsenide", to be published in the Proceedings of Lord Rayleigh Centenery Symposium on Rayleigh Waves, sponsored by Frank Prize Funds and Royal Inst. of England, July 14-18, 1985

8.57 A.F. Slobodnik: GaAs acoustic-surface-wave propagation losses at 1000 MHz. Electron. Lett. 8, 307-309 (June 1972)

8.58 M.R. Melloch, R.S. Wagers: Propagation loss of the acoustic pseudosurface wave on (ZXt) 45° GaAs. Appl. Phys. Lett. 43, 1008-1009 (Dec. 1983)

8.59 R.T. Webster: 1.5 GHz GaAs surface acoustic wave delay lines, IEEE Trans. Microwave Theory Tech. MIT-33, 824-827 (SEpt. 1985)

8.60 R. Ulrich, R.J. Martin: Geometrical optics in film light guides. Appl. Opt. 10, 2077 (1971)

8.61 C.C. Righini, V. Russo, S. Sottini, G. Toraldo de Francia: Geodesic lenses for guided optical waves. Appl. Opt. 12, 1477 (1973)

8.62 F. Zernike: Luneberg Lens for optical waveguide use. Opt. Commun. 12, 379 (1974)

8.63 D.B. Anderson, R.L. Davis, J.T. Boyd, R.R. August: Comparison of optical waveguide lens technologies. IEEE J. QE-13, 275 (1977)

8.64 P.K. Tien, S. Riva-Sangeverino, R.J. Martin, G. Smolinsky: Two-layered construction of integrated optical circuits and formation of thin-film prism, lenses, and reflectors. Appl.Phys. Lett. 24, 547 (1974)

8.65 E. Spiller, J.S. Harper: High-resolution lenses for optical waveguides. Appl. Opt. 13, 2105 (1974)

8.66 C.M. Verber, D.W. Vahey, V.E. Wood: Focal properties of geodesic waveguide lenses. Appl. Phys. Lett. 28, 514 (1976)

8.67 B. Chen, E. Marom, A. Lee: Geodesic lenses in single-mode LiNbO$_3$ waveguides. Appl. Phys. Lett. 31, 263 (1977)

8.68 W.H. Southwell: Inhomogeneous optical waveguide lens analysis. J. Opt. Soc. Am. 67, 1004-1010 (1977)

8.69 S.K. Yao, D.E. Thomson: Chirp-grating lens for guided wave optics. Appl. Phys. Lett. 33, 635 (1978)
G.I. Hatakoshi, S.I.. Tanaka: Grating lenses for integrated optics. Opt. Lett. 2, 142 (1978)

8.70 B. Chen, O.G. Ramer: Diffraction-limited geodesic lens for integrated optic circuit. IEEE J. QE-15, 853 (1979)

8.71 D. Mergerian, E.C. Malarkey, R.P. Pautienus, J.C. Bradley: Diamond machined geodesic lenses in LiNbO$_3$. Proc. SPIE 176, 85 (1979)

8.72 P. Mottier, S. Valetter: Integrated Fresnel lens on thermally oxidized silicon substrate. Appl. Phys. Lett. **20**, 1630 (1981)

8.73 T. Suhara, K. Kobayashi, J. Koyama: Graded-index Fresnel lenses for integrated optics. Appl. Phys. Lett. **21**, 1966 (1982)

8.74 C. Warren, S. Forouhar, W.S.C. Chang, S.K. Yao: Double ion exchanged chirp grating lens in lithium nibate waveguide. Appl. Phys. Lett. **43**, 424 (1983)

8.75 G.C. Righini, V. Russo, S. Sottini, G. Toraldo de Trancia: Thin film geodesic lens. Appl. Opt. **11**, 1442 (1972)

8.76 Q. Li, C.S. Tsai, S. Sottini, C.C. Lee: Acousto-optic interaction in a LiNbO$_3$ spherical waveguide. 1984 Topical Meeting on Integrated and Guided-Wave Optics, Kissimmee, FL, *Technical Digest*, IEEE Cat. No. 84CH1997-6 pp.TuB2-1 to 4

8.77 Q. Li, C.S. Tsai, S. Sottini, C.C. Lee: Light propagation and acousto-optic interaction in a LiNbO$_3$ spherical waveguide. Appl. Phys. Lett. **46**, 707-709)1985)

8.78 H.I. Smith, F.J. Bachner, N. Efremow: A high-yield photolithographic technique for surface wave devices. J. Electrochem. Soc. **188**, 821 (1971)

8.79 W. Chen, P. Le, C.S. Tsai: Acoustooptic interaction between optical whistling gallery waves and surface acoustic waves in a spherical surface (unpublished)

8.80 Q. Li, C.S. Tsai: Wideband acousto-optic Bragg diffraction in a LiNbO$_3$ spherical waveguide and its applications (unpublished)

Subject Index

319